# Student Manual

for

# Mathematics for Business Decisions Part 2

and

# Calculus for Business Decisions

Library of Congress Catalog Card Number 2007935279

ISBN 978-0-88385-749-6

Printed in the United States of America

Current printing (last digit):
10 9 8 7 6 5 4 3 2 1

# Student Manual

for

# Mathematics for Business Decisions Part 2

and

# Calculus for Business Decisions

**Release 1.5**
**Second Edition**

**David Williamson, Marilou Mendel,**
**Julie Tarr, and Deborah Yoklic**

*Published and Distributed by*
*The Mathematical Association of America*

## CLASSROOM RESOURCE MATERIALS

Classroom Resource Materials is intended to provide supplementary classroom material for students—laboratory exercises, projects, historical information, textbooks with unusual approaches for presenting mathematical ideas, career information, etc.

*101 Careers in Mathematics, 2nd edition* edited by Andrew Sterrett

*Archimedes: What Did He Do Besides Cry Eureka?,* Sherman Stein

*Calculus Mysteries and Thrillers,* R. Grant Woods

*Combinatorics: A Problem Oriented Approach,* Daniel A. Marcus

*Conjecture and Proof,* Miklós Laczkovich

*A Course in Mathematical Modeling,* Douglas Mooney and Randall Swift

*Cryptological Mathematics,* Robert Edward Lewand

*Differential Geometry and its Applications*, John Oprea

*Elementary Mathematical Models,* Dan Kalman

*Environmental Mathematics in the Classroom,* edited by B. A. Fusaro and P. C. Kenschaft

*Essentials of Mathematics: Introduction to Theory, Proof, and the Professional Culture,* Margie Hale

*Exploratory Examples for Real Analysis*, Joanne E. Snow and Kirk E. Weller

*Fourier Series*, Rajendra Bhatia

*Geometry From Africa: Mathematical and Educational Explorations,* Paulus Gerdes

*Historical Modules for the Teaching and Learning of Mathematics* (CD), edited by Victor Katz and Karen Dee Michalowicz

*Identification Numbers and Check Digit Schemes,* Joseph Kirtland

*Interdisciplinary Lively Application Projects,* edited by Chris Arney

*Inverse Problems: Activities for Undergraduates,* Charles W. Groetsch

*Laboratory Experiences in Group Theory,* Ellen Maycock Parker

*Learn from the Masters,* Frank Swetz, John Fauvel, Otto Bekken, Bengt Johansson, and Victor Katz

*Mathematical Connections: A Companion for Teachers and Others*, Al Cuoco

*Mathematical Evolutions,* edited by Abe Shenitzer and John Stillwell

*Mathematical Modeling in the Environment,* Charles Hadlock

*Mathematics for Business Decisions Part 1: Probability and Simulation* (electronic textbook), Richard B. Thompson and Christopher G. Lamoureux

*Mathematics for Business Decisions Part 2: Calculus and Optimization* (electronic textbook), Richard B. Thompson and Christopher G. Lamoureux

*Math Made Visual: Creating Images for Understanding Mathematics*, Claudi Alsina and Roger B. Nelsen

*Ordinary Differential Equations: A Brief Eclectic Tour,* David A. Sánchez

*Oval Track and Other Permutation Puzzles*, John O. Kiltinen

*A Primer of Abstract Mathematics,* Robert B. Ash

*Proofs Without Words,* Roger B. Nelsen

*Proofs Without Words II,* Roger B. Nelsen

*A Radical Approach to Real Analysis*, 2nd edition, David M. Bressoud

*Real Infinite Series*, Daniel D. Bonar and Michael Khoury, Jr.

*She Does Math!,* edited by Marla Parker

*Solve This: Math Activities for Students and Clubs,* James S. Tanton

*Student Manual for Mathematics for Business Decisions Part 1: Probability and Simulation,* David Williamson, Marilou Mendel, Julie Tarr, and Deborah Yoklic

*Student Manual for Mathematics for Business Decisions Part 2 and Calculus for Business Decisions,* David Williamson, Marilou Mendel, Julie Tarr, and Deborah Yoklic

*Teaching Statistics Using Baseball,* Jim Albert
*Topology Now!*, Robert Messer and Philip Straffin
*Understanding our Quantitative World*, Janet Andersen and Todd Swanson
*Writing Projects for Mathematics Courses: Crushed Clowns, Cars, and Coffee to Go*, Annalisa Crannell, Gavin LaRose, Thomas Ratliff, Elyn Rykken

MAA Service Center
P.O. Box 91112
Washington, DC 20090-1112
1-800-331-1MAA    FAX: 1-301-206-9789

# Acknowledgements

Initial funding for this manual was provided by Pima Community College's Educational Technology Grant.

The files ***Graphing.xls***, ***Integrating.xls***, and ***Differentiating.xls*** were used in this manual to create many of the graphs and provided solutions to various problems. These files are in the course materials for *Mathematics for Business Decisions Part II: Calculus and Optimization* by Dr. Richard B. Thompson Department of Mathematics University of Arizona and Dr. Christopher G. Lamoureux Department of Finance University of Arizona.

| Definition | Computation | | Plot Interval | |
|---|---|---|---|---|
| Formula for $f(x)$ | $x$ | $f(x)$ | $a$ | $b$ |
| 0.25 | | 0.25 | 0 | 20 |

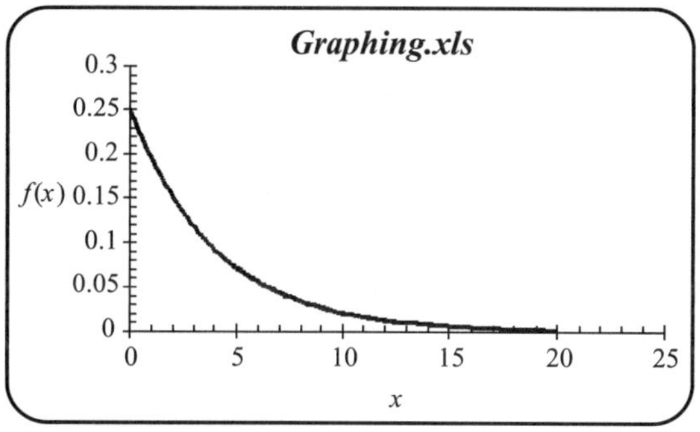


| Definition | Computation | | Plot Interval | | Integration Interval | | $\int_a^b f(x)\,dx$ |
|---|---|---|---|---|---|---|---|
| Formula for $f(x)$ | $x$ | $f(x)$ | $A$ | $B$ | $a$ | $b$ | |
| 0 | | 0 | 30 | 34 | 32 | 32.5 | 0.3196 |

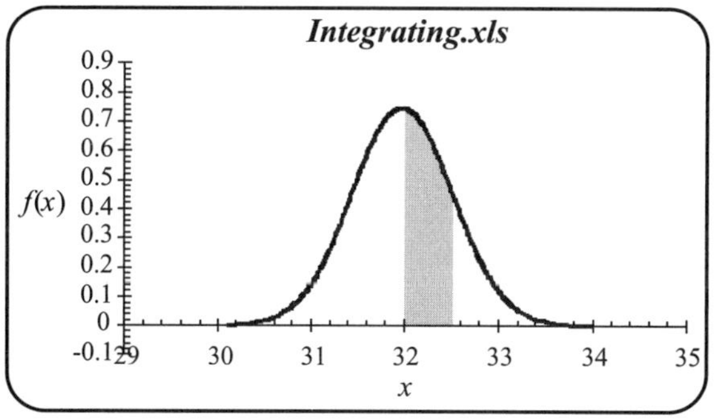


| Definition | Computation | | | Plot Interval | | Increment |
|---|---|---|---|---|---|---|
| Formula for $f(x)$ | $x$ | $f(x)$ | $f'(x)$ | $a$ | $b$ | $h$ |
| 5 | | 5 | 2.000 | 0 | 5 | 0.000001 |

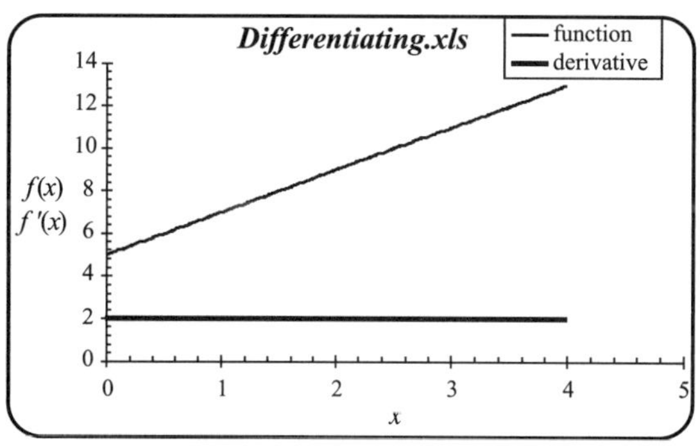

# Contents

# Mathematics for Business Decisions

# Students and the Workforce

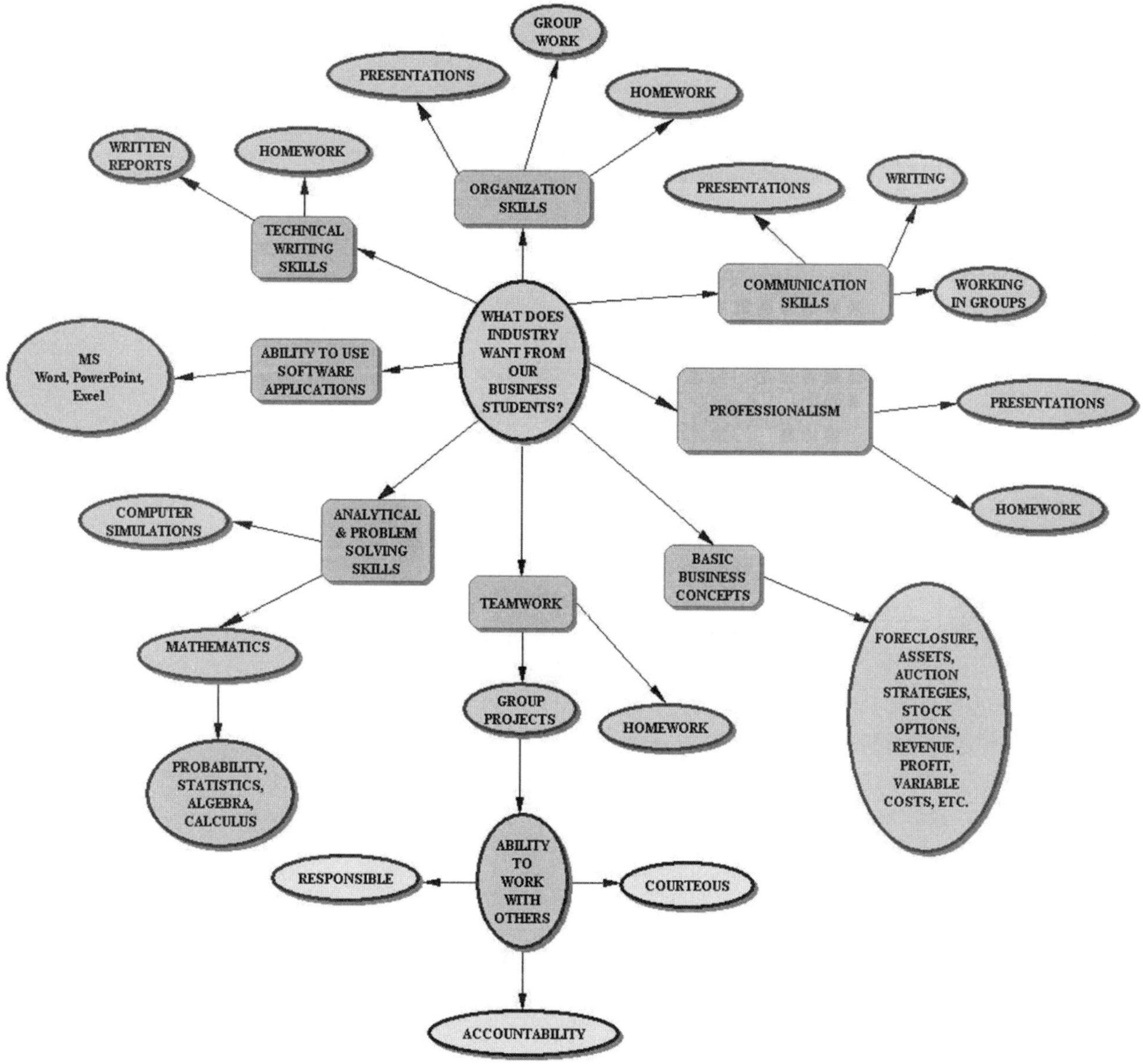

## Suggestions from Former Business Mathematics Students: What It Takes to be Successful in this Class

- ❑ "Have a very thorough understanding of *Excel*."
- ❑ "Work on the project as you go along, so you won't have to rush at the end and you can see how the lecture materials apply to the project. This helps greatly as you write your paper and make your PowerPoint presentation."
- ❑ "Stay on top of the reading."
- ❑ "Warn students to be prepared to spend a lot of time outside of the classroom working on the team projects."
- ❑ "Set aside regular group meeting times each week."
- ❑ "Make sure that everyone in the group understands the material before you move on to something new."
- ❑ "Working with members of your own group can be helpful in solving problems. Attempting this class on your own is too difficult."
- ❑ "Try to get along with your teammates. The projects and the class will be a lot easier if you work well with them."
- ❑ "Develop good computer, software, and Internet skills."
- ❑ "Attend all lectures and see the teacher at the appointment times. Don't leave questions unanswered."
- ❑ "The homework is very complicated, although working with my group made it more understandable and less complicated."
- ❑ "The material is not easy — ask for help from other students or your instructor."
- ❑ "You need to have a flexible schedule. There will be many times you will need to meet with your group."
- ❑ "Look at the examples in PowerPoint. It helped me understand how to do some of the problems using *Excel*."
- ❑ "The student should know that going into this class, there would also be times that you meet outside of class. I found it hard to meet with a student that worked full time."
- ❑ "The quizzes, tests, and projects have to do with using *Excel*, so you need to have some knowledge on how to use *Excel*."
- ❑ "Reading all the assigned texts for the week is crucial."

- ❑ "Do not leave everything to the last minute. In this class you can not procrastinate."
- ❑ "Collaborate as a group and pull your own weight."
- ❑ "Adjust your project to what you have learned in the lectures."
- ❑ "This class is very confusing and you need to get help as soon as you are lost."
- ❑ "Homework takes a long time. Do not think you can get the homework done an hour before class."
- ❑ "Put your project together well before it is due so you can get it reviewed by the instructor."
- ❑ "Never allow yourself to feel overwhelmed. The time allowed to complete each project is ample, just do not save it until the end."
- ❑ "Understanding the material is the single most important aspect to the course. Taking advantage of the other minds in your group as well as making a point to see your instructor during office hours will benefit you the most."
- ❑ "Stay on top of the reading."
- ❑ "Group projects should not be put off until the last minute. There are lots of points that need to be covered and they should be done well in advance."
- ❑ "Finish the homework at least one class before it is due. That way you will be able to ask your teacher questions that are pertinent to the work."
- ❑ "Practice the *Excel* problems. This class is heavily reliant on many different *Excel* functions. Your comprehension with *Excel* may mean the difference between a passing and failing grade on the tests and presentations."
- ❑ "While homework can be overwhelming and time consuming, you are cautioned against dividing up the problems among group members. This is because many quiz and test problems are similar to the homework."
- ❑ "This class moves very fast and you will probably get behind before you even realize it. Seek help as soon as you need it."
- ❑ "Do not overflow your schedule with work and school because it can become dangerous."
- ❑ "This is a fast moving class. Do your homework in groups so you can ask questions if you need help."
- ❑ "Before meeting with other members of your group to discuss the homework, have everyone do it individually. This way you will be able to compare your answers rather than just relying on those of one person."
- ❑ "Make sure that everyone in the group participates and knows how to do each homework problem. Then everyone will be prepared for the test and giving the oral reports."
- ❑ "Practice giving your oral reports several times before you actually do them. Also, think of potential questions that your teacher could ask, and make sure everyone in your group knows how to answer them."

- "When working on the projects periodically check your data with the professor and work on it from day one. Waiting until the week before it is due can cause problems."
- "Practice your oral reports so that you can give it without needing to read from anything — you will be less likely to read from an index card or the computer screen. Eye contact is important."

## Time Management: A Note from the Instructor

This is not a typical math class. In addition to learning definitions and procedures and working math problems, this course involves group projects, oral presentations, written reports, using *Excel* to solve math problems, and the use of *PowerPoint* and the equation editor in *Word*, not to mention an electronic text. All of this requires more of you than a traditional math course. The year you take this course, you will need to make sure you have time in your schedule for these additional demands, particularly that you have time to meet outside of class with your group.

# Having a Successful Team

So you've been assigned to Team 2, and the only other person you know in the class has been assigned to Team 1. And, oh, those other people in Team 2! They can't find a time to meet, this one is always on her cell phone, that one has to rush off to work, no one has any idea what the instructor was talking about in class. Oh, you just **know** this is going to be a disaster!

Well, that depends. Depends on **you**, that is. The most important factor in the success of your team is you: your attitude, your expectations, your efforts, your understanding. So here are some ideas to help your group be the best it can be.

## Rule #1: Have a Positive Attitude

Go in with a positive attitude, ready to contribute your best to the group. All you really have control over is your own attitude and your own contribution. If everyone has a great attitude, the rest will be that much easier.

What characterizes someone with a "good attitude"? We all know a good attitude when we see one. Someone with a good attitude is curious, eager, interested, well mannered, prepared, helpful, respectful, focused, involved, open-minded, active, a good listener and a team player. Someone with a positive attitude has enthusiasm and optimism for all kinds of ideas. Someone with a positive attitude comes to all group meetings, definitely gets in touch with others when he or she cannot (for a very good reason), has done the readings and the work before the group meets. Have a positive attitude, have a good group.

## Rule #2: Be a Scout: Be Prepared

Come to all meetings prepared. Read the electronic text. Work through the examples in the text. If your group works on homework sets together, start all assignments before the group meeting. Go over the ***Focus*** sections and try to figure out how they apply to your group's data. Make a list of questions to bring to the group.

## Rule #3: Keep an Open Mind toward Other Group Members

Expect a lot from yourself; expect little from the others. If you expect the people around you to be something they cannot live up to, you will be disappointed, confused and even angry.

Each person has her/his own work and communication styles, which may be very different from yours. You may have to work with someone you just don't get along with. You have to figure out how to get past whatever clashes come up. Keep in mind that in the "real world" your boss will decide whom you work with. Learn to adapt and make the best of every situation.

Take advantage of each person's strengths. Work together to make up for other's weaknesses. Even that person who seems like a "drag" on the rest of group may turn out to have the best ideas, or the best writing skills, or even be the best mediator. You'll never know if you write them off from the first day.

No one person in the group — not the person with the best grades, nor the person who understands the quickest, nor the person with the best computer skills — no one person has all the information, skills or resources needed for the highest possible quality solution and presentation.

## Rule #4: Look at Group Work as an Opportunity

Approach the requirement to work in a group as an opportunity to learn from others. When you teach others, you learn more yourself. Group work is an opportunity to get the best work from each person. One person in the group cannot succeed unless everyone in the group succeeds. You've heard the old saw: the sum is more than its parts. Make that happen!

## Rule #5: Nip Conflict in the Bud

Group conflicts often stem from different expectations group members have of each other. At the first meeting, prepare a list of ground rules. These rules might include some of the following: come to meetings on time, come to meetings prepared, turn off cell phones at meetings, outline problems before the group meeting, assign tasks for each group member. Add your own rules to this list! Have *each person* sign the rules.

If — or rather, when — conflict arises, the most important thing to do is **talk the situation over soon**. Talk with the person or persons you have the conflict with, not with others who will inevitably pick sides or with those not even involved. First, though, a couple of minutes of cool-down time might be wise. Whatever you do, don't let things go on too long.

If you don't like someone else's behavior or attitude, talk with them about it in a non-accusatory way: "I don't like it when you don't come prepared because then we have to spend time on what the rest of us already know. It would be a better use of our time together solving the problem." Or, "It is hard for me when you come late because I have only an hour and we need all the time together we can manage." Or, "I'm willing to try to help you with this, but when I can't help,

it would be better for us to go see the instructor than to get angry with me for also not understanding."

Initiate resolution. Be the first person to say you're sorry. Even when you think the other person is wrong, it's not a bad thing to say "I'm sorry you feel that way" or "I'm sorry I offended you in that way."

If you can't agree, then agree to disagree. Compromise: let's do slide #6 your way, and slide #8 my way. The bottom line is: be flexible.

## Rule #6: Get Help Before It's Too Late

If there is a concept or homework problem no one in the group understands fully, go to the instructor or a tutor. The instructor might not go over that concept again in class. If you don't know it now, it's sure to show up on an exam, and you still won't know it.

If there is a conflict or problem the group cannot resolve yourselves, go see the instructor before it gets impossible to solve, before the hurt feelings and resentments cannot be turned around. Your instructor may want to meet with you individually and then as a group to try to resolve the conflict.

You are probably in this group for the duration of this project, so it's up to you to make the best of it. It's not helpful to be an ostrich and hope the problems will go away if you ignore them. It is helpful to be that kid with the entrepreneurial stand: when life hands you lemons, make lemonade.

## Bibliography


Feenstra, Kristin. "Resolving Conflict in Friendships." *IamNext.* 1 October 2002. <www.iamnext.com/people/tips/conflict.html>.

Felder, R. M. <felder@eos.ncsu.edu> and R. Brent. "Effective Strategies for Cooperative Learning."*Active and Cooperative Learning.* 2001. North Carolina State University. 1 October 2002. <www2.ncsu.edu/unity/lockers/users/f/felder/public/Papers/CLStrategies(JCCCT).pdf>

Godshall, Leslie. "Surviving Group Projects for Life." *IamNext.* 1 October 2002. <www.iamnext.com/academics/inclass/group.html>.

Johnson, Roger T. <johns009@umn.edu> and David W. Johnson. "Cooperative Learning." *Cooperative Learning Center at the University of Minnesota.* 19 June 2002. University of Minnesota. 1 October 2002. <www.clcrc.com/pages/cl.html>.

O'Keefe, Dr. Barbara <b-okeefe@uiuc.edu>. "Teamworks Module: Problem Solving; Lesson 4: Creative Problem-Solving." *Teamworks: Skills for Colalborative Work.* The Sloan Center for Asynchronous Learning Environments (SCALE). 1 October 2002. <www.vta.spcomm.uiuc.edu/PSG/psg14-ov.html>.

# Group Presentations

The scenario is that your group is asked to answer questions X and Y for the project.

How would you go about presenting the solution?

Before presenting in front of an audience make sure your group has completed the following tasks:

- **GROUP ORGANIZATION**

  Who will be doing what part of the presentation?

  - **Introduction**
    - State the project objectives.
    - Give any background information.
    - Define the project terms.
    - State the assumptions made.
  - **Steps used to solve questions X & Y**
    - Explain any mathematics used to solve the questions.
    - Explain the *Excel* functions used in the project.
    - Explain why you are doing specific tasks.
    - Use graphs or visual aids.
    - Present solution X.
    - Repeat the steps to answer question Y.

It may take several steps and calculations to reach the final result. Make sure you cover all the steps and calculations.

Remember how you felt lost and confused when you started this project. Your audience will possibly feel that way if your presentation is not clear and concise, or you skip steps.

If you are not sure about any aspect of your part of the presentation, do not mention it at all because if you do say anything incorrect, you will most likely be asked a question about it.

Now that you have answered all the questions and explained in detail the processes involved for obtaining your solution, your audience has most likely forgotten some, if not all, of your solutions. So it is best to summarize your results and give closure to your presentation.

- **Summary**
  - Restate the solution for X.
  - Restate the solution for Y.
  - Ask the audience if they have any questions.

Before you present in front of your audience make sure the group has practiced together.

- **GROUP PRACTICE**
  - **Practice the entire presentation**
    - Decide who will control the slides.
    - Try to balance the amount of time each person presents.
    - Keep practicing until the presentation goes smoothly.
      - Remember this is a **group** presentation and it only takes one person to make your group look bad.

You should also practice your part of the presentation on your own.

Make sure each group member knows what the other members are talking about. For example, if Steve is talking about how he used *Excel* to solve question X, then all the members should know how to solve question X. Remember you can be asked a question at the end of your presentation about something other than your part of the presentation.

Be prepared for computer problems. The projector may not work or the computer could freeze up. Make sure you have a backup plan. It is a good idea to print your slides so you have a hard copy.

Now that the presentation is organized, your group has practiced and is ready to go, you need to know how to present the material. We need to think about professionalism.

- **PROFESSIONALISM**
  - **Proper attire**
    - No jeans, T-shirts, shorts, sandals, tennis shoes, etc.

Imagine that you are working for a corporation and representing this company. The company wants you to represent them in a professional manner and the first place to start is how you dress.

There is a time and place when it is appropriate to dress up, and when presenting in front of an audience is one of them. Also, the way you dress does not represent your intelligence, but it does reflect some characteristics of your personality and how others may judge you.

For example, laser vision surgery is becoming very popular for near-sighted people, and you are interested in learning more about the procedure, risks involved, recovery time, costs, etc. You have signed up for two seminars.

The first seminar has about 200 potential patients in the audience. The doctor comes out wearing jeans, tennis shoes, and a T-shirt that is not tucked in. The T-shirt is also wrinkled. Think about your first impression.

The second seminar is just as full of potential patients and the doctor is wearing slacks, dress shoes, and a nice shirt. Compare your first impression of this doctor to the first one.

With all other factors considered equal, such as cost, doctor's experience, and the presentation, which doctor would you choose?

An analogy would be going into a restaurant that was not clean. Right away you might leave or at least think that the kitchen is also dirty.

**Bottom Line** Part of your grade depends on professionalism, and one part of professionalism is how you dress for your presentation. Right now your grade is the only thing being affected; however, later in your career your job or promotion may be affected.

Another part of professionalism is organization. We have already discussed how to organize the group presentation, so what can you do to prepare for your part?

The answer to this question is simple. PRACTICE! PRACTICE! PRACTICE!

- **Organization**
  - Proofread the PowerPoint slides.
    - Make sure there are no spelling or mathematical errors.
      - Your presentation will not look good if there are errors in your slides. Especially miscalculations, incorrect assumptions, and definitions.
    - Have at least two group members proofread the slides.
      - Would you turn in a writing assignment without proofreading it?
  - Practice your part of the presentation for a family member or friend and ask for criticism.
  - Practice by yourself.
    - Pretend you are speaking to an audience.
    - Mentally go through your presentation.
  - Make cue cards to help guide you through the presentation.
    - Do not read directly from the cue cards.
    - The cue cards should only be used to keep you on track or to help you remember something.
  - Understand the material you are presenting.
    - Get help if you are unclear about any aspect of your presentation.
  - Present your material systematically.

    - Each part of your presentation should build upon the previous part.
    - There should be a smooth transition when changing speakers.

So far your group has organized and practiced the presentation. You have decided what to wear and practiced your part. At this point you should feel confident about the presentation.

The final step for a professional presentation is the delivery.

- **Presentation delivery**
  - Minimize the number of times you turn your back to the audience.
    - Use your cue cards so you do not have to look at the screen.
  - Face your audience when you speak.
    - Maintain eye contact with the audience.
      - Do not speak to the projector screen.
  - Avoid nervous habits.
    - Do not speak to the audience with your hands in your pockets.
    - Avoid scratching your head, playing with your hair or jewelry, excessive pacing, etc.
  - Do not read directly from the *PowerPoint* slides or cue cards.

    - Since you have practiced, you should be able to explain all the concepts, definitions, and calculations without reading from the cue cards or slides.
    - If there is a long definition, quote, or you completely go blank, then you can read from the cue cards.
  - *PowerPoint* slides should contain bulleted pieces of information.
    - Avoid lengthy definitions or calculations. This accomplishes two things:
      - First, it does not clutter up the slides with a lot of text.
      - Second, it prevents you from reading directly from the slides.
  - Speak clearly.
    - Enunciate all words and sentences.
    - Do not use empty filler words such as "uh," "um," or "ok."
    - Avoid speaking in a monotone voice.
      - You may put some of your audience asleep.
    - Do not speak too quickly.
      - You will confuse your audience and eventually lose their attention.
  - Avoid standing in front of the projection screen.
    - While you are presenting to the audience they will be reading from the slides.
  - Format the slides so they are easy to read and follow.
    - Use colors that are clear.
      - Some colors that might look clear on your personal computer may not show up well on the projector.

    - The only way to check this is to view the slides on the projection screen before your group presents. Again, this is a part of being prepared.
    - Do not use dark backgrounds with dark lettering.
- Do not link to large *Excel* files from *PowerPoint.*
  - Use the cut & paste special in *Excel* to show only a part of the spreadsheet.
    - All of the spreadsheet is not necessary and it would be difficult for the audience to follow.
  - This prevents the computer from locking up or needing lengthy load times.
    - Points will be deducted from your presentation if this occurs. Remember you have a time limit on the presentation.
- Show graphs and calculations when necessary.

Use the KISS principal— **Keep It Simple Stupid**
Some of the best presentations are the ones that are kept simple and well organized.

The objective of the project is to answer questions X and Y. In order for your audience to understand the process used to answer questions X and Y, the presentation should contain all objectives, definitions, calculations, and graphs. All of these elements should be presented in a logical and clear manner so it is easy for your audience to follow and comprehend the way in which your group has solved the project objectives.

# Check List for Group Presentations

## Group Organization

- ❑ Identify each group member's part in the presentation.
- ❑ Decide who will control the slides while you are presenting.

## Organize the power point slides

### Introduction

- ❑ Identify the project objectives.
- ❑ Include any business background information.
- ❑ Define the project terms.
- ❑ State any project assumptions.

### Solution process

- ❑ Mathematics used.
- ❑ *Excel* functions used to perform the mathematics.
- ❑ Rationale for your entire solution process.
- ❑ Graphs or visual aids.
- ❑ Present solution.
- ❑ Does your answer make sense?
- ❑ Repeat the process to answer other questions.

## Summary

- ❑ Restate all your results.
- ❑ Ask audience if they have any questions.

## Professionalism

### Organization

- ❑ Practice as a group as well as individually.
- ❑ Use cue cards as a guide.
- ❑ Understand material that is being presented.
- ❑ Present the material in a logical order.
- ❑ Minimize the amount of time you turn your back to the audience.
- ❑ Minimize reading directly from the slides or cue cards.
- ❑ *PowerPoint* slides should contain bulleted pieces of information.
- ❑ Slides should be easy to read.
- ❑ Do not link large *Excel* files into *PowerPoint.*
- ❑ Dress appropriately.

# Technical Writing

The technical reports you will be writing in this course will involve detailed explanations as to how your group solved the project. The solutions to the projects involve computer simulation, direct computation, and mathematical concepts and assumptions. The writing style for a technical document is different than the writing style for a non-technical document. To help improve your reports, some guidelines on technical writing can be found at

http://www.technical-writing-course.com. This is an excellent resource that is short and easy to following with many examples.

# Technology Tips & Tricks
## (Featuring Microsoft *Office* 2003)

## Basic Microsoft *PowerPoint* Features

### Getting Started

- Click 

.
- Move the pointer to "All Programs" and wait for the submenu to appear.
- Move the pointer to the Microsoft Office folder and select Microsoft Office PowerPoint 2003.
- Wait for Microsoft *PowerPoint* to open. Your window should look similar to the window below. On the right-hand side, click the "X" to the right of "**Getting Started**" to make more room. You can also disable this menu from being automatically displayed by pulling down the "Tools" menu, selecting "Options…," and choosing the "View" tab. Uncheck the "Startup Task Pane."

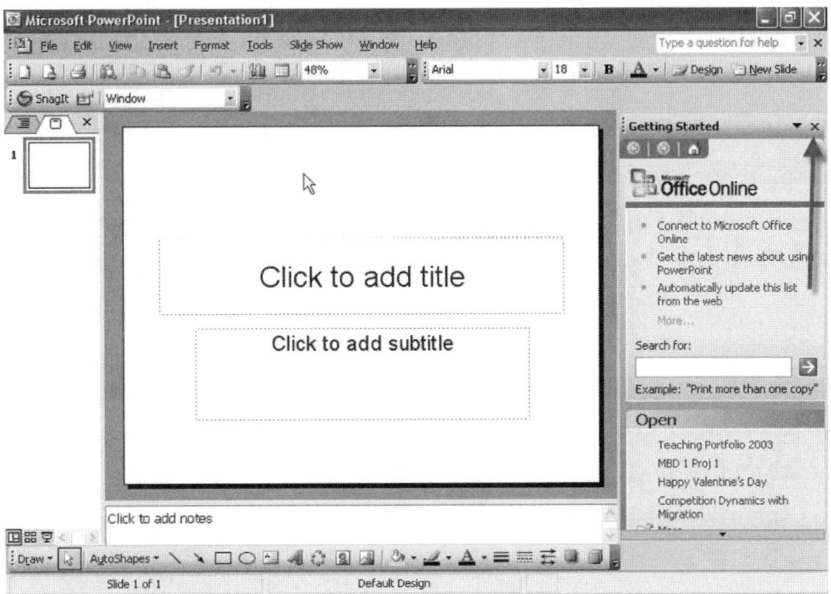


### Creating a Basic Presentation

- From the *PowerPoint* toolbar, pull down the "Format" menu and select "Slide Design…"

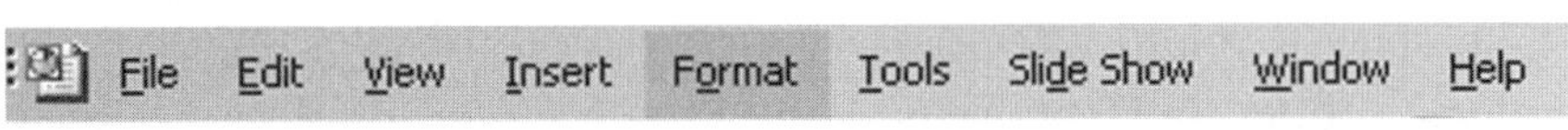

- Choose your design template from the selection that appears on screen. You may also browse for a template if you have purchased additional templates from Microsoft or another vendor.
- Once you have selected a template, pull down the "Format" menu again and this time select "Slide Layout..."
- Select a slide layout for your first slide. You can change your layout for each slide or select "Duplicate Slide" from the Insert pull-down menu for more copies of the same layout.
- Add text (just click and type!) to your slides.
- Don't forget to save your work frequently.

## Adding Pictures and Changing the Format or Background

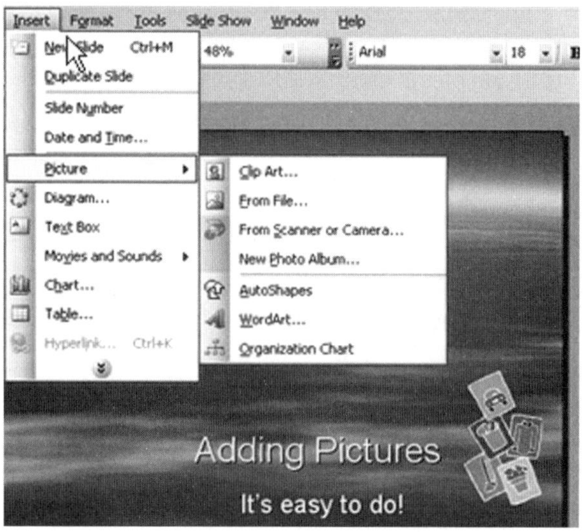


- To add pictures, pull down the "Insert" menu and select "Picture."
  - Notice that you can add clip art pictures as well as pictures that you have saved as files (such as bitmap, gif, or jpeg files).
  - Also note that you can select a layout that has a pre-set spot for a picture or you can insert a picture essentially anywhere you want using the drag and drop method.
- The "Format" pull-down menu allows you to change the appearance of the slide.
  - You can change font, background, and line spacing.
  - You can alter slide layouts and slide color schemes.
- Note that changes can be made to take effect at many levels.
- To change the look of all of the slides at once, use the "Slide Master" under the "View" pull-down menu.

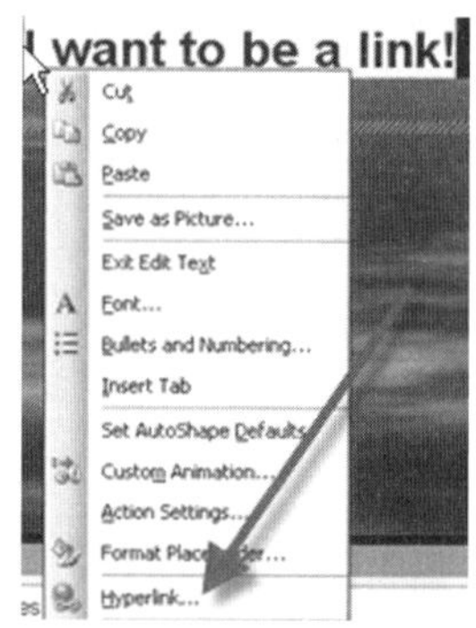


## Adding Links

- Links can be made to existing files, to the Internet, from one slide to another within a *PowerPoint* presentation, or to new documents.
- Select the object or text that you wish to become a hyperlink.
- From the "Insert" pull-down menu, select "Hyperlink."
- Browse for a file or URL, select another slide in the presentation, create a new document, or enter an email address. **Make sure that all linked files are accessible from where you run the presentation.**
- Warning! Linking large Microsoft *Excel* files can freeze the presentation or take several minutes to load. This is quite embarrassing, especially during timed presentations. If you have *Excel* graphs or

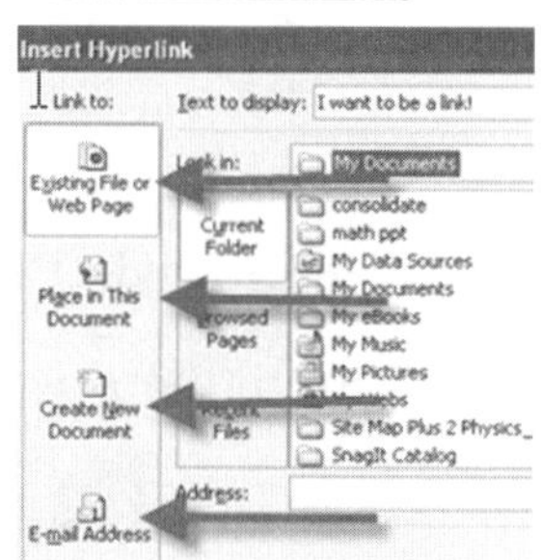

clips that you wish to show, then you should use "Paste Special" under the "Edit" menu to paste them in as pictures or bitmaps instead of linked files.

## Customizing Animation

- Select the object that you wish to animate.
- From the "Slide Show" pull-down menu, select "Custom Animation."
- Click "Add Effect" in the "Custom Animation" task pane.
- Once you have selected your effect, additional options for that particular effect, such as speed, direction, and order of animation, will be available.
- You can also specify whether you want the animation to occur automatically, after a preset amount of time, or only by mouse-click.
- Warning! Too many animations can make your professional presentation look more like a cheap commercial. Although visuals can spruce up a dry presentation, you should be judicious in your selection of visuals and animations. Remember — not every slide needs something whipping in or wiping out!
- Also, for the most part, it is best to avoid sound effects. Canned clapping noises may work well in sitcoms, but they usually fall flat in professional presentations.

## Printing Options

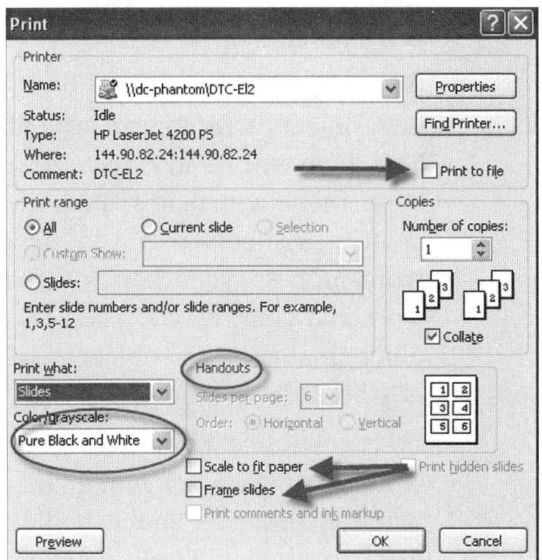


- You can print all of your slides or just a selection.
- You can print one slide per page or several slides per page or a handout on which your audience can take notes.
- You can print in color, black and white, or grayscale. Printing in black and white can save ink and is useful for people with inkjet printers.
- There are numerous other options that you can explore as well, such as framing your slides, scaling them to fit an entire page, and only printing selected slides.

## Miscellaneous Pointers

- At the bottom left of your *PowerPoint* work area, you should see a set of icons: . The first icon selects the "normal" view of your work area. This is the view that you want when you are building your slides. The second icon selects the "slide sorter" view. This is useful for organizing your slides, just click, drag, and move a slide wherever you would like it to appear. The third icon selects the slide show view. This allows you to preview your presentation.

- You can add charts by using the "Add Chart" option under the "Insert" menu, but it is probably better to build your charts using Microsoft *Excel*. Use the "Paste Special" option under the "Edit" menu to insert your charts as bitmaps (as opposed to linked files) — remember, you don't want to be standing up in front of an expectant audience with a frozen *PowerPoint* presentation.
- You can add equations in the same way you add them in Microsoft *Word*: Insert-Object-Microsoft Equation, but if you have already typed your equations in *Word*, then cutting and pasting will save time.

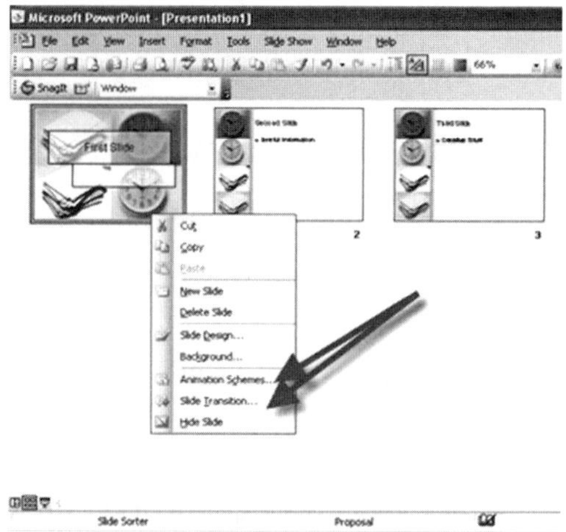

## Useful Microsoft *Word* Features

### Using the Microsoft *Equation Editor*

- The Microsoft *Equation Editor* is accessed through the "Insert" pull-down menu by selecting "Object." Once the "Object" window appears, scroll down and select "Microsoft Equation" (whichever version is available).

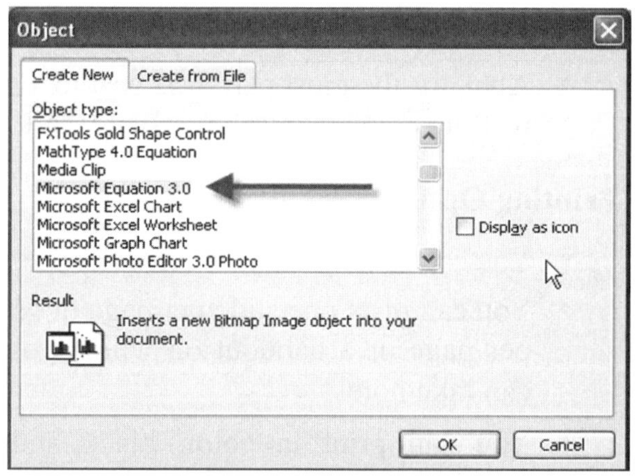


- After you click "OK" or press Enter, two new objects will appear in your working area. One will be an *Equation Editor* pull-down menu with many categories of mathematics symbols that you may need to word-process your homework and your reports. The other smaller empty object is where the symbols will appear once selected.

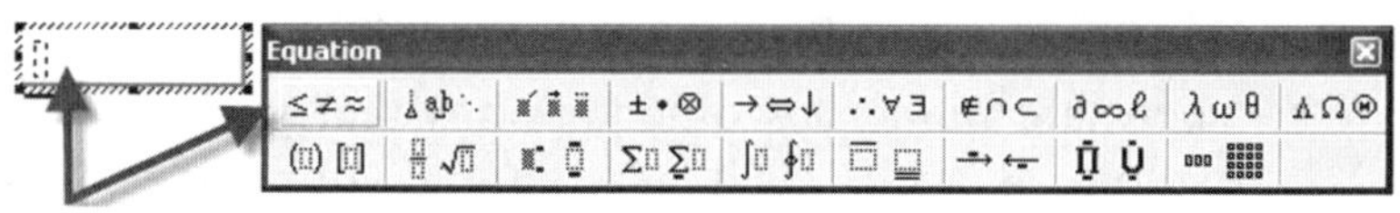


- Notice that when you turn on the *Equation Editor*, the usual *Word* toolbars at the top of your window disappear and a smaller toolbar for use with the *Equation Editor* appears. If you want to align multiple equivalent equations by lining up the equality signs, then the "Format" menu is the pull-down menu for you. If you want to insert regular text in a mathematical argument but do not want to exit the *Equation Editor*, then try the "Style" menu and select "Text"; pull it back down and reselect "Math" to return to the editor's default style. "Other" under the "Size" pull-down menu will allow you to change the size of your math-type font.

- To save time, you can customize your Microsoft *Word* toolbar to include the Microsoft *Equation Editor* icon. Under the "View" menu on the standard toolbar, select "Toolbars" and scroll all the way down to the bottom to select "Customize." A pop-up window will appear. Select the "Command" tab and "Insert" under the "Categories" option. You will have to scroll down quite a bit under the "Commands" option to find the "Equation Editor" icon (the square root of alpha). Just drag and drop the icon onto your toolbar and you are good to go! Now, you only have to click that icon to run the *Equation Editor*.

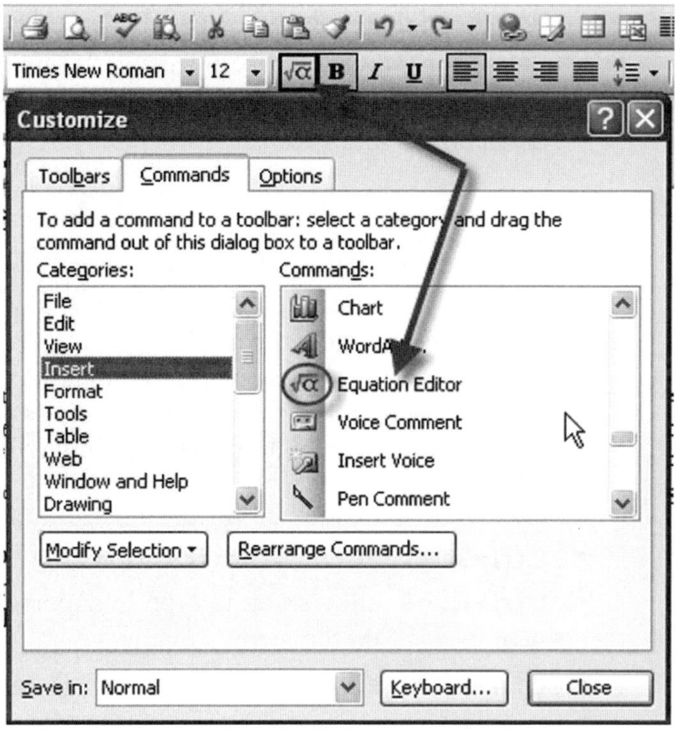


## Two Special Cases

- Piecewise functions can be a pain! This is especially true when it comes time to word-process them. Suppose that you want to word-process the following piecewise defined function:

$$f(x)=\begin{cases} x, & \text{if } x>0 \\ x^2, & \text{if } x\le 0 \end{cases}.$$

A simple way to proceed is as follows:

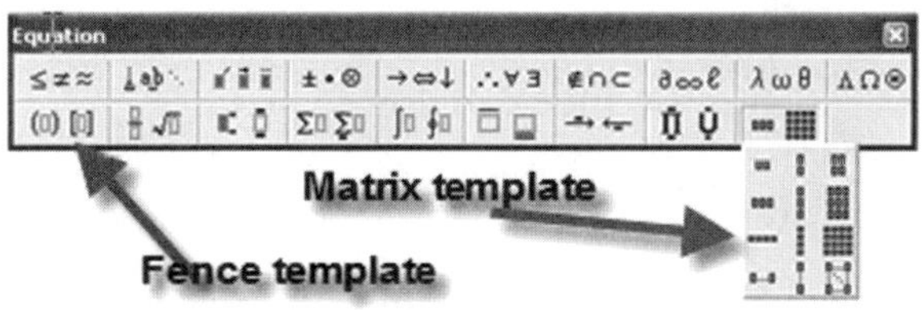


  - Call up the equation editor.
  - Type "$f(x)=$" using the keyboard.
  - Select the single left bracket from the "Fence templates" (the pull-down menu on the far bottom left).
  - Select the 2-by-2 matrix from the "Matrix templates" on the far right.
  - Insert the four entries, $x$, $x^2$, if $x>0$, and if $x\le 0$, each in their own cell.
  - Voilà! You have a perfectly aligned piecewise function.
- Sample spaces can also provide word-processing challenges. Suppose that you want to word-process the sample space given below.

$$\begin{Bmatrix} (1,1) & (1,2) & (1,3) \\ (2,1) & (2,2) & (2,3) \\ (3,1) & (3,2) & (3,3) \end{Bmatrix}$$

  - Call up the equation editor.

- Select the double brackets, "{}", from the "Fence templates."
- Select the 3-by-3 matrix from the "Matrix templates" on the far right. (You can construct higher than a 4-by-4 matrix by selecting the bottom-corner rectangular matrix of unspecified size.)

### Hotkeys

- Sometimes pull-down menus can be tedious to use. You must interrupt your keyboard usage, find your mouse, click here, click there, …, sigh. Here are some shortcuts to avoid the pull-down menus in the *Equation Editor* for some commonly used objects (Note: **CTRL+Character** indicates that you should simultaneously depress both the control key and that character):
    - **Ctrl+F** inserts the fraction builder.
    - **Ctrl+/** inserts the in-line slash fraction builder.
    - **Ctrl+9** inserts parentheses.
    - **Ctrl+H** allows you to type in superscripts (**Tab** puts you back on the main line, as do the arrow keys).
    - **Ctrl+L** allows for subscripts (**Tab** puts you back).
    - **Ctrl+R** inserts a square root builder.
    - **Ctrl+I** inserts an definite integral builder.
- Here are a few more useful shortcuts in which you first simultaneously depress the Control key and "T", let go, and then depress another character. This is represented as **Ctrl+T, character.**
    - **CTRL+T, S** inserts the summation notation builder.
    - **CTRL+T, M** inserts a $3 \times 3$ matrix.
    - **CTRL+T, U** inserts an underscript, which is very useful with limit notation.
- Look for more shortcuts by visiting the Equation Editor Help menu and typing "shortcuts" into the index dialogue box.

### Templates

- When creating reports, working from a pre-structured template can save time. Under the "File" menu, select "New" to bring up an option window on the right-hand-side of your workspace. Select either "Templates on Office Online" or "On my computer…" for free report templates to help you get started with professional looking reports.
- Make sure to modify the report somewhat to give it your team's personal touch.

## Useful Microsoft *Excel* Features

### General Helpful Hints

- Highlighting a cell reference in a formula and pressing the **F4** function key will automatically render the reference absolute; i.e., **F4** will change a cell reference such as A1 to \$A\$1.

- If you have set the calculation process to manual calculation (through the "Tools" pull-down menu and "Options…" calculation tab), then press the **F9** function key to instigate calculations. You should also uncheck "recalculate before saving" so *Excel* does not change the values.

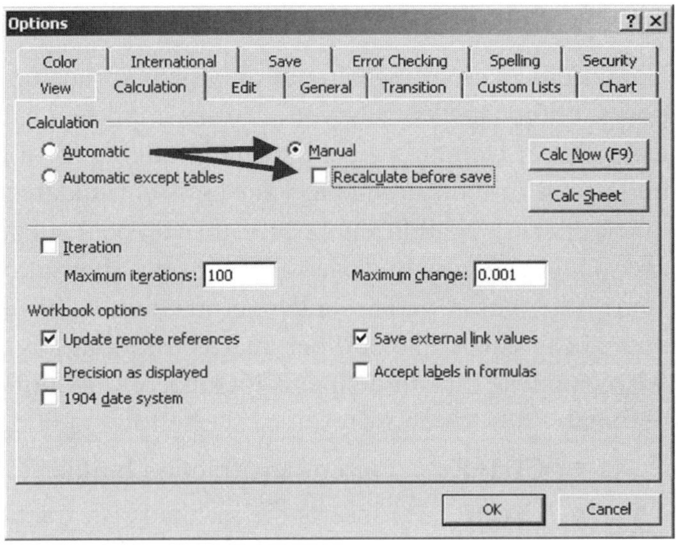


- Use the **CTRL + SHIFT** and the arrow keys (as well as the END and HOME keys) to help select large ranges of data instead of scrolling with the mouse, since it is often hard to control the scrolling speed.
- The **Go To** command (under the "Edit" pull down menu) is also useful for navigating within large spreadsheets.
- Quickly generating a column of values with a prescribed start and stop point will be useful for many problems in Mathematics for Business Decisions! The "Fill" sub-menu under the "Edit" pull-down menu is used for this. Click "Series" and a dialogue box will appear. Enter the information shown to the right to create a column of numbers starting at one and ending at five thousand, in steps of size one.

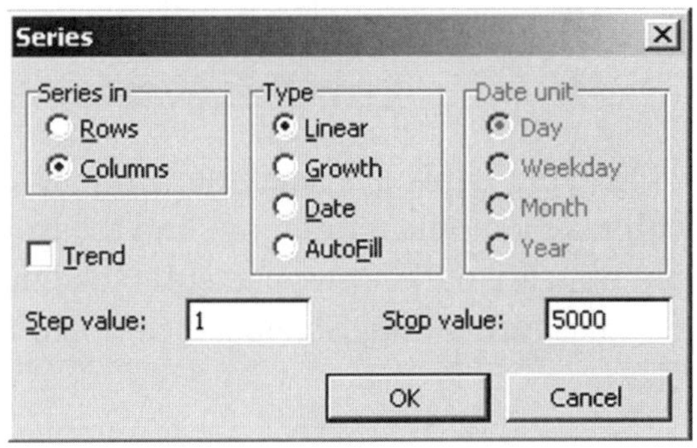


- Right-clicking an object usually gives options designed particularly for that object.
- Right-click a chart and select *Chart Options* to change the chart's title, gridlines, axes, data labels, and move the legend. Select *Format Chart Area* to format the background area and font used. *Source Data* allows you to change the range of data selected, add more graphs, or detail the legend. *Chart Window* is nice if you want a full-page printout of your graph.
- Right-click the plot area itself to format the plot area; the curve to format the curve or add a trend line to a scatter plot; the axis to change the scale, etc.
- Order of operation is just as important when using Microsoft *Excel* as it is with any calculation tool. You may know that $\frac{x+1}{x+8}$ means divide the sum $x+1$ by the sum $x+8$ but, in *Excel*, you enter formulas *in-line*. Thus $x+1/x+8$ will be interpreted by *Excel* as $x+(1/x)+8$ or $x+\frac{1}{x}+8$. Use parentheses to avoid such misinterpretations. $\frac{x+1}{x+8}$ should be entered as $(x+1)/(x+8)$
- Similarly, $5^{x+1}$ typed in-line becomes $5\^(x+1)$ *not* $5\^x+1$. The latter expression is equivalent to $5^x+1$.

## Conditional Formatting

Conditional formatting can not only help spruce up your Microsoft *Excel* files for reporting purposes, but it can also help you identify important calculations. For example, suppose that you are using *Excel* to monitor complicated budgets for a department that you are managing. As your units send in their numbers, your assistant updates your files and sends them over to you for a quick check. Wouldn't it be nice if overspent accounts were automatically formatted in red and accounts dangerously close to going over their allocated funds were automatically formatted in some other color, perhaps a dangerous looking shade of pink? Suppose that Division #1's budget is not to exceed \$5,000 per quarter and that Division #5 is not to exceed \$10,000 per quarter. Also, suppose that all divisions together should not exceed \$25,000. per quarter and \$100,000.00 per year. You have just received the following budget update:

| **Department Budget** | | | | | |
|---|---|---|---|---|---|
| | 1st Quarter | 2nd Quarter | 3rd Quarter | 4th Quarter | Division Totals |
| Division#1 | \$4,998.00 | \$4,789.00 | \$5,023.00 | \$4,920.00 | \$19,730.00 |
| Division#2 | \$6,200.00 | \$6,298.00 | \$6,100.00 | \$5,987.00 | \$24,585.00 |
| Division#3 | \$1,852.00 | \$1,997.00 | \$1,856.00 | \$1,945.00 | \$7,650.00 |
| Division#4 | \$1,978.00 | \$2,100.00 | \$2,200.00 | \$2,178.00 | \$8,456.00 |
| Division#5 | \$9,900.00 | \$9,999.00 | \$9,789.00 | \$9,745.00 | \$39,433.00 |
| Quarter Totals | \$24,928.00 | \$25,183.00 | \$24,968.00 | \$24,775.00 | \$99,854.00 |

Granted, this is a small department and an extremely simplified budgetary view, but it would still take a moment or two to find the trouble-makers. This manual is not in color, so both red and hot pink are out, but we can still make the problems quickly stand out automatically through conditional formatting. Shoppers, compare the above boring and unformatted chart, to the following chart:

| **Department Budget** | | | | | |
|---|---|---|---|---|---|
| | 1st Quarter | 2nd Quarter | 3rd Quarter | 4th Quarter | Division Totals |
| Division#1 | \$4,998.00 | \$4,789.00 | ***\$5,023.00*** | \$4,920.00 | \$19,730.00 |
| Division#2 | \$6,200.00 | \$6,298.00 | \$6,100.00 | \$5,987.00 | \$24,585.00 |
| Division#3 | \$1,852.00 | \$1,997.00 | \$1,856.00 | \$1,945.00 | \$7,650.00 |
| Division#4 | \$1,978.00 | \$2,100.00 | \$2,200.00 | \$2,178.00 | \$8,456.00 |
| Division#5 | \$9,900.00 | \$9,999.00 | \$9,789.00 | \$9,745.00 | \$39,433.00 |
| Quarter Totals | \$24,928.00 | ***\$25,183.00*** | \$24,968.00 | \$24,775.00 | \$99,854.00 |

Okay, it is not that much more exciting, but the dollar amounts that have exceeded their respective allocations are more easily detected. Conditional formatting allows the user to specify formatting properties dependent upon conditions specified on the contents of a cell. In the above example, the cells containing Division #1's quarterly totals were formatted with a bold italic font and grey fill if the dollar amount exceeded the limit of \$5,000 for the division. Similarly, the quarter total cells were formatted in the same way if the dollar amount exceeded the limit of \$25,000.

Closer inspection shows that, on at least a couple of occasions, certain divisions came dangerously close to their limits. We can add additional criteria to highlight these dangerously close situations.

| **Department Budget** | | | | | |
|---|---|---|---|---|---|
| | 1st Quarter | 2nd Quarter | 3rd Quarter | 4th Quarter | Division Totals |
| Division#1 | $4,998.00 | $4,789.00 | $5,023.00 | $4,920.00 | $19,730.00 |
| Division#2 | $6,200.00 | $6,298.00 | $6,100.00 | $5,987.00 | $24,585.00 |
| Division#3 | $1,852.00 | $1,997.00 | $1,856.00 | $1,945.00 | $7,650.00 |
| Division#4 | $1,978.00 | $2,100.00 | $2,200.00 | $2,178.00 | $8,456.00 |
| Division#5 | $9,900.00 | $9,999.00 | $9,789.00 | $9,745.00 | $39,433.00 |
| Quarter Totals | $24,928.00 | $25,183.00 | $24,968.00 | $24,775.00 | $99,854.00 |

Now cells that contain dollar amounts within $50 of their respective limits stand out as well. For your spreadsheets, you should probably use colors as opposed to patterns. Cell formatting is easily accomplished by performing the following steps:

- Under the "Format" pull-down menu, select "Conditional formatting…"; you may have to use the menu expansion arrows at the bottom of the menu.
- The dialogue box below will appear. Enter the desired conditions with regard to the contents of the cell, click "Format…", chose your conditional formatting options, and click "OK".
- As in the above example, you may have more than one condition for which you want to impose a conditional formatting protocol. Notice the "Add>>" button at the bottom of the dialogue box? Yes, it really is as easy as that! Add multiple conditions using the same or different associated formatting options.

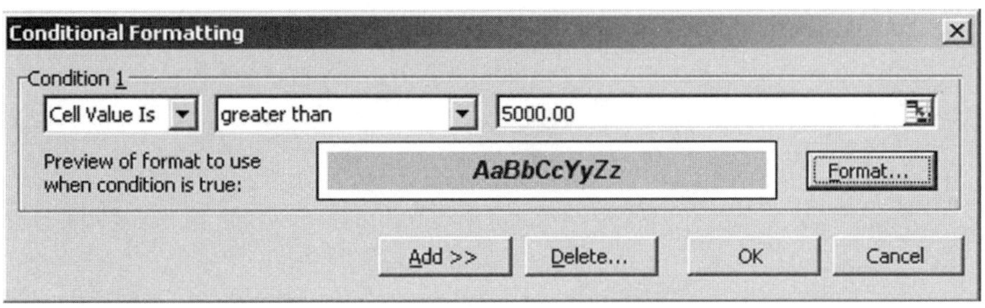


- Note that the full set of formatting options is not available through conditional formatting, but you should have enough options to make the very good, very bad, and somewhat worrisome values all stand out.

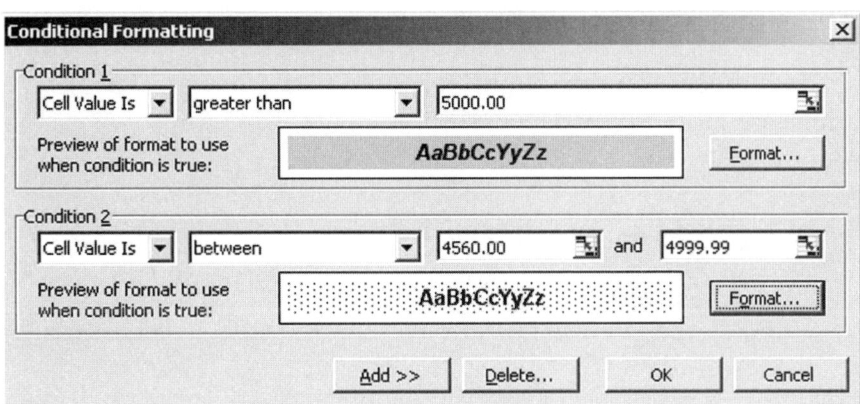

# Technology Tips & Tricks
## (Featuring Microsoft *Office* 2007)

## Basic Microsoft *PowerPoint* Features

### Getting Started

- Click 

.
- Move the pointer to **All Programs** and wait for the submenu to appear.
- Move the pointer to the Microsoft Office folder and select Microsoft Office PowerPoint 2007.
- Wait for Microsoft *PowerPoint* to open. Your window should look similar to the window below.

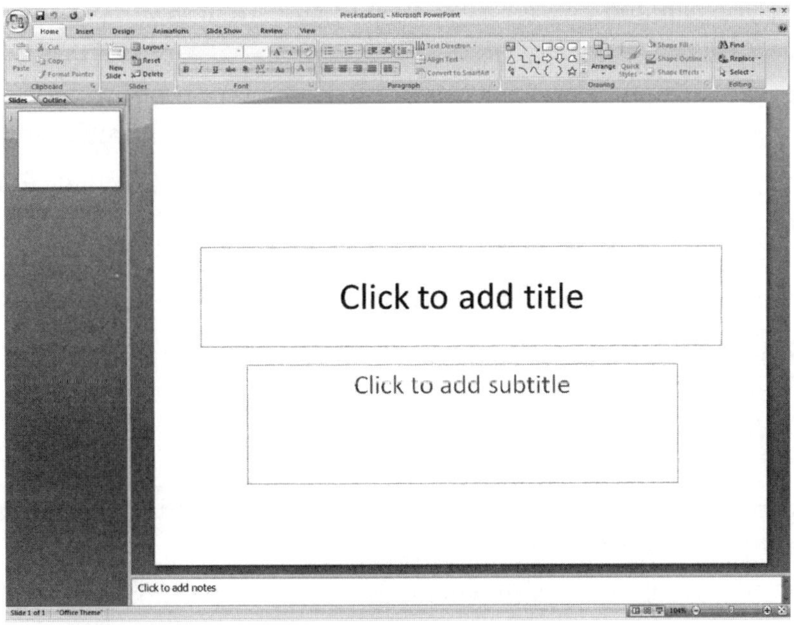


### Creating a Basic Presentation

- Click the **Design** tab and choose a theme from the selection that appears in the **Themes** group.

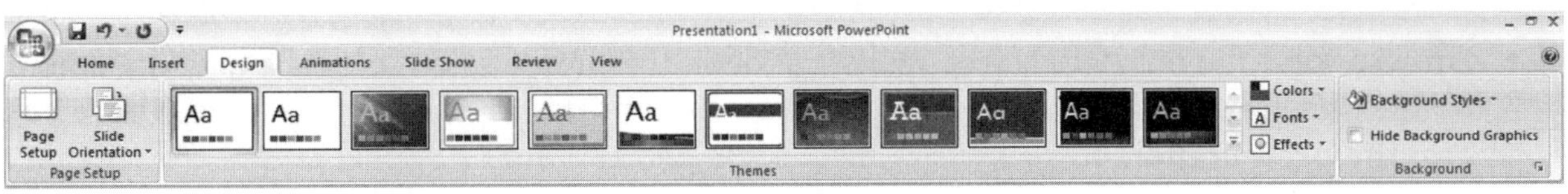

You may also download additional themes from Microsoft or browse for a previously saved theme.

- Once you have selected a theme, click the **Home** tab and then click **Layout**.

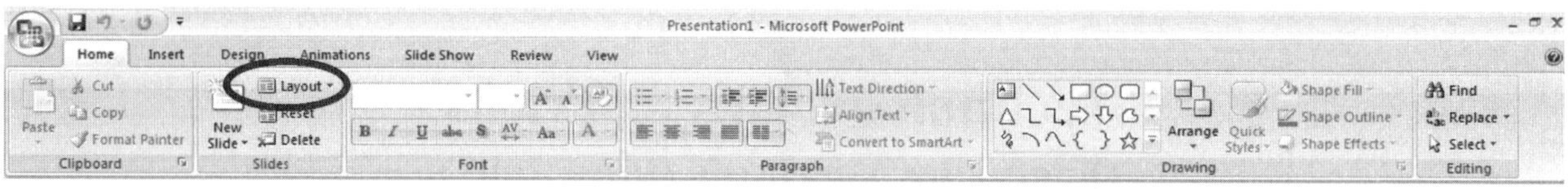


- Select a slide layout for your first slide. You can change your layout for each slide or duplicate a slide for more copies of the same layout.
- Add text (just click and type!) to your slides.
- Don't forget to save your work frequently.

## Adding Pictures and Changing the Format or Background

- To add pictures, click the **Insert** tab and then click **Picture**.

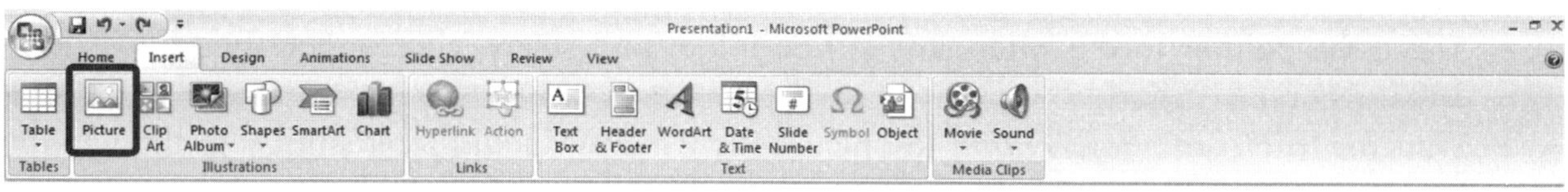


- Notice that you can add clip art as well as pictures that you have saved as files (such as a bitmap, gif, or jpeg files).
- Also note that you can select a layout that has a pre-set spot for a picture or you can insert a picture essentially anywhere you want using the drag and drop method.
- Commands on the **Home** tab allow you to change the slide layout, font, and line spacing; commands on the **Design** tab allow you to change the slide color scheme and background style.
    - Note that changes can be made to take effect at many levels.
    - To change the look of all of the slides at once, use the **Slide Master** command on the **View** tab.

## Adding Links

- Links can be made to existing files, to the Internet, from one slide to another within a *PowerPoint* presentation, or to new documents.
- Select the object or text that you wish to become a hyperlink.
- Click the **Insert** tab and then click **Hyperlink**.

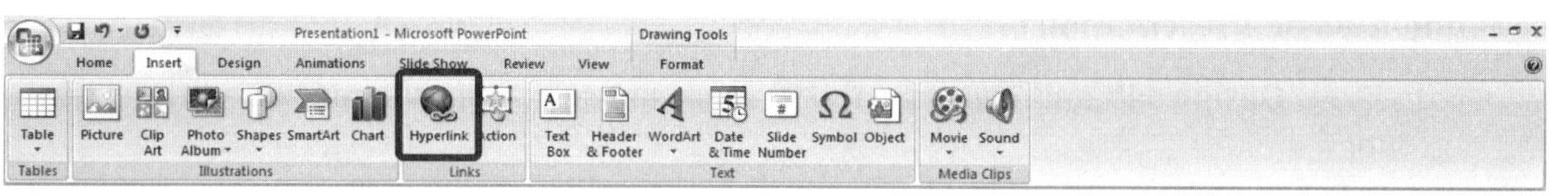


- Browse for a file or URL, select another slide in the presentation, create a new document, or enter an email address. **Make sure that all linked files are accessible from where you run the presentation.**

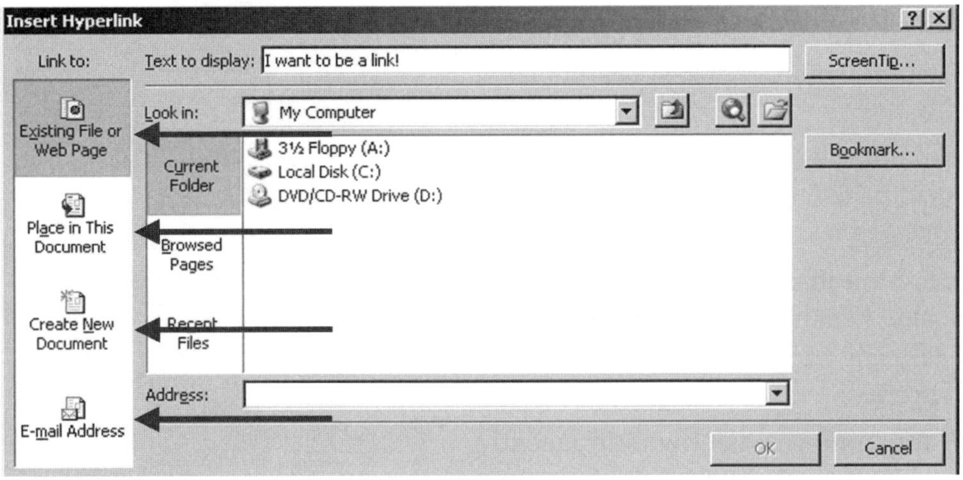


- Warning! Linking large Microsoft *Excel* files can freeze the presentation or take several minutes to load. This is quite embarrassing, especially during timed presentations. If you have *Excel* graphs or clips that you wish to show, then you should use the **Paste Special** command on the **Home** tab to paste them in as pictures or bitmaps instead of linked files.

### Customizing Animation

- Select the object that you wish to animate.
- Click the **Animations** tab and then click **Custom Animation**.

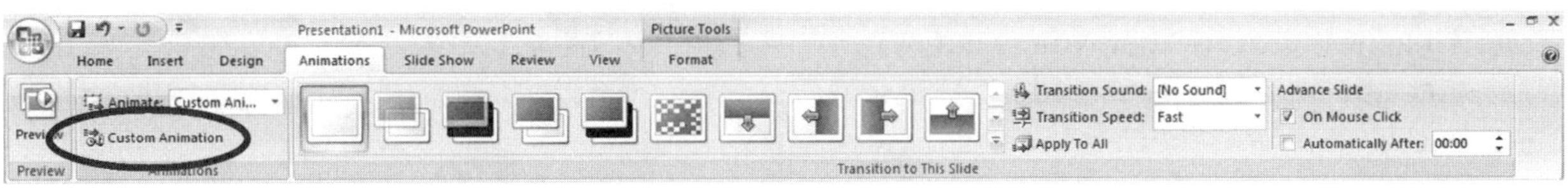

- Click **Add Effect** in the **Custom Animation** task pane.
- Once you have selected your effect, additional options for that particular effect, such as speed, direction, and order of animation, will be available.
- You can also specify whether you want the animation to occur automatically, after a preset amount of time, or only by mouse-click.
- Warning! Too many animations can make your professional presentation look more like a cheap commercial. Although visuals can spruce up a dry presentation, you should be judicious in your selection of visuals and animations. Remember — not every slide needs something whipping in or wiping out!
- Also, for the most part, it is best to avoid sound effects. Canned clapping noises may work well in sitcoms, but they usually fall flat in professional presentations.

### Printing Options

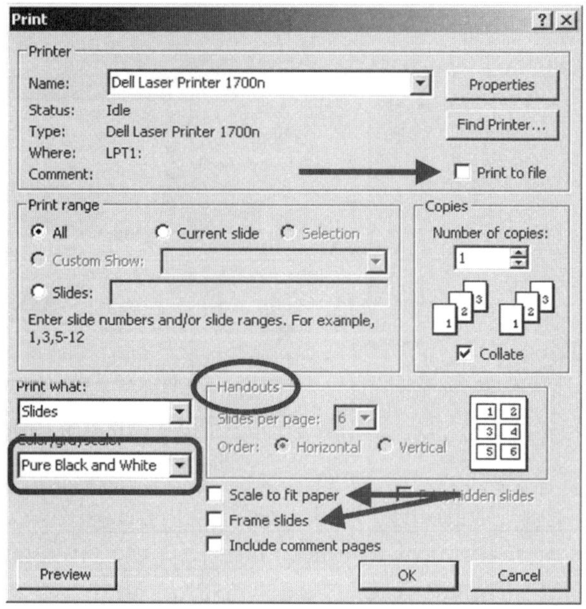


- You can print all of your slides or just a selection.
- You can print one slide per page or several slides per page or a handout on which your audience can take notes.
- You can print in color, black and white, or grayscale. Printing in black and white can save ink and is useful for people with inkjet printers.
- There are numerous other options that you can explore as well, such as framing your slides, scaling them to fit an entire page, and only printing selected slides.

### Miscellaneous Pointers

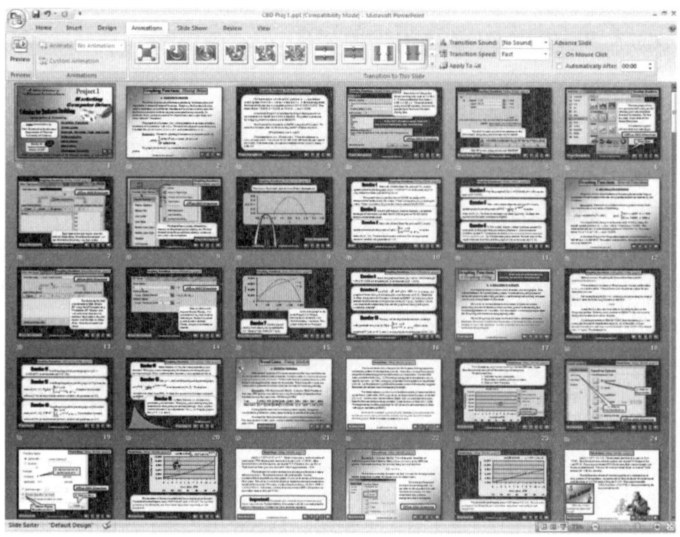

- At the bottom right of your *PowerPoint* work area, you should see a set of icons: . The first icon selects the "normal" view of your work area. This is the view that you want when you are building your slides. The second icon selects the "slide sorter" view. This is useful for organizing your slides, just click, drag, and move a slide wherever you would like it to appear. The third icon selects the slide show view. This allows you to preview your presentation.
- You can create charts in *PowerPoint*, but it is probably better to use Microsoft *Excel*. Use the **Paste Special** command on the **Home** tab to insert your charts as bitmaps (as opposed to linked files) — remember, you don't want to be standing up in front of an expectant audience with a frozen *PowerPoint* presentation.
- You can add equations by copying and pasting equations that you have created in Microsoft *Word*.

## Useful Microsoft *Word* Features

### Using the Microsoft *Equation Builder*

- The Microsoft *Equation Builder* is accessed from the **Insert** tab by clicking **Equation**.

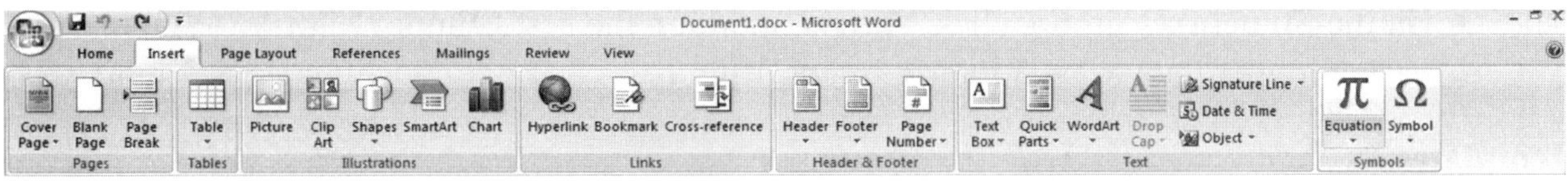


- On the **Design** tab, under **Equation Tools**, you will find many symbols and structures that you will need to word-process your homework and your reports.

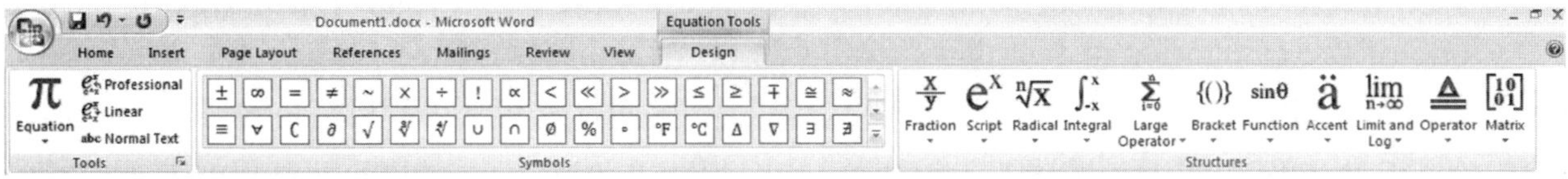


The object that appears in your document is where the symbols will appear once selected.

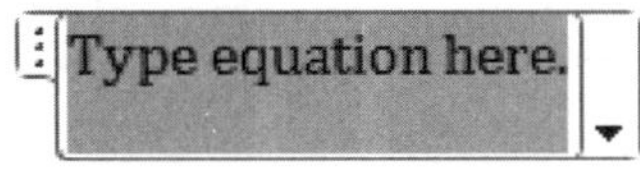


**Two Special Cases**

- Piecewise functions can be a pain! This is especially true when it comes time to word-process them. Suppose that you want to word-process the following piecewise defined function:

$$f(x)=\begin{cases} x, & \text{if } x>0 \\ x^2, & \text{if } x\le 0 \end{cases}.$$

A simple way to proceed is as follows:

  - Launch the equation builder.
  - Type "$f(x)=$" using the keyboard.
  - In the **Structures** group, click **Bracket** and then the single left bracket.
  - In the **Structures** group, click **Matrix** and then the $2\times 2$ Empty Matrix.
  - Insert the four entries, $x$, $x^2$, if $x>0$, and if $x\le 0$, each in their own cell.
  - Voilà! You have a perfectly aligned piecewise function.

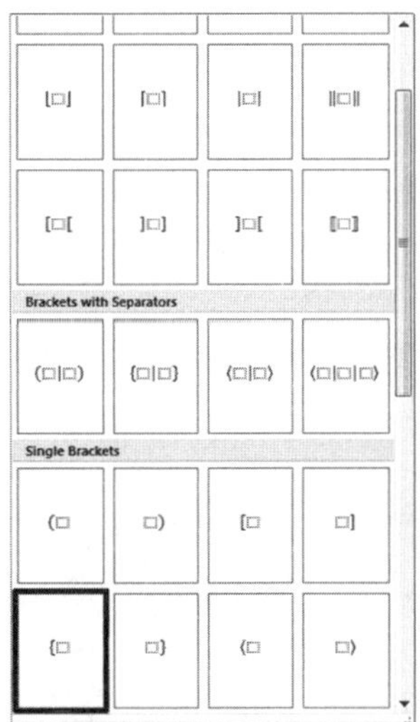


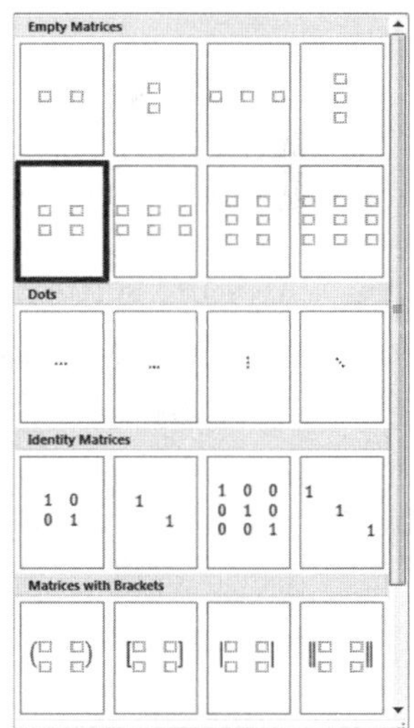


- Sample spaces can also provide word-processing challenges. Suppose that you want to word process the sample space given below.

$$\begin{Bmatrix} (1,1) & (1,2) & (1,3) \\ (2,1) & (2,2) & (2,3) \\ (3,1) & (3,2) & (3,3) \end{Bmatrix}$$

  - Launch the equation builder.
  - In the **Structures** group, click **Bracket** and then the double brackets, "{}".

- In the **Structures** group, click **Matrix** and then the 3 × 3 Empty Matrix. (You can construct higher than a 3 × 3 matrix by inserting rows and columns into a 3 × 3 matrix.)

### Templates

- When creating reports, working from a pre-structured template can save time. Click the **Microsoft Office Button** and then click **New** to view options that can be used to create a document from a template. Select either **Installed Templates** or one of the categories of templates available on **Microsoft Office Online** for free report templates to help you get started with professional looking reports.
- Make sure to modify the report somewhat to give it your team's personal touch.

## Useful Microsoft *Excel* Features

### General Helpful Hints

- Highlighting a cell reference in a formula and pressing the **F4** function key will automatically render the reference absolute; i.e., **F4** will change a cell reference such as A1 to $A$1.
- If you have set the calculation process to manual calculation (through the **Calculation Options** command on the **Formulas** tab), then press the **F9** function key to instigate calculations.

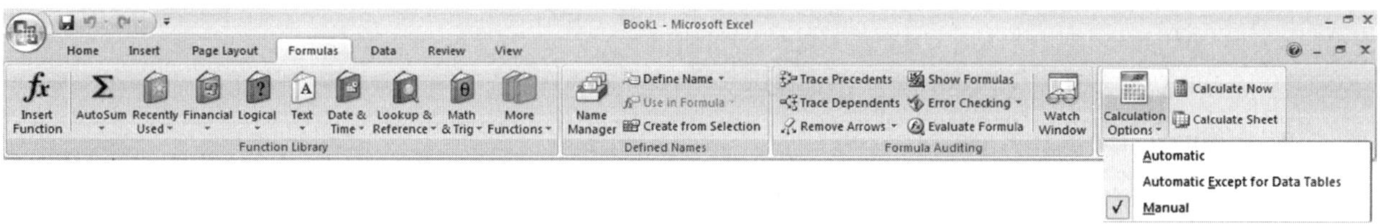


- Use the **CTRL + SHIFT** and the arrow keys (as well as the END and HOME keys) to help select large ranges of data instead of scrolling with the mouse, since it is often hard to control the scrolling speed.
- The **Find & Select** command on the **Home** tab is also useful for navigating within large spreadsheets.

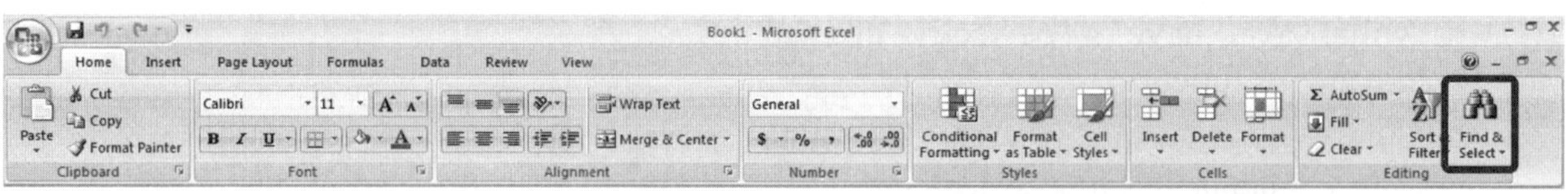


- Quickly generating a column of values with a prescribed start and stop point will be useful for many problems in Mathematics for Business Decisions! The **Fill** command on the **Home** tab is used for this. Click **Series** and a dialog box will appear. Enter the information shown to the right to create a column of numbers starting at one and ending at five thousand, in steps of size one.

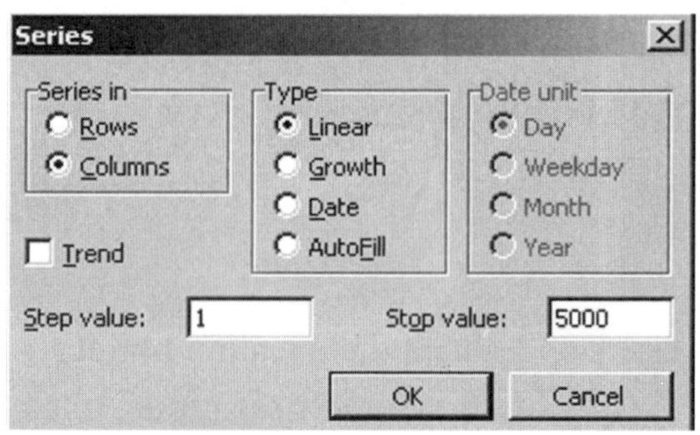

- Right-clicking an object usually gives options designed particularly for that object.
- Right-click a chart and click the appropriate command on the **Mini toolbar** to format the background, border, and font.
- Right-click the plot area itself to format the plot area; the curve to format the curve or to add a trend line to a scatter plot; the axis to change the scale; etc.
- Order of operation is just as important when using Microsoft *Excel* as it is with any calculation tool. You may know that $\frac{x+1}{x+8}$ means divide the sum $x+1$ by the sum $x+8$ but, in *Excel*, you enter formulas *in-line*. Thus $x+1/x+8$ will be interpreted by *Excel* as $x+(1/x)+8$ or $x+\frac{1}{x}+8$. Use parentheses to avoid such misinterpretations. $\frac{x+1}{x+8}$ should be entered as $(x+1)/(x+8)$
- Similarly, $5^{x+1}$ typed in-line becomes 5^$(x+1)$ *not* 5^$x+1$. The latter expression is equivalent to $5^x+1$.

## Conditional Formatting

Conditional formatting can not only help spruce up your Microsoft *Excel* files for reporting purposes, but it can also help you identify important calculations. For example, suppose that you are using *Excel* to monitor complicated budgets for a department that you are managing. As your units send in their numbers, your assistant updates your files and sends them over to you for a quick check. Wouldn't it be nice if overspent accounts were automatically formatted in red and accounts dangerously close to going over their allocated funds were automatically formatted in some other color, perhaps a dangerous looking shade of pink? Suppose that Division #1's budget is not to exceed \$5,000 per quarter and that Division #5's budget is not to exceed \$10,000 per quarter. Also, suppose that all divisions together should not exceed \$25,000 per quarter and \$100,000 per year. You have just received the following budget update:

| **Department Budget** | | | | | |
|---|---|---|---|---|---|
| | 1st Quarter | 2nd Quarter | 3rd Quarter | 4th Quarter | Division Totals |
| Division#1 | $4,998.00 | $4,789.00 | $5,023.00 | $4,920.00 | $19,730.00 |
| Division#2 | $6,200.00 | $6,298.00 | $6,100.00 | $5,987.00 | $24,585.00 |
| Division#3 | $1,852.00 | $1,997.00 | $1,856.00 | $1,945.00 | $7,650.00 |
| Division#4 | $1,978.00 | $2,100.00 | $2,200.00 | $2,178.00 | $8,456.00 |
| Division#5 | $9,900.00 | $9,999.00 | $9,789.00 | $9,745.00 | $39,433.00 |
| Quarter Totals | $24,928.00 | $25,183.00 | $24,968.00 | $24,775.00 | $99,854.00 |

Granted, this is a small department and an extremely simplified budgetary view, but it would still take a moment or two to find the trouble-makers. This manual is not in color, so both red and hot pink are out, but we can still make the problems quickly stand out automatically through conditional formatting. Shoppers, compare the above boring and unformatted chart, to the following chart:

| **Department Budget** | | | | | |
|---|---|---|---|---|---|
| | 1st Quarter | 2nd Quarter | 3rd Quarter | 4th Quarter | Division Totals |
| Division#1 | $4,998.00 | $4,789.00 | ***$5,023.00*** | $4,920.00 | $19,730.00 |
| Division#2 | $6,200.00 | $6,298.00 | $6,100.00 | $5,987.00 | $24,585.00 |
| Division#3 | $1,852.00 | $1,997.00 | $1,856.00 | $1,945.00 | $7,650.00 |
| Division#4 | $1,978.00 | $2,100.00 | $2,200.00 | $2,178.00 | $8,456.00 |
| Division#5 | $9,900.00 | $9,999.00 | $9,789.00 | $9,745.00 | $39,433.00 |
| Quarter Totals | $24,928.00 | ***$25,183.00*** | $24,968.00 | $24,775.00 | $99,854.00 |

Okay, it is not that much more exciting, but the dollar amounts that have exceeded their respective allocations are more easily detected. Conditional formatting allows the user to specify formatting properties dependent upon conditions specified on the contents of a cell. In the above example, the cells containing Division #1's quarterly totals were formatted with a bold italic font and grey fill if the dollar amount exceeded the limit of $5,000 for the division. Similarly, the quarter total cells were formatted in the same way if the dollar amount exceeded the limit of $25,000.

Closer inspection shows that, on at least a couple of occasions, certain divisions came dangerously close to their limits. We can add additional criteria to highlight these dangerously close situations.

| **Department Budget** | | | | | |
|---|---|---|---|---|---|
| | 1st Quarter | 2nd Quarter | 3rd Quarter | 4th Quarter | Division Totals |
| Division#1 | **$4,998.00** | $4,789.00 | ***$5,023.00*** | $4,920.00 | $19,730.00 |
| Division#2 | $6,200.00 | $6,298.00 | $6,100.00 | $5,987.00 | $24,585.00 |
| Division#3 | $1,852.00 | $1,997.00 | $1,856.00 | $1,945.00 | $7,650.00 |
| Division#4 | $1,978.00 | $2,100.00 | $2,200.00 | $2,178.00 | $8,456.00 |
| Division#5 | $9,900.00 | **$9,999.00** | $9,789.00 | $9,745.00 | $39,433.00 |
| Quarter Totals | $24,928.00 | ***$25,183.00*** | **$24,968.00** | $24,775.00 | $99,854.00 |

Now cells that contain dollar amounts within $50 of their respective limits stand out as well. For your spreadsheets, you should probably use colors as opposed to patterns. Cell formatting is easily accomplished by performing the following steps:

- Click the **Home** tab, then click **Conditional Formatting**, then click **Highlight Cells Rules**, and then click **More Rules**.

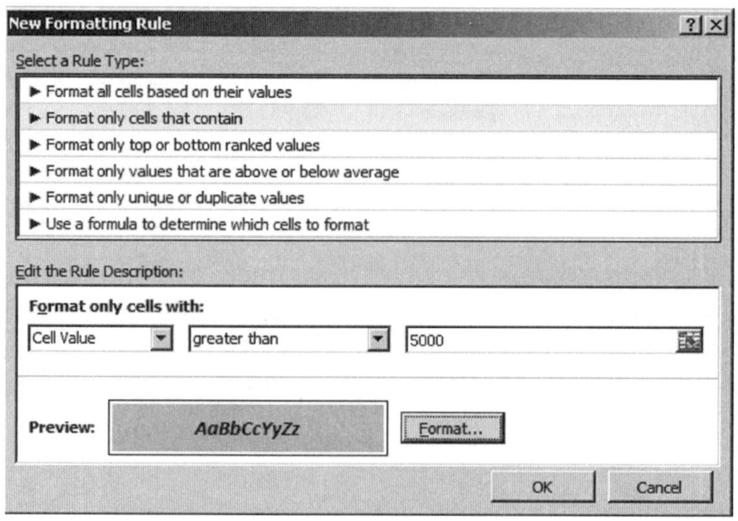


- The dialogue box to the right will appear. Enter the desired conditions with regard to the contents of the cell, click **Format**, chose your conditional formatting options, and then click **OK**.
- As in the above example, you may have more than one condition for which you want to impose a conditional formatting protocol. Repeat the previous steps to add another condition using the same or different associated formatting options.

- Note that the full set of formatting options is not available through conditional formatting, but you should have enough options to make the very good, very bad, and somewhat worrisome values all stand out.

# Project 1: Marketing Computer Drives

## Business Background

One of the most important decisions made by companies is the production decision—determining how many units of each product to produce and sell. This decision is affected by the structure of the market in which the firm operates. In particular, the structure of the market determines the degree of control that companies have over prices.

Perfectly competitive markets are characterized by a large number of buyers and sellers, easy entry and exit, and a totally undifferentiated product. Because the number of participants is large and because the product produced by any one company is indistinguishable from that produced by any other company, individual companies have little or no control over price. If a company attempts to increase the price of the product, then that company will lose customers to its competitors and will therefore lose revenue. There is also no incentive to decrease the price of the product, because a reduction in price will also result in the loss of potential revenue. The focus of the production decision for companies operating in this type of market is on output.

Monopolies are characterized by the presence of only one seller and significant barriers to entry. Such deterrents may include control over vital raw materials or specialized labor, economies of scale, product differentiation, and legal restrictions. If crucial raw materials and skilled labor are not available to prospective competitors, then those companies are prohibited from entering the market. High fixed costs must be spread over a large number of units in order for the venture to be profitable. Thus, it may not be economical for competitors to enter the market. If the perceived quality of a company's product is high due to, say, extensive advertising, then it may be difficult for other companies to compete. Legal restrictions such as patent protection also serve to restrict competition. Such protection is typically granted in order to encourage investment in capital intense product development. Due to the lack of competition, a monopolist has considerable freedom in making pricing and output decisions.

The demand curve for a product shows the relationship between the price of the product and the quantity demanded. A monopolist is faced with a downward sloping demand curve—the quantity demanded decreases as the price increases. The particular shape of the demand curve is a function of consumer preferences, income, prices of other goods, etc. One technique that is used to forecast the demand for a new product is test marketing. The company introduces the product in a few selected markets in order to examine variations in the product, in the marketing strategy, and in prices. The results are then used to predict national sales. This technique is extremely useful, albeit expensive and time-consuming.

When deciding how many units of a product to produce and sell, cost must be considered as well as demand. The cost curve shows the relationship between the total cost of producing a product and the number of units produced. Total cost is comprised of fixed costs and variable costs. Fixed costs include the costs of the production facility and the equipment and are incurred whether or not any units are produced. Variable costs include the costs of raw materials and direct labor and are incurred on a per unit basis.

When a new product is developed, the production and engineering departments typically provide estimates of these costs for various levels of production. Generally the total variable cost of production increases as the number of units produced increases, and therefore, so does the total cost of production.

The profit received from the production and sale of a product is equal to the difference between revenue, the product of price and quantity sold, and costs. One of the primary objectives of any for-profit organization is to maximize profit. It is with this objective in mind that the production decision is made.

Please visit `http://www.quickmba.com` for additional information about the production decision.

## Graphing Functions

(Featuring Microsoft *Excel* 2003)

This section will specifically help you to graph functions using Microsoft *Excel*. In addition you will learn how to graph functions using the utility, ***Graphing.xls***, which is one of your course files.

### *Basic Functions*

Using Microsoft *Excel* to create graphs is fun and easy to do. In the following example you will see how to graph simple functions.

**Example 1:** Plot the quadratic function $f(x) = -0.5x^2 + 1$ over the interval $[-4, 4]$.

**SOLUTION**

*Step 1*: Create a table in *Excel* with the same rows and columns as shown to the right. Column A contains the $x$-values (inputs) and Column B contains the corresponding values of $f(x)$ (outputs).

| | A | B |
|---|---|---|
| 1 | $x$ | $f(x)$ |
| 2 | -4 | -7 |
| 3 | -3 | -3.5 |
| 4 | -2 | -1 |
| 5 | -1 | 0.5 |
| 6 | 0 | 1 |
| 7 | 1 | 0.5 |
| 8 | 2 | -1 |
| 9 | 3 | -3.5 |
| 10 | 4 | -7 |

*Step 2*: Plot the function. To open the **Chart Wizard**, pull down the **Insert** menu and select **Chart**, or simply click the **Chart Wizard** button on the **Standard Toolbar**. In the **Chart type** section of the first page of the **Chart Wizard**, click **XY (Scatter)**; and in the **Chart sub-type** section, click the subtype *Scatter with data points connected by smoothed Lines without markers*. Then click **Next** to advance to the second page of the **Chart Wizard**.

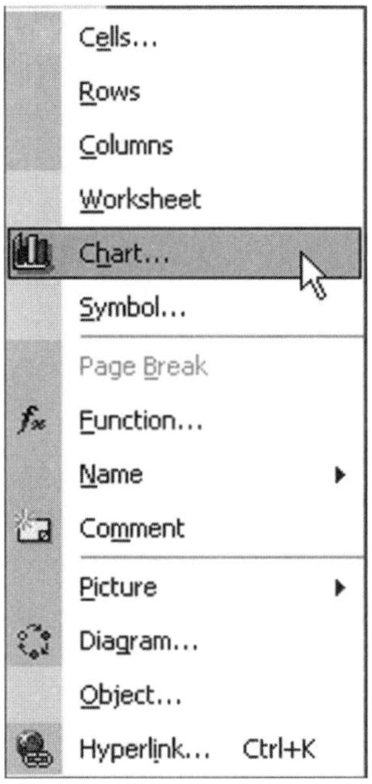


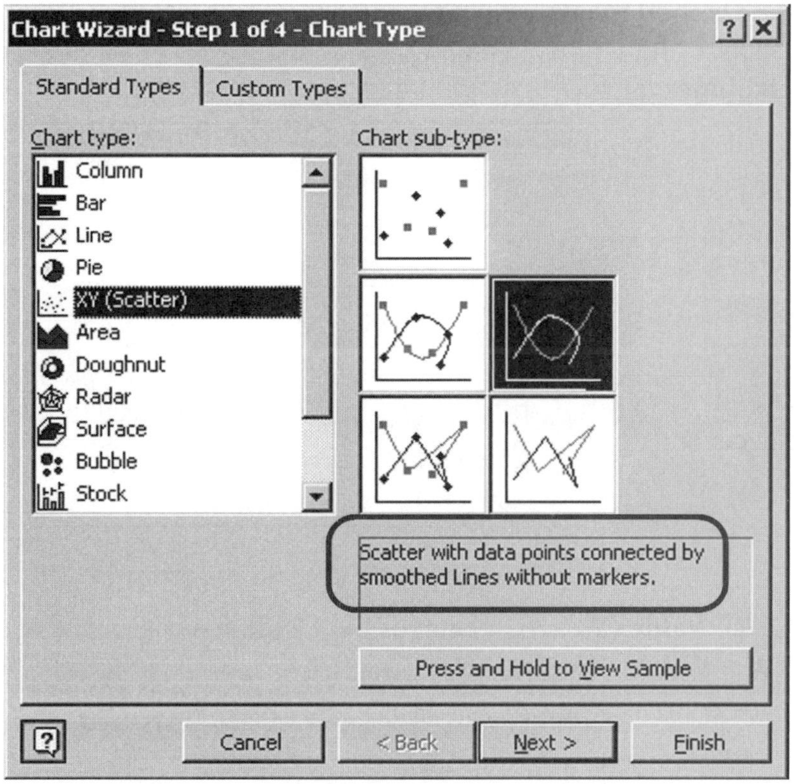


At this point, you need to indicate which cells contain the $x$-values (inputs) and which cells contain the values of $f(x)$ (outputs). To define the outputs, click in the **Data Range** box and select *Cells B2:B10*. Then click the **Series** tab. To define the inputs, click in the **X Values** box and select *Cells A2:A10*. Then click **Next** to advance to the third page of the **Chart Wizard**.

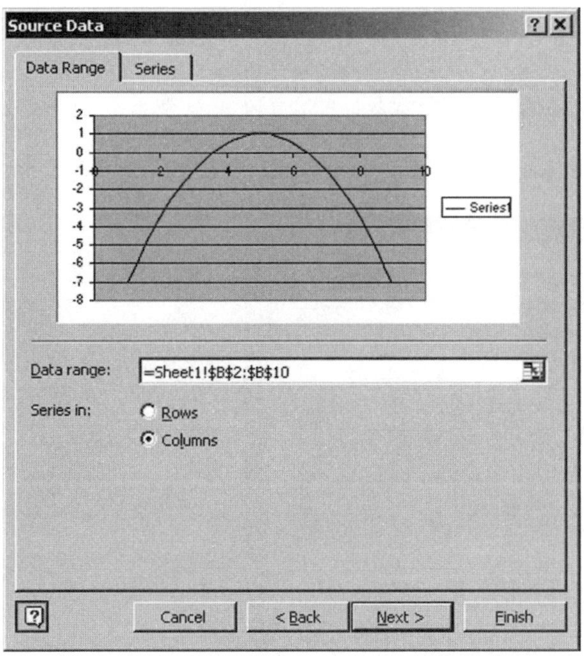


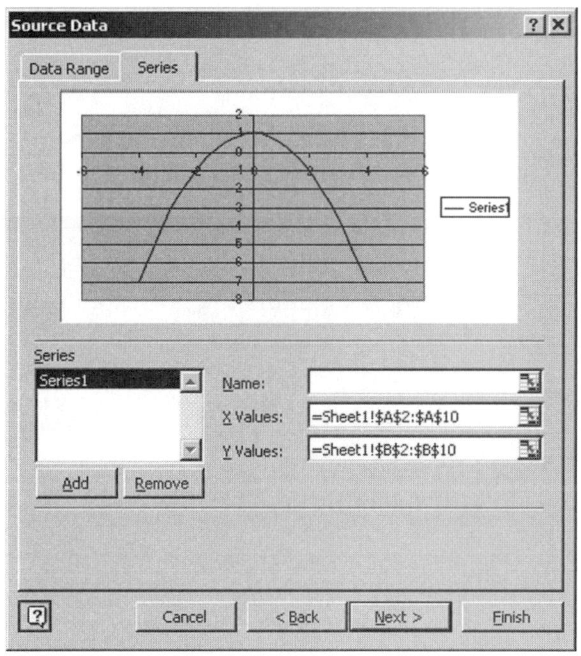

On the **Titles** tab, enter the title QUADRATIC FUNCTION in the **Chart title** box and the labels $x$ and $y$ in the **Value (X) axis** and **Value (Y) axis** boxes, respectively. Then click the **Gridlines** tab.

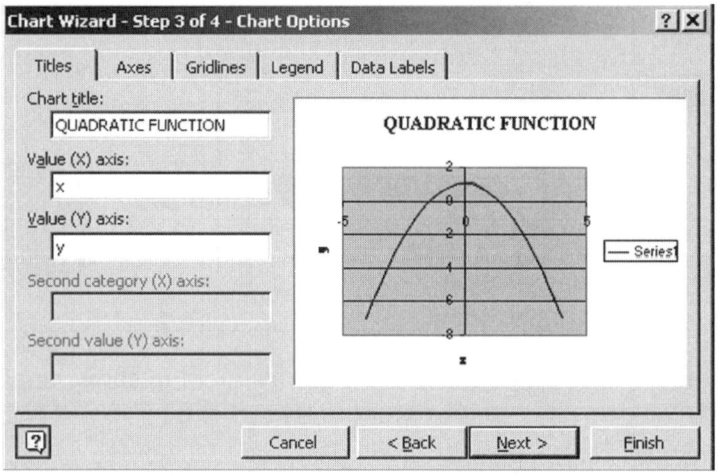


In the **Value (Y) axis** section of the **Gridlines** tab, uncheck the **Major gridlines** check box. Then click the **Legend** tab.

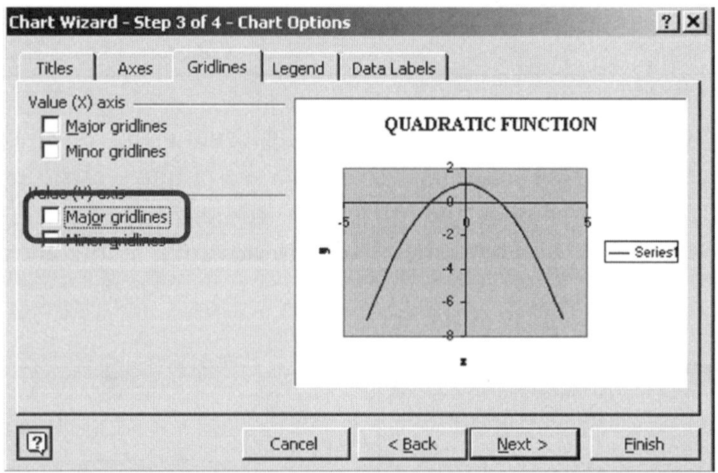


Uncheck the **Show legend** check box and click **Finish**.

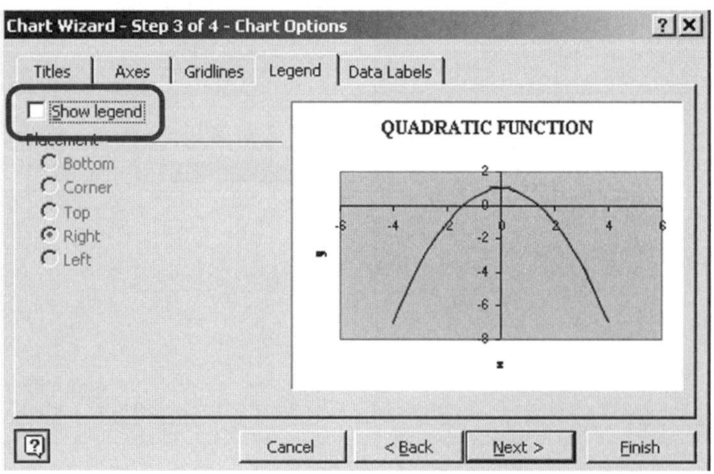

Your graph should look similar to the one shown below.

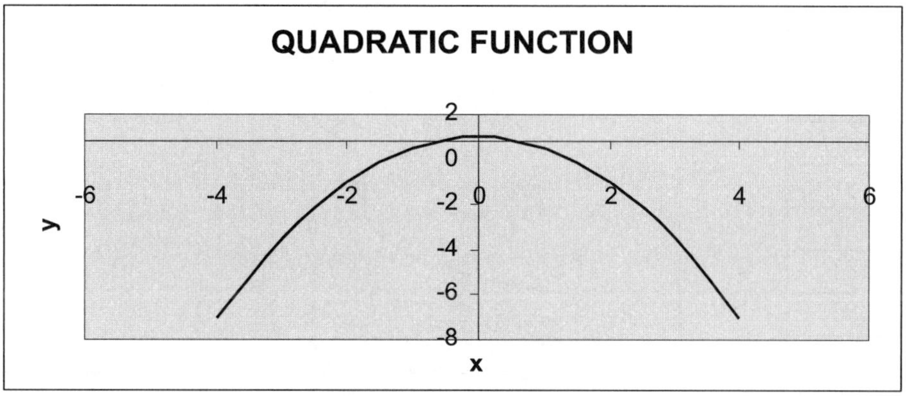


The number of points plotted in **Example 1** was sufficient to show an accurate representation of the function. However, there are many functions that require a large number of data points in order to obtain an accurate representation of the function. The next two graphs are of the same function, $f(x)=\sqrt[3]{x+1}$, and are plotted over the same interval, [–5, 5].

In **GRAPH 1**, 11 equally spaced points are plotted. To compute the difference between the $x$-coordinates of consecutive points, the left endpoint of the interval [–5, 5] is subtracted from the right endpoint, and the result is divided by 10.

$$\frac{5-(-5)}{10}=1$$

| $x$ | $f(x)$ |
|---|---|
| –5.0 | –1.59 |
| –4.0 | –1.44 |
| –3.0 | –1.26 |
| –2.0 | –1.00 |
| –1.0 | 0.00 |
| 0.0 | 1.00 |
| 1.0 | 1.26 |
| 2.0 | 1.44 |
| 3.0 | 1.59 |
| 4.0 | 1.71 |
| 5.0 | 1.82 |

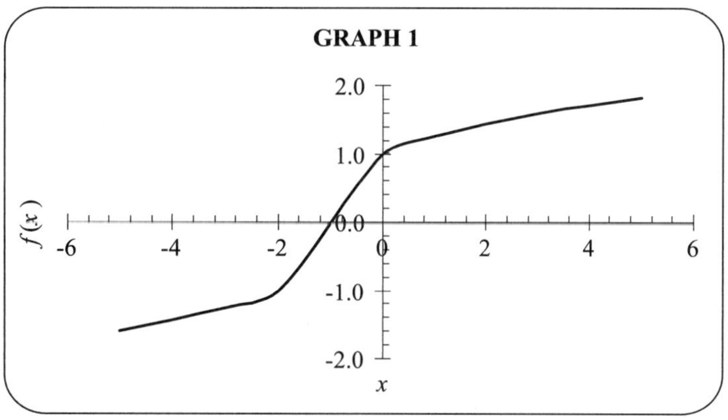


In **GRAPH 2**, 201 equally spaced points are plotted. To compute the difference between the $x$-coordinates of consecutive points, the left endpoint of the interval [–5, 5] is subtracted from the right endpoint, and the result is divided by 200.

$$\frac{5-(-5)}{200}=0.05$$

| $x$ | $f(x)$ |
|---|---|
| −5.00 | −1.59 |
| −4.95 | −1.58 |
| −4.90 | −1.57 |
| −4.85 | −1.57 |
| ⋮ | ⋮ |
| 4.85 | 1.80 |
| 4.90 | 1.81 |
| 4.95 | 1.81 |
| 5.00 | 1.82 |

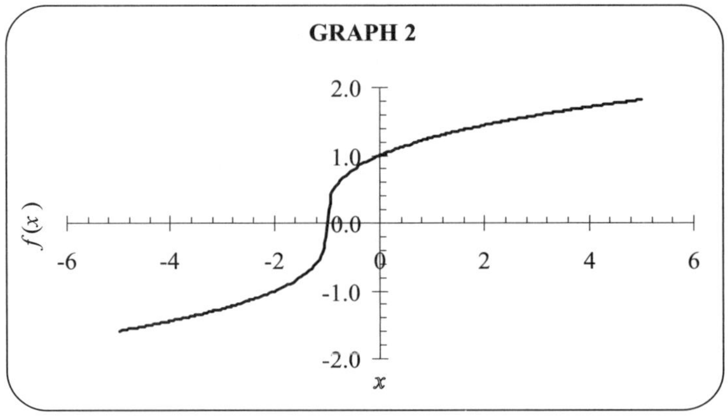


**GRAPH 2** is more accurate than **GRAPH 1**. This is because more points were plotted in **GRAPH 2** than in **GRAPH 1**. As the number of points plotted increases, the more accurate is the graph of the function. However, there are exceptions to this rule. One exception is a linear function (straight line). It doesn't matter whether two points are plotted or one thousand points are plotted; the graph will look the same. Another exception is a quadratic function (parabola). It doesn't matter whether 10 points are plotted or 200 points are plotted; the graph will look the same.

More complicated functions require more data points to obtain an accurate representation of the function. If you are not sure whether or not the function you are plotting would require a large number of data points, it is better to use more data points than fewer; plotting 500 points for any function usually will ensure an accurate graph.

The two previous graphs **GRAPH 1** and **GRAPH 2**, were plotted over the interval [−5, 5], but the interval was divided into a different number of subintervals for each graph. Looking back at these examples, you can see how this was achieved. Any interval $[a, b]$ can be divided into $n$ subintervals of equal width by dividing the width of the interval by $n$. Let $\Delta x$ represent the width of each subinterval. Then

$$\Delta x = \frac{b-a}{n}.$$

This gives a total of $n+1$ data points. The $x$-coordinate of first point will be $a$, and the $x$-coordinate of the last point (the $n+1$ point) will be $b$.

**Example 2**: Suppose that a function is to be plotted over the interval [−7, 7].

a. What is the width of each subinterval if the interval is divided into 400 subintervals of equal width?
b. How many data points will be plotted?
c. What is the $x$-coordinate of the second data point?
d. What is the $x$-coordinate of the second to last data point?

**SOLUTION**

a. $\Delta x = \dfrac{7-(-7)}{400} = \dfrac{14}{400} = 0.035.$

b. The total number of data points is $n+1$ where $n=400$, so 401 data points will be plotted.
c. The $x$-coordinate of the second data point is equal to the sum of the $x$-coordinate of the first data point and the width of the first subinterval; $-7 + 0.035 = -6.965$.
d. The $x$-coordinate of the second to last data point is equal to the difference between the $x$ coordinate of the last data point and the width of the last subinterval; $7 - 0.035 = 6.965$.

## *Piecewise Functions*

A piecewise function consists of two or more functions, each of which is defined on a different part of the domain. An example is given below.

$$f(x)=\begin{cases} x^2+1 & \text{if } -1\le x\le 1 \\ 2 & \text{elsewhere} \end{cases}$$

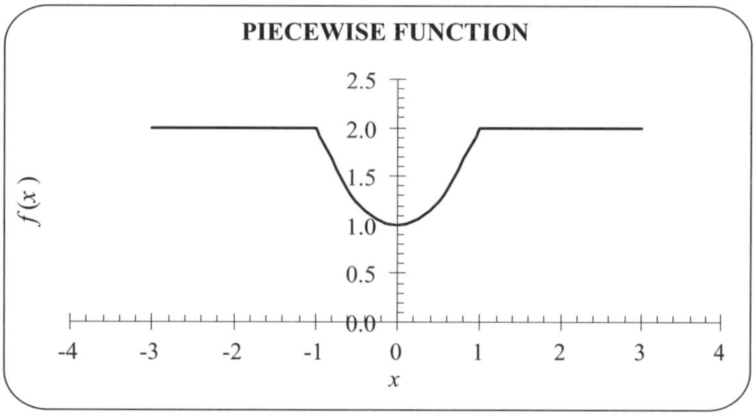


As you can see, when $x$ is between $-1$ and 1 the graph is a parabola, and when $x$ is less than $-1$ or greater than 1 the graph is a horizontal line.

Graphing piecewise functions in *Excel* requires the use of logical operators (*if*, *and*, *or*, *then*). Note that you will not see the logical operator *then* in *Excel* but it is implied with the use of a comma. The *Excel* formula shown below can be read as follows: ***If*** $-1$ is less than or equal to $x$ ($x$ is represented by the cell reference) ***and*** $x$ is less than or equal to 1, ***then*** $f(x)=x^2+1$, ***else*** $f(x)=2$.

```
IF(AND(-1<=A2,A2<=1,),A2^2+1,2)
```

↑ ***and*** ↑ ***then*** ↑ ***else***

**Example 3**: Plot the cumulative distribution function, $F_X(x)$, over the interval [–10, 40] using 501 equally spaced points.

$$F_X(x)=\begin{cases} 0 & \text{if } x<0 \\ 1-e^{-0.25x} & \text{if } x\ge 0 \end{cases}$$

**SOLUTION**

$$\Delta x = \frac{40-(-10)}{501-1} = \frac{50}{500} = 0.1$$

| $x$ | $F_X(x)$ |
|---|---|
| −10.0 | 0 |
| −9.9 | 0 |
| −9.8 | 0 |
| −9.7 | 0 |
| −9.6 | 0 |
| −9.5 | 0 |
| ⋮ | ⋮ |
| 39.6 | 0.999950 |
| 39.7 | 0.999951 |
| 39.8 | 0.999952 |
| 39.9 | 0.999953 |
| 40.0 | 0.999955 |

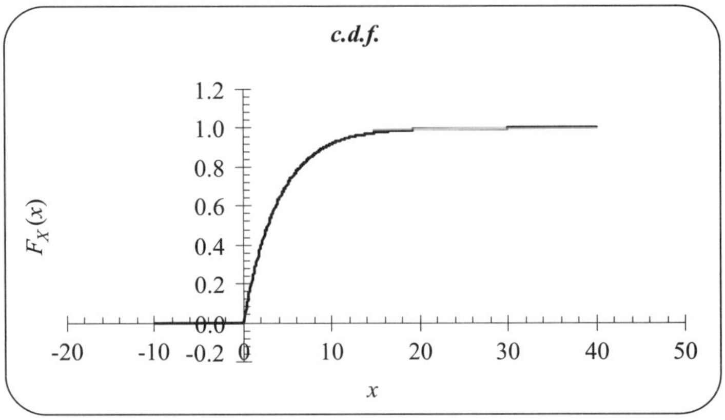


**Example 4**: Plot the probability density function, $f_X(x)$, over the interval [−10, 40] using 501 equally spaced points.

$$f_X(x) = \begin{cases} 0 & \text{if } x < 0 \\ 0.25e^{-0.25x} & \text{if } x \geq 0 \end{cases}$$

**SOLUTION**

$$\Delta x = \frac{40-(-10)}{501-1} = \frac{50}{500} = 0.1.$$

| $x$ | $f_X(x)$ |
|---|---|
| -10.0 | 0 |
| -9.9 | 0 |
| -9.8 | 0 |
| -9.7 | 0 |
| -9.6 | 0 |
| -9.5 | 0 |
| ⋮ | ⋮ |
| 39.6 | 1.25E-05 |
| 39.7 | 1.22E-05 |
| 39.8 | 1.19E-05 |
| 39.9 | 1.16E-05 |
| 40.0 | 1.13E-05 |

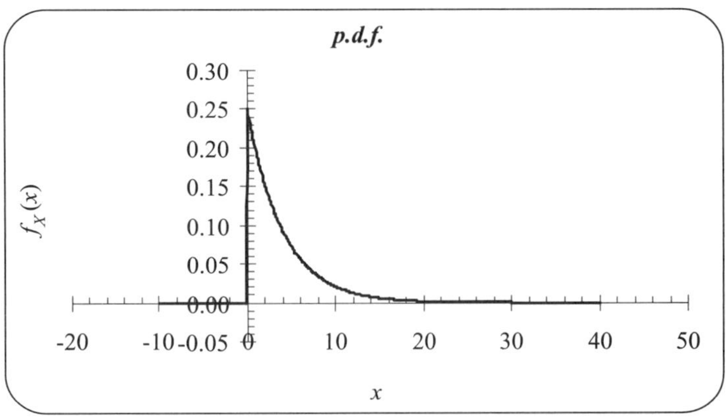

When plotting a piecewise function, unwanted vertical lines will appear when there are vertical gaps between parts of the graph. An unwanted vertical line is shown above in the graph of $f_X(x)$. This vertical line is not part of the graph and it needs to be removed. The removal of the vertical line will be covered in ***Advanced Formatting***.

## *Multiple Functions*

There are times when you may want to plot two functions on one graph. Specifically, in the ***Marketing Project***, you will be asked to plot both the cost function and the revenue function on one graph.

**Example 5:** Plot the functions $f(x) = 10$ and $g(x) = 0.75x + 5$ over the interval [0, 10].

**SOLUTION**

*Step 1*: Create a table in *Excel* with the same rows and columns as shown below. Column A contains the $x$-values, and Columns B and C contain the corresponding values of $f(x)$ and $g(x)$, respectively.

| | A | B | C |
|---|---|---|---|
| 1 | $x$ | $f(x)$ | $g(x)$ |
| 2 | 0 | 10 | 5 |
| 3 | 1 | 10 | 5.75 |
| 4 | 2 | 10 | 6.5 |
| 5 | 3 | 10 | 7.25 |
| 6 | 4 | 10 | 8 |
| 7 | 5 | 10 | 8.75 |
| 8 | 6 | 10 | 9.5 |
| 9 | 7 | 10 | 10.25 |
| 10 | 8 | 10 | 11 |
| 11 | 9 | 10 | 11.75 |
| 12 | 10 | 10 | 12.5 |

*Step 2*: Plot the functions. To open the **Chart Wizard**, pull down the **Insert** menu and select **Chart**, or simply click the **Chart Wizard** button on the **Standard Toolbar**. In the **Chart type** section of the first page of the **Chart Wizard**, click **XY (Scatter)**; and in the **Chart sub-type** section, click the subtype *Scatter with data points connected by smoothed Lines without markers*. Then click **Next** to advance to the second page of the **Chart Wizard**.

To define the outputs for the first function, click in the **Data Range** box and select *Cells B2:B12*. Then click the **Series** tab. To name the series, click in the **Name** box and enter the name $f(x)$. To define the inputs for the first function, click in the **X Values** box and select *Cells A2:A12*. Then click **Add**. To name the new series, click in the **Name** box and enter the name $g(x)$. To define the inputs for the second function, click in the **X Values** box and select *Cells A2:A12*. To define the outputs for the second function, click in the **Y Values** box and select *Cells C2:C12*. Then click **Next** to advance to the third page of the **Chart Wizard**.

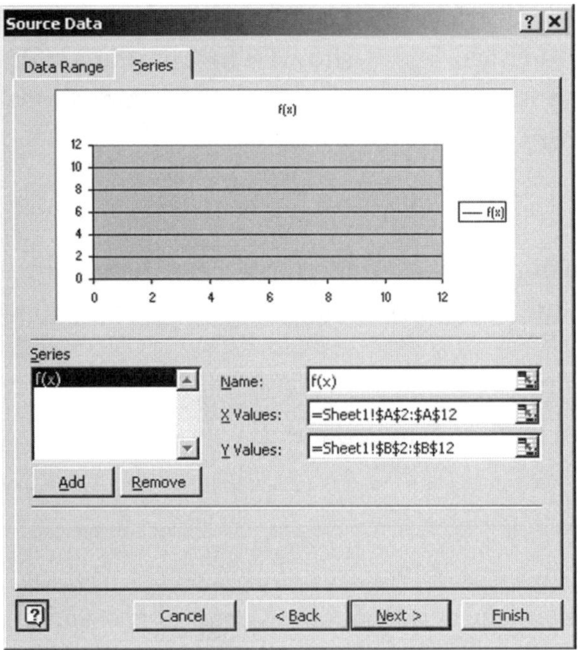


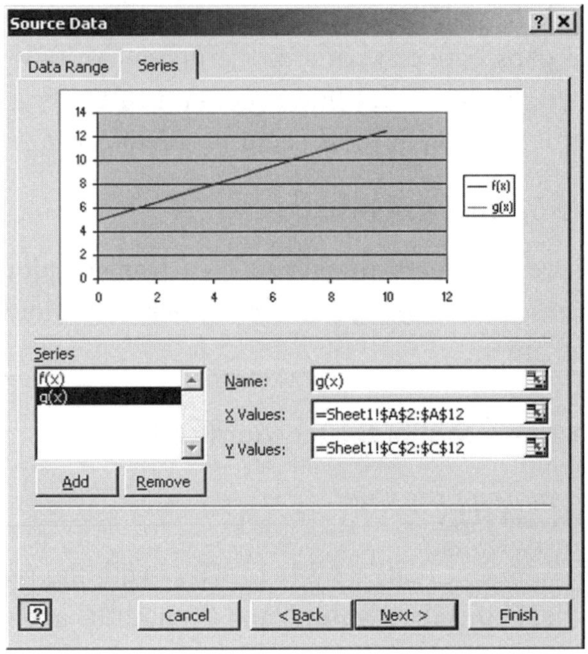


On the **Titles** tab, enter the title MULTIPLE FUNCTIONS in the **Chart title** box and the labels $x$ and $y$ in the **Value (X) axis** and **Value (Y) axis** boxes, respectively. Then click the **Gridlines** tab. In the **Value (Y) axis** section of the **Gridlines** tab, uncheck the **Major gridlines** check box. Then click **Finish**.

Your graph should look similar to the one shown below.

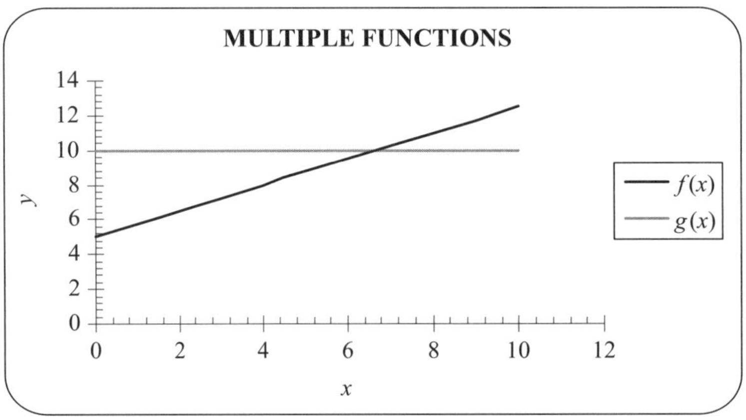


## *Advanced Formatting*

It is easy to change the appearance of any chart element. Just right-click the element you want to modify and select the appropriate option from the shortcut menu that appears.

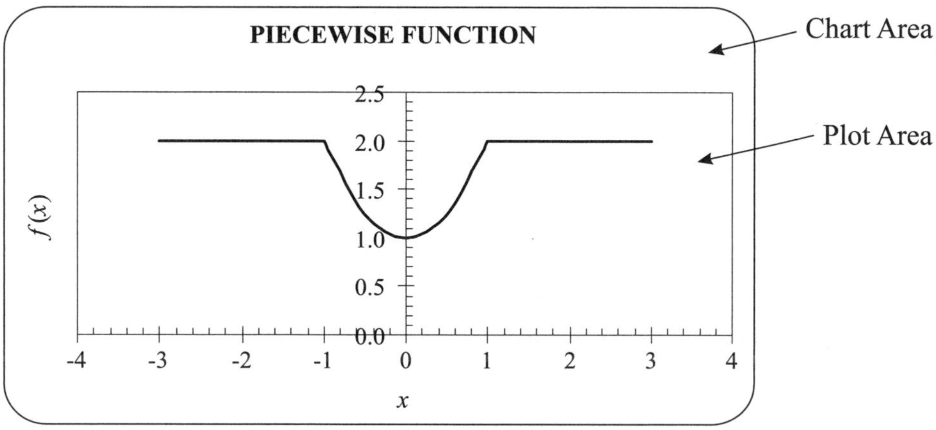


*Formatting the Chart Area*

To change the appearance of the chart area, right-click the chart area and select **Format Chart Area** from the shortcut menu that appears.

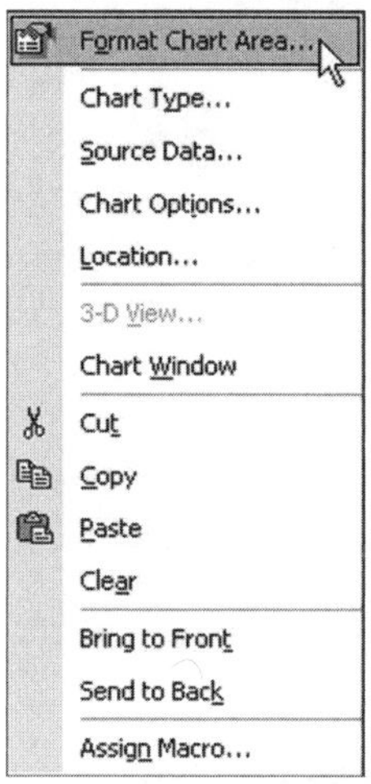


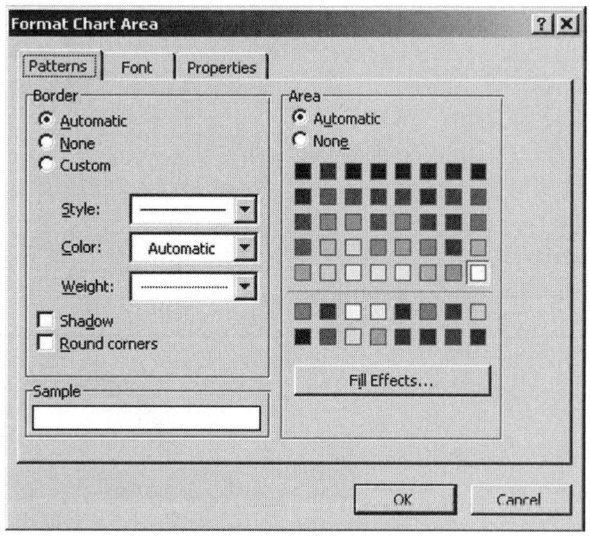


In the **Border** section of the **Patterns** tab, you can change the style, color, and weight of the chart area border. You can also add a shadow or round the corners of the chart area border. In the **Area** section of the tab, you can change the color of the chart area and add a gradient or texture. On the **Font** tab, you change the font, font style, and size of the text and numbers in the chart.

Note that changing the attributes of the text and numbers in the chart area will affect all text and numbers in the chart — this includes the chart title, axis titles, and tick mark labels. To change the font, font style, or size of any title individually, right-click the title you want to change and select the appropriate option from the shortcut menu that appears.

You can also change the font, font style, or size of any part of a title. For example, you can format the $X$ in the titles $F_X(x)$ and $f_X(x)$ as a subscript. To do this, click the axis title and highlight the $X$. Then right-click the axis title and select **Format Axis Title** from the shortcut menu that appears. Check the **Subscript** check box and click **OK**.

**Example 6**: Format the graph from **Example 1** as follows:

a. Change the weight of the chart area border to maximum thickness.
b. Round the corners of the chart area border.
c. Change the chart area font to Times New Roman.
d. Change the chart area font size to 12.
e. Change the chart title font style to bold.
f. Change the font style of the *x*-axis title to italic.
g. Change the font style of the *y*-axis title to italic.

## SOLUTION

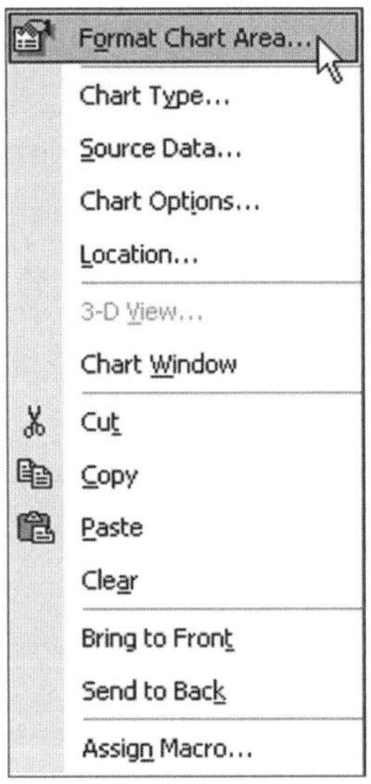


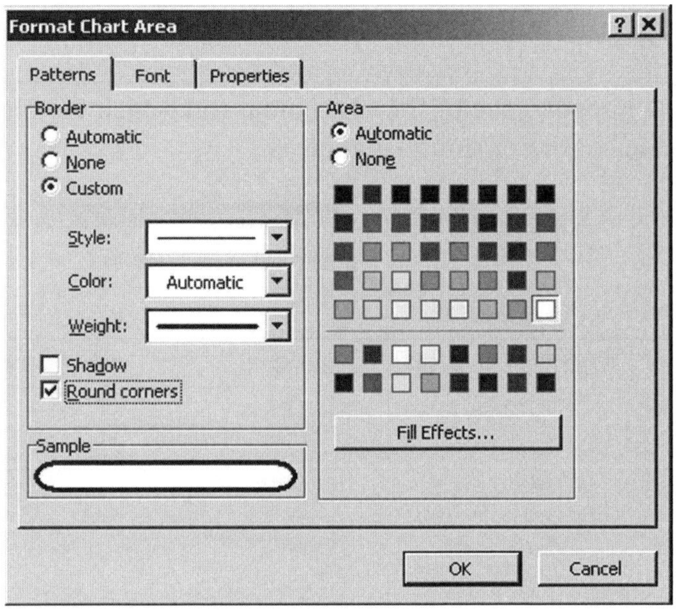


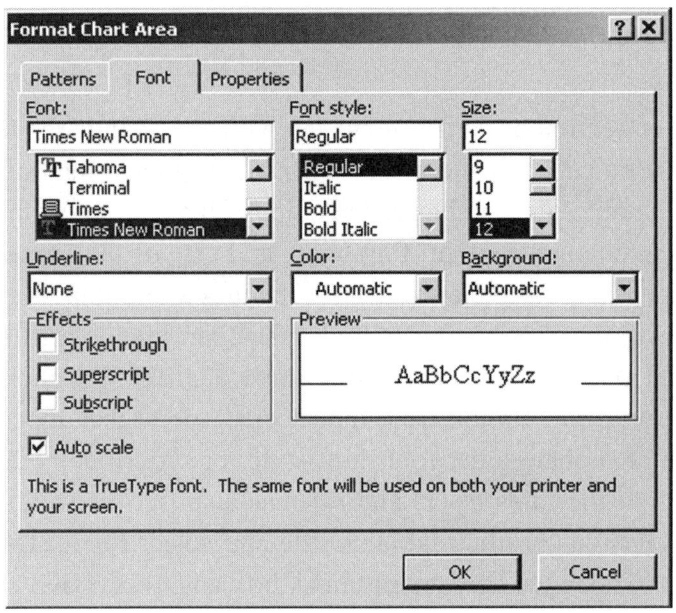

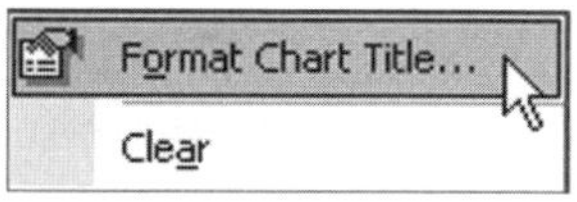
Format Chart Title...
Clear

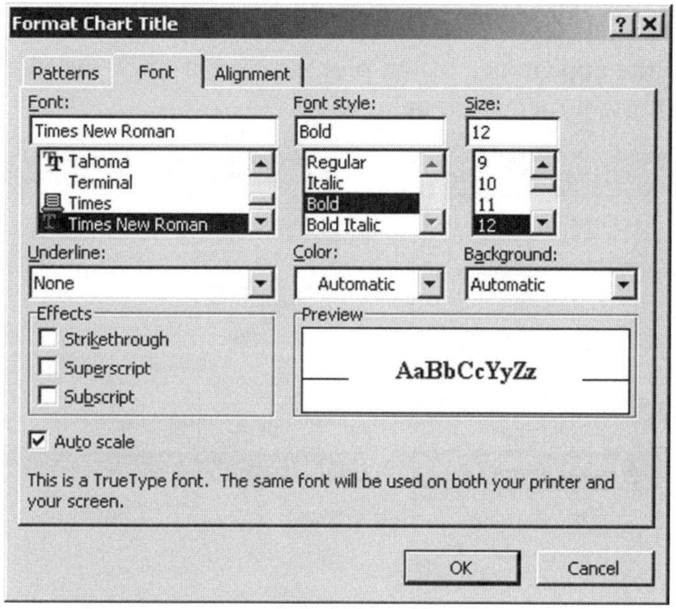
Format Chart Title
Patterns
Font
Alignment
Font:
Times New Roman
Tahoma
Terminal
Times
Times New Roman
Font style:
Bold
Regular
Italic
Bold
Bold Italic
Size:
12
9
10
11
12
Underline:
None
Color:
Automatic
Background:
Automatic
Effects
Strikethrough
Superscript
Subscript
Preview
AaBbCcYyZz
Auto scale
This is a TrueType font. The same font will be used on both your printer and your screen.
OK
Cancel

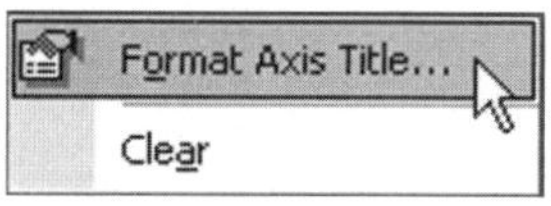
Format Axis Title...
Clear

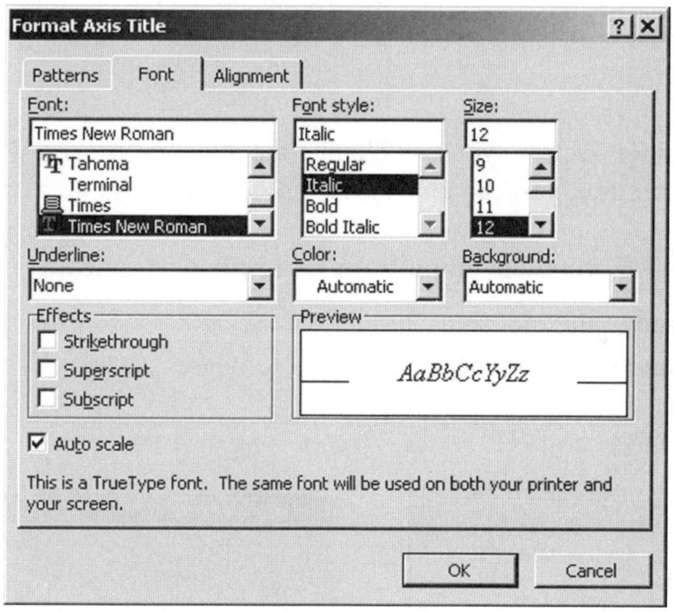
Format Axis Title
Patterns
Font
Alignment
Font:
Times New Roman
Tahoma
Terminal
Times
Times New Roman
Font style:
Italic
Regular
Italic
Bold
Bold Italic
Size:
12
9
10
11
12
Underline:
None
Color:
Automatic
Background:
Automatic
Effects
Strikethrough
Superscript
Subscript
Preview
AaBbCcYyZz
Auto scale
This is a TrueType font. The same font will be used on both your printer and your screen.
OK
Cancel

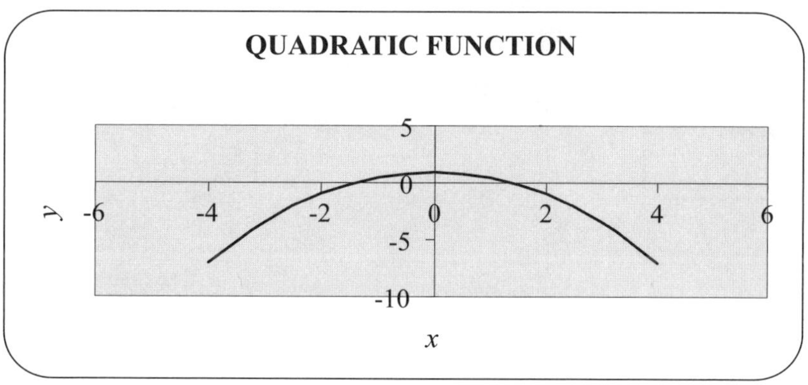
QUADRATIC FUNCTION
5
0
-5
-10
-6
-4
-2
0
2
4
6
y
x

*Formatting the Plot Area*

To change the appearance of the plot area, right-click the plot area and select **Format Plot Area** from the shortcut menu that appears.

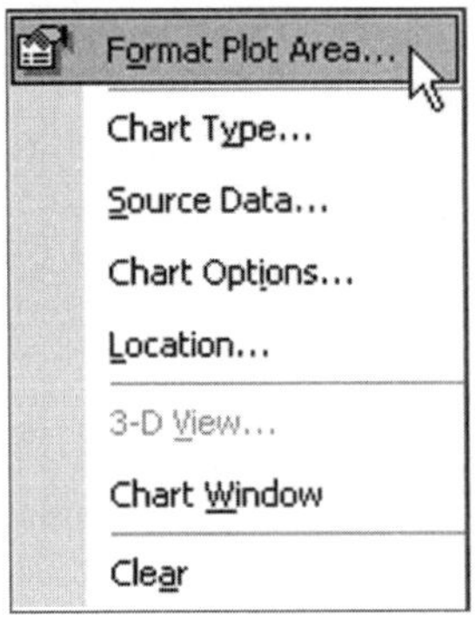


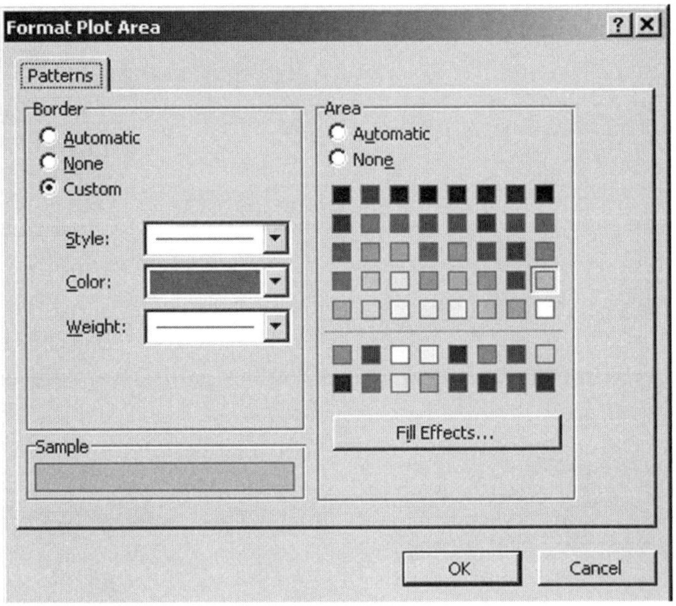


In the **Border** section, you can change the style, color, and weight of the plot area border, and in the **Area** section, you can change the color of the plot area and add a gradient or texture.

**Example 7**: Format the graph from **Example 6** as follows:

a. Remove the plot area border.
b. Change the color of the plot area to white.

**SOLUTION**

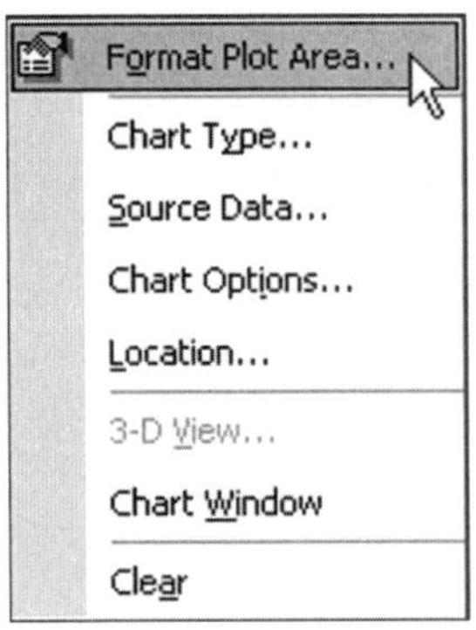


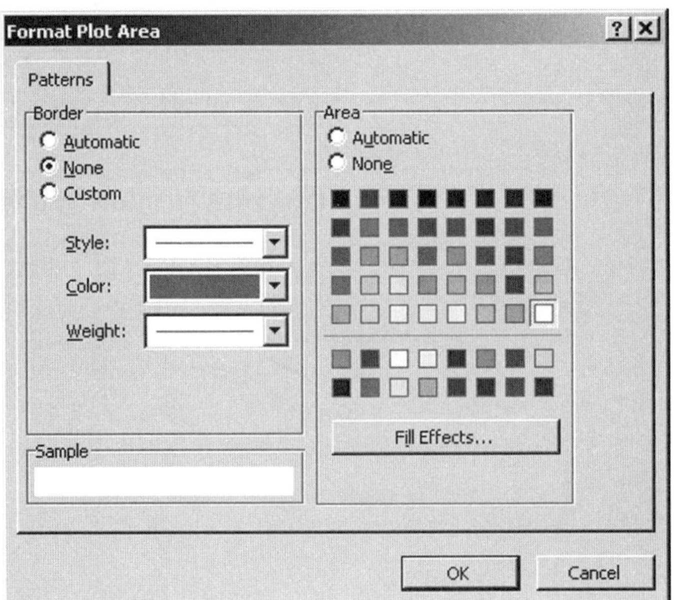

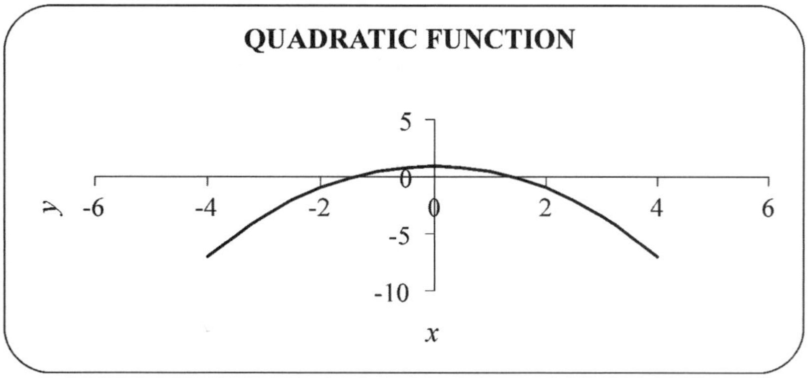


*Formatting the Data Series*

To change the appearance of the line that is plotted in the graph, right-click the line and select **Format Data Series** from the shortcut menu that appears.

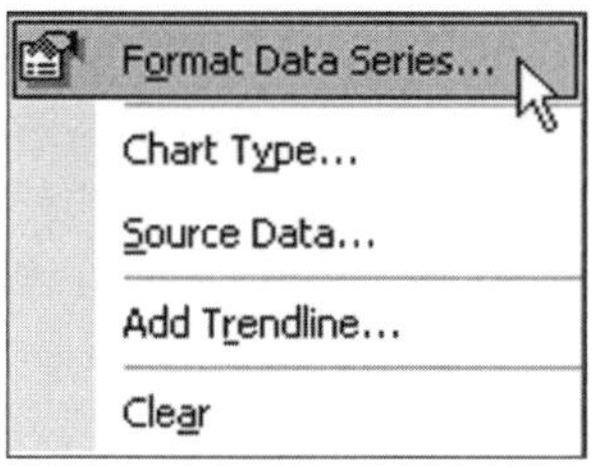


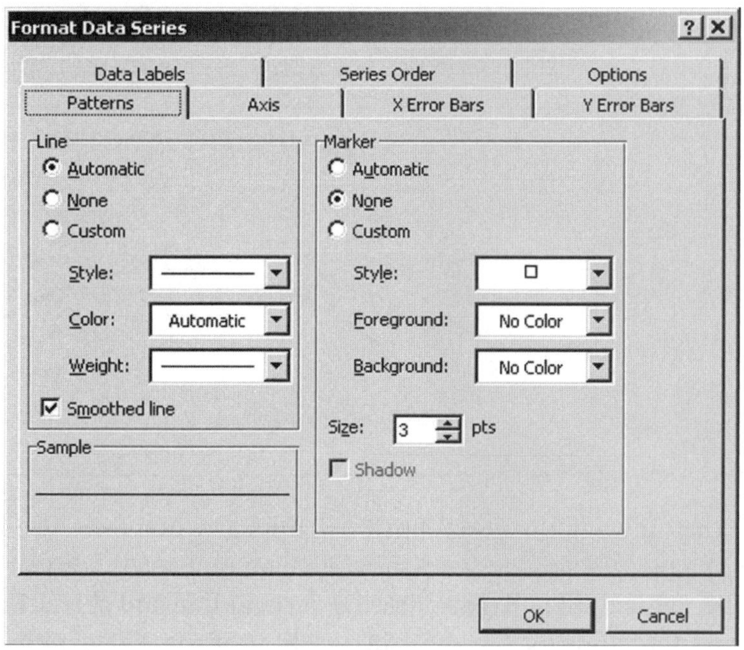


On the **Patterns** tab, you can change the style, color, and weight of the line.

**Example 8**: Change the weight of the line that is plotted in the graph from **Example 7** to maximum thickness.

## SOLUTION

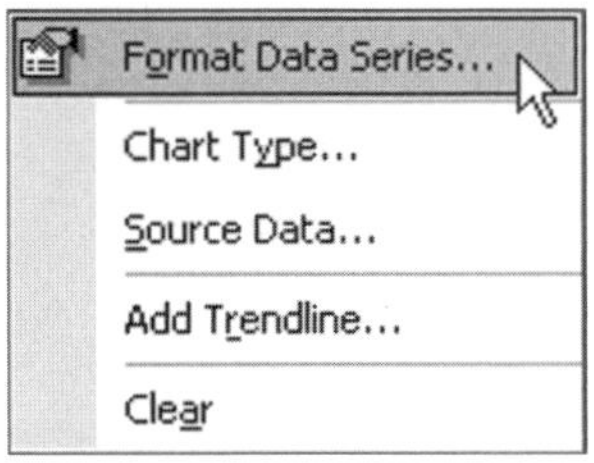


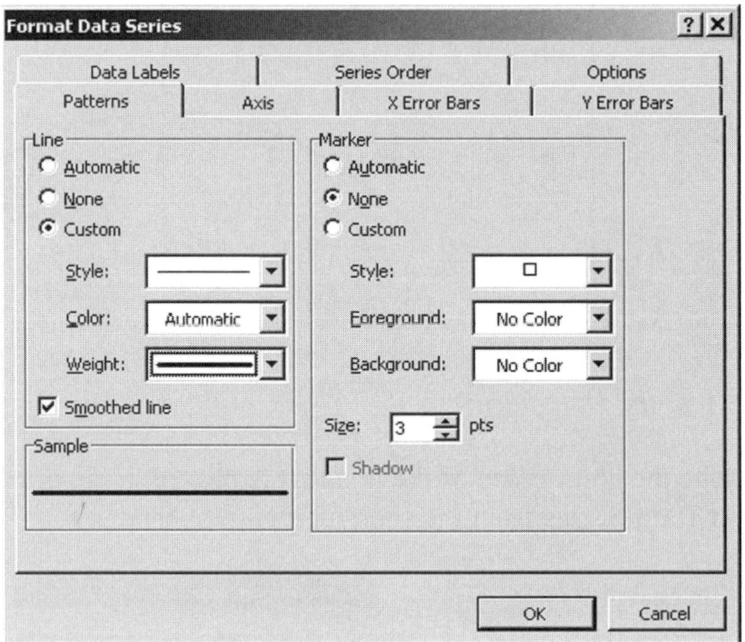


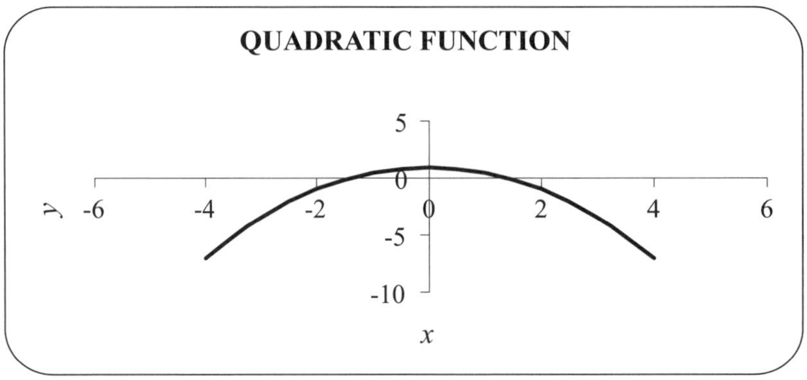


Recall that unwanted vertical lines will appear when there are vertical gaps between parts of the graph of a piecewise function. To remove an unwanted vertical line, click the data series and then click the vertical line. Right-click the vertical line and select **Format Data Point** from the shortcut menu that appears. In the **Line** section of the **Patterns** tab, select the **None** option button. Then click **OK**.

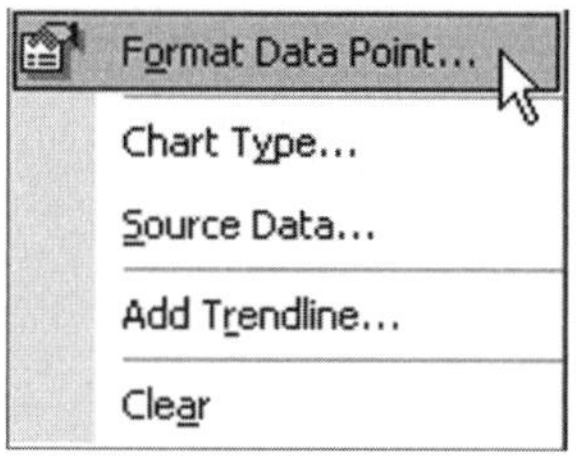


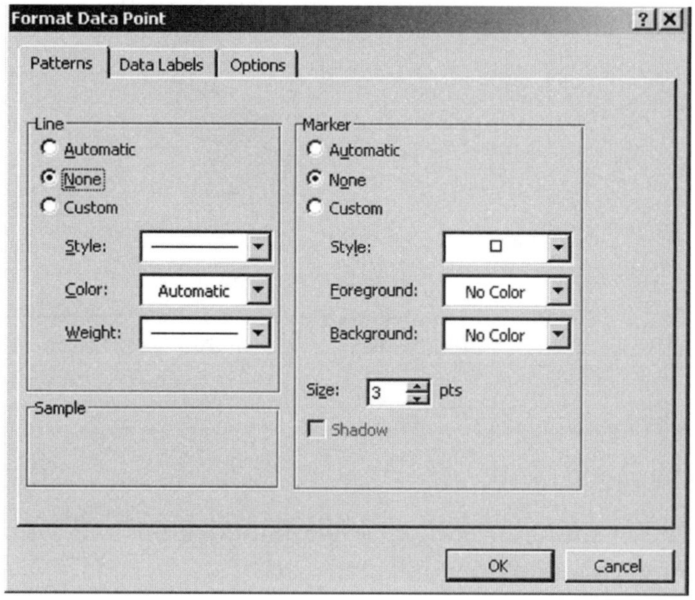


**Example 9:** Remove the unwanted vertical line from the graph from **Example 4**.

**SOLUTION**

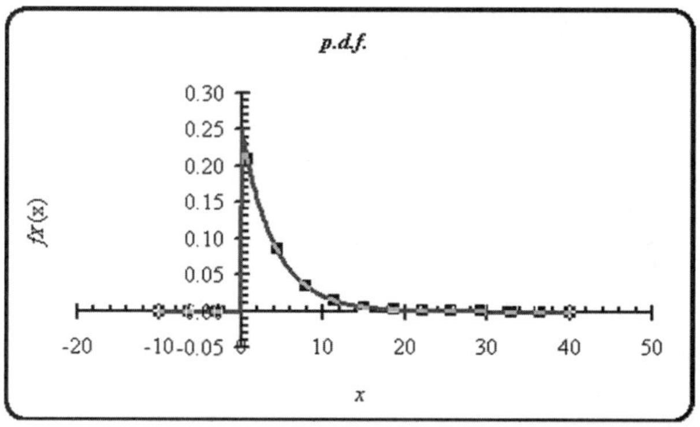


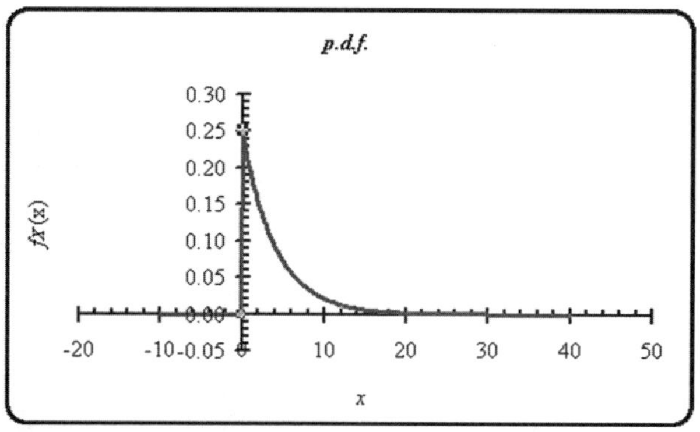


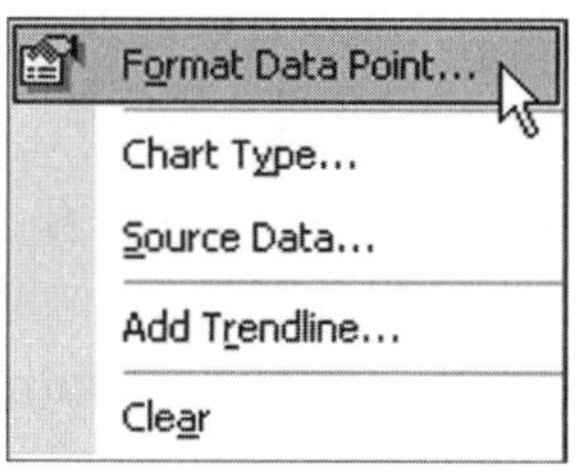


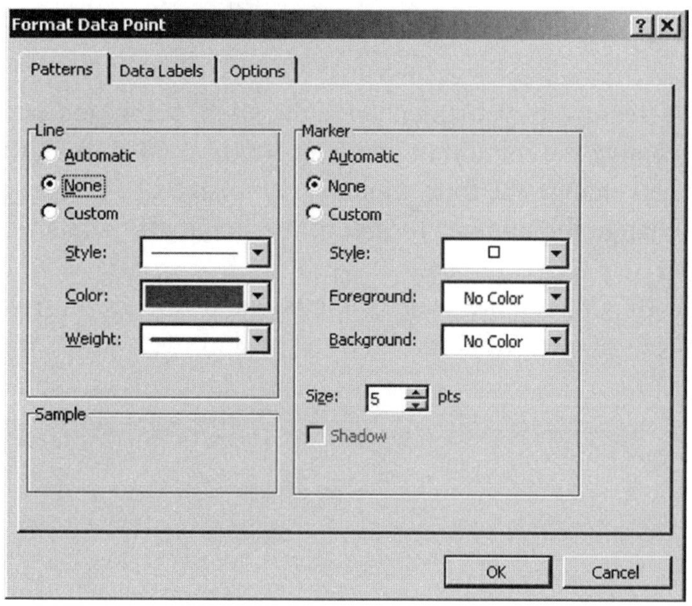

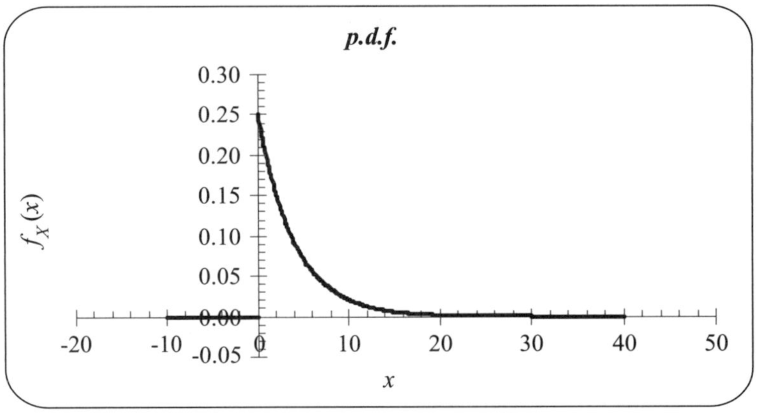


*Formatting the Axes*

To change the appearance of an axis, right-click the axis you want to modify and select **Format Axis** from the shortcut menu that appears.

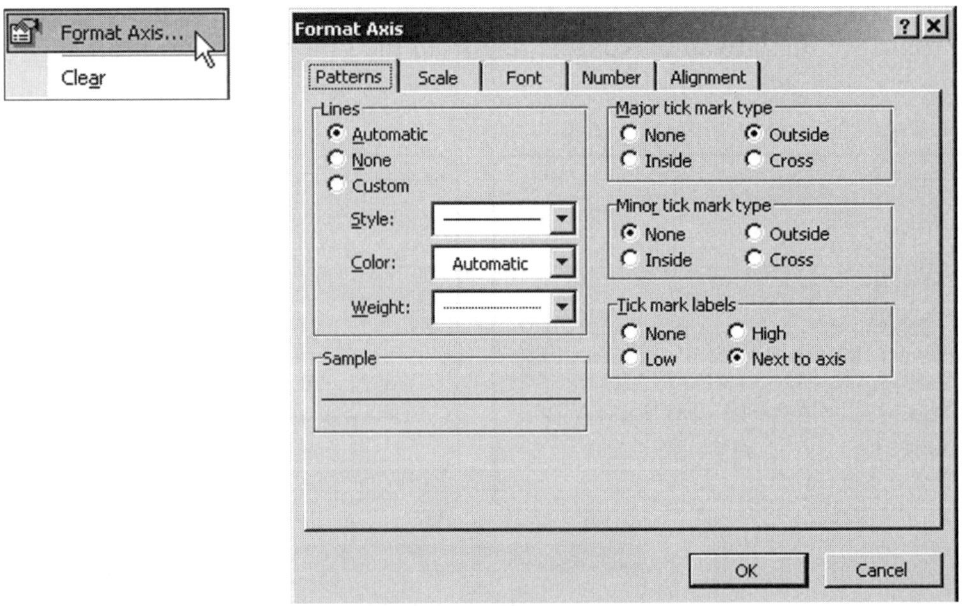


On the **Patterns** tab, you can change the style, color, and weight of the axis. On the **Scale** tab, you can change the minimum value, maximum value, major unit, and minor unit. On the **Font** tab, you can change the font, font style, and size of the tick mark labels. On the **Number** tab, you can change the number format of the tick mark labels.

**Example 10**: Change the scale of the $y$-axis of the graph from **Example 8** so that it has a minimum of $-8$, a maximum of 2, a major unit of 2, and a minor unit of 1.

**SOLUTION**

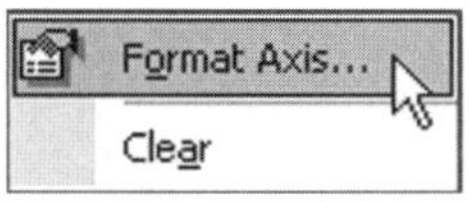


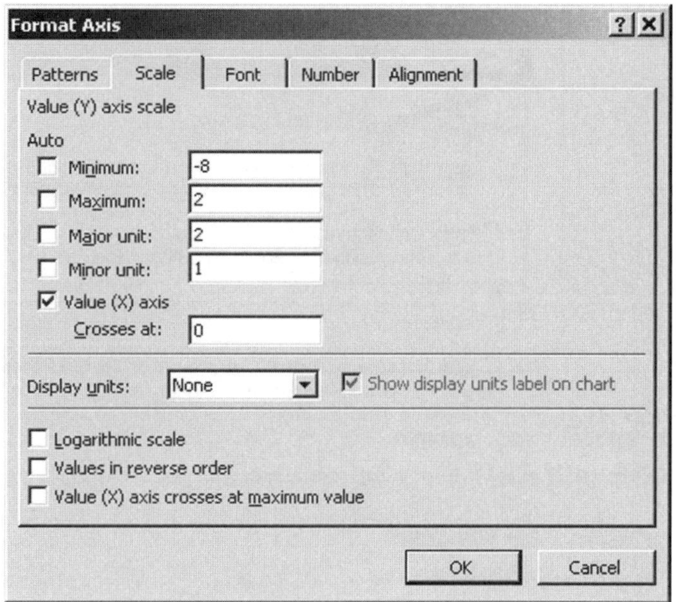


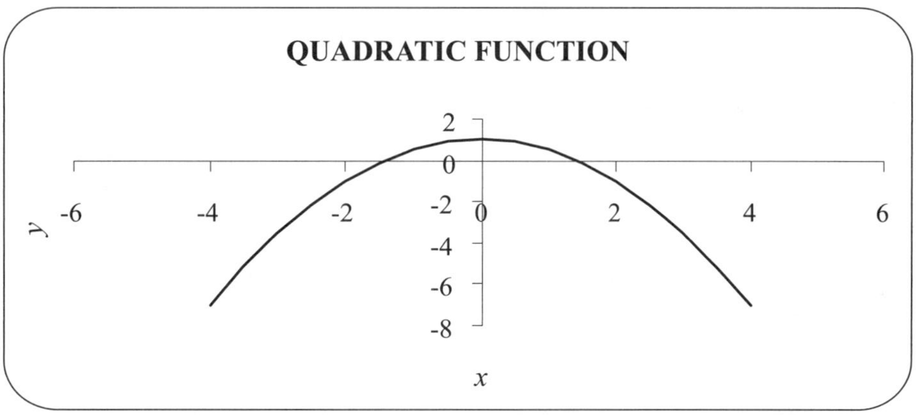


## *Using Graphing.xls*

***Graphing.xls*** is a utility found in your course files. Before you actually begin to use ***Graphing.xls***, a few important issues need to be addressed.

First, if the security alert shown below appears when you open ***Graphing.xls***, you will need to change the macro security level.

Pull down the **Tools** menu, point to **Macro**, and select **Security**. Select the **Medium** or **Low** option button and click **OK**.

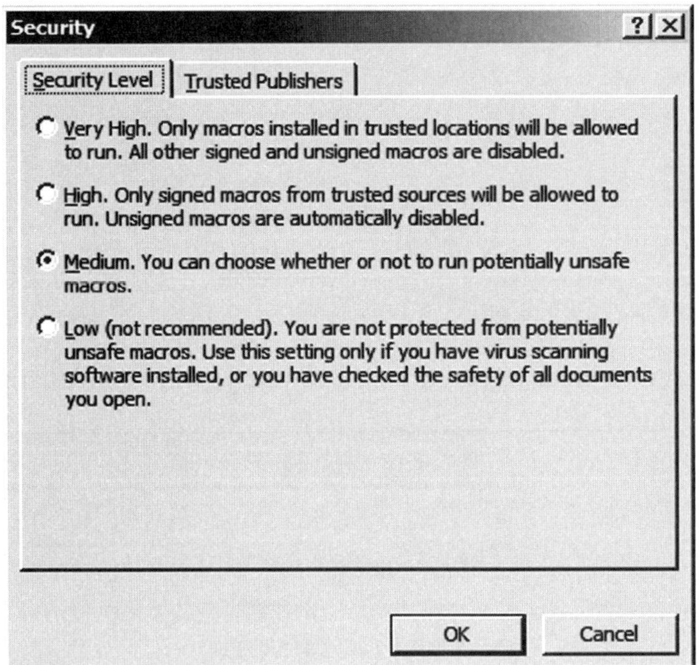


If you select Medium, every time you open a file containing a macro, you will be prompted to enable or disable the macro; if you select Low, the macro will automatically be enabled.

Second, you need to make certain that *Excel* is set for automatic computation. This is the default calculation setting. However, if for some reason *Excel* is set for manual calculation, pull down the **Tools** menu and select **Options**. Then click the **Calculation** tab. In the **Calculation** section of the **Calculation** tab, select the **Automatic** option button and click **OK**.

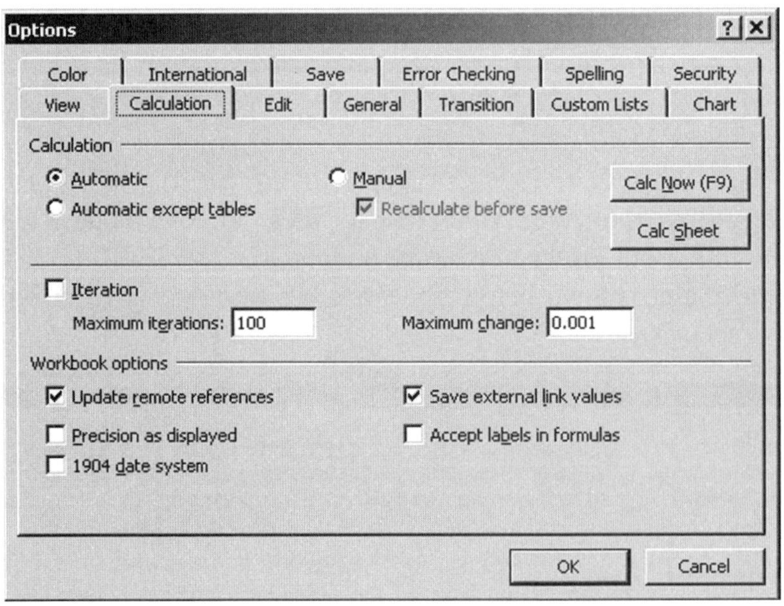


Finally, when you enter formulas in ***Graphing.xls***, always call the variable $x$.

***Graphing.xls*** is easy to use. Simply enter the formula in *Cell C18* and the endpoints of the plot interval in *Cells H18* and *I18*. Then click the **Graph** button to run the macro. Note that if you change the formula in *Cell C18*, you need to rerun the macro.

| | A | B | C | D | E | F | G | H | I | J | K | L |
|---|---|---|---|---|---|---|---|---|---|---|---|---|
| 16 | | Definition | | | Computation | | | Plot Interval | | | Constants | |
| 17 | | Formula for $f(x)$ | | | $x$ | $f(x)$ | | $a$ | $b$ | | $s$ | |
| 18 | | | = | | | = | | | | | $t$ | |
| 19 | | | | | | | | | | | $u$ | |
| 20 | | | | | | | | | | | $v$ | |
| 21 | | | | | | | | | | | $w$ | |

Initially, there is no need to enter anything in the **Computation** or **Constants** cells.

**Example 11**: Use ***Graphing.xls*** to plot the quadratic function $f(x) = x^2 + 4x + 1$ over the interval $[-10, 10]$.

**SOLUTION**

Enter the formula = *x*^2+4**x*+1 in *Cell C18*, and the values –10 and 10 in *Cells H18* and *I18*, respectively. Then click the **Graph** button to run the macro.

| | A | B | C | D | E | F | G | H | I |
|---|---|---|---|---|---|---|---|---|---|
| 16 | | Definition | | | Computation | | | Plot Interval | |
| 17 | | Formula for $f(x)$ | | | $x$ | $f(x)$ | | $a$ | $b$ |
| 18 | | | 1 | | | 1 | | -10 | 10 |

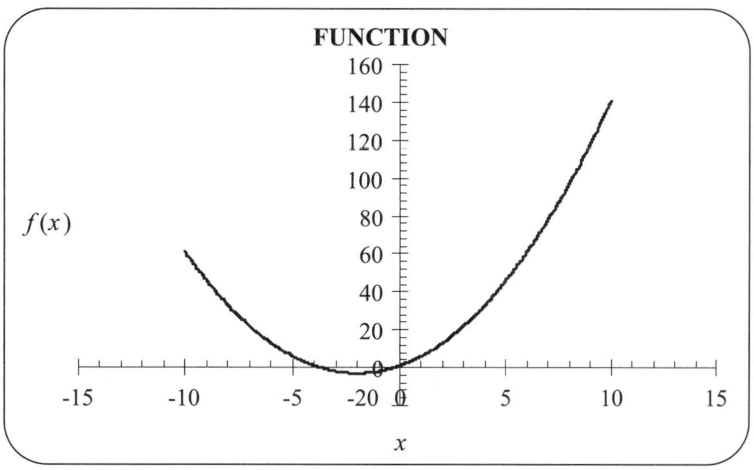


The number 1 in *Cell C18* is the value of the function at $x = 0$. The same is true for the 1 in *Cell F18*. If you wanted to evaluate the function at any other value of $x$, you would enter that value of $x$ in *Cell E18*. That is, if you entered 2 in *Cell E18*, the number 13 would appear in *Cell F18* since $f(2) = 13$.

The formula for the function in **Example 11** contains a constant. If you wanted to see the effects of changing the constant without changing the formula every time, you would use the

**Constants** cells. To do this, you would enter the formula in *Cell C18* using any of the variables *s, t, u, v,* or *w* in place of the constant. If you decided to use *t* as the constant, you would enter the formula =x^2+4*x+t in *Cell C18* and a value for *t* in *Cell L18*. Then you would click the **Graph** button to run the macro.

| | A | B | C | D | E | F | G | H | I | J | K | L |
|---|---|---|---|---|---|---|---|---|---|---|---|---|
| 16 | | Definition | | | Computation | | | Plot Interval | | | Constants | |
| 17 | | Formula for $f(x)$ | | | $x$ | $f(x)$ | | $a$ | $b$ | | $s$ | |
| 18 | | | 50 | | | 50 | | -10 | 10 | | $t$ | 50 |
| 19 | | | | | | | | | | | $u$ | |
| 20 | | | | | | | | | | | $v$ | |
| 21 | | | | | | | | | | | $w$ | |

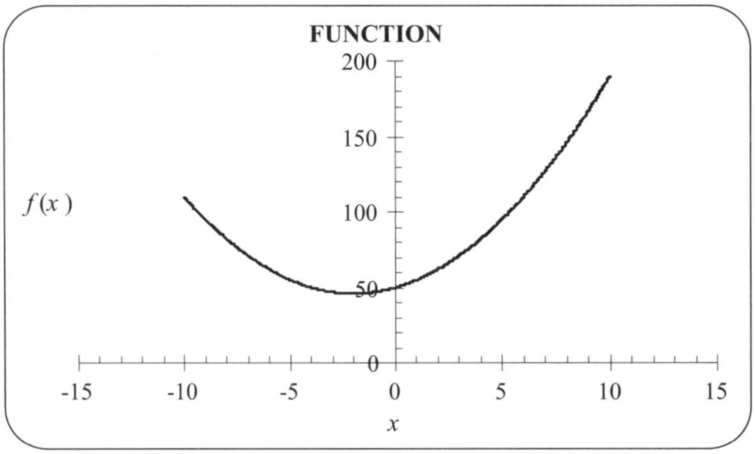


**Example 12**: Use ***Graphing.xls*** to plot the function $f(x) = u^x + s$ over the interval $[-4, 4]$ with the following conditions:

a. $s = 5$, $u = 1$; find $f(2)$.
b. $s = -50$, $u = 2$; find $f(-2)$.
c. $s = 40$, $u = 3$; find $f(3.71)$.

**SOLUTION**

a.

| Definition | |
|---|---|
| Formula for $f(x)$ | |
| | 6 |

| Computation | |
|---|---|
| $x$ | $f(x)$ |
| 2 | 6 |

| Plot Interval | |
|---|---|
| $a$ | $b$ |
| -4 | 4 |

| Constants | |
|---|---|
| $s$ | 5 |
| $t$ | |
| $u$ | 1 |
| $v$ | |
| $w$ | |

Formula for $f(x)$: =u^x+s

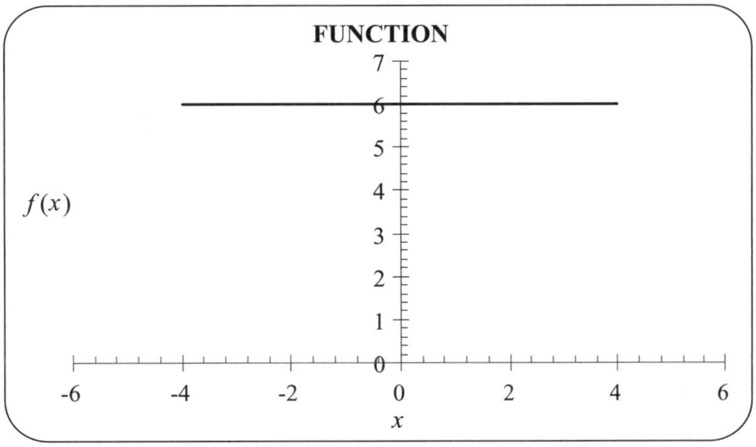


b.

**Definition**

| Formula for $f(x)$ |
|---|
| -49 |

**Computation**

| $x$ | $f(x)$ |
|---|---|
| -2 | -49.75 |

**Plot Interval**

| $a$ | $b$ |
|---|---|
| -4 | 4 |

**Constants**

| | |
|---|---|
| $s$ | -50 |
| $t$ | |
| $u$ | 2 |
| $v$ | |
| $w$ | |

Formula for $f(x)$: =u^x+s

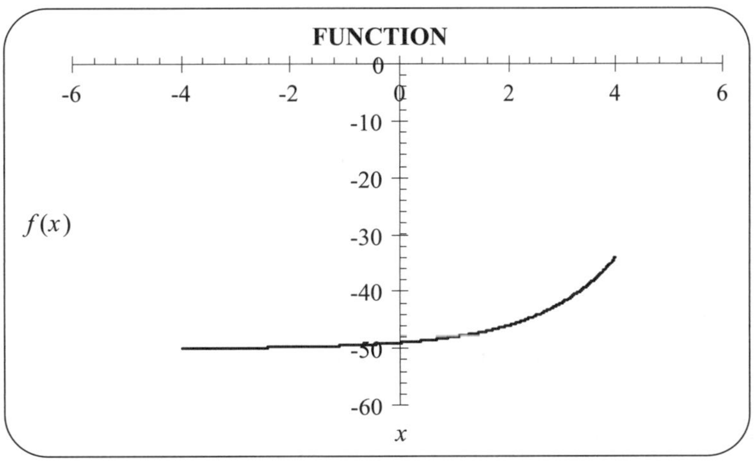


c.

**Definition**

| Formula for $f(x)$ |
|---|
| 41 |

**Computation**

| $x$ | $f(x)$ |
|---|---|
| 3.71 | 98.90062 |

**Plot Interval**

| $a$ | $b$ |
|---|---|
| -4 | 4 |

**Constants**

| | |
|---|---|
| $s$ | 40 |
| $t$ | |
| $u$ | 3 |
| $v$ | |
| $w$ | |

Formula for $f(x)$: =u^x+s

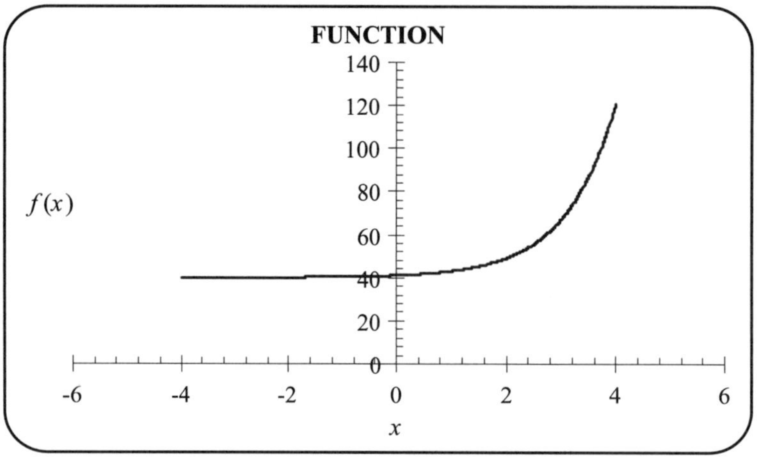


Notice that all of the values used for $u$ in **Example 12** were greater than zero. When $u$ is less than zero, *Excel* will not correctly interpret the formula used in **Example 12**. Look what happens if $s = 10$ and $u = -2$.

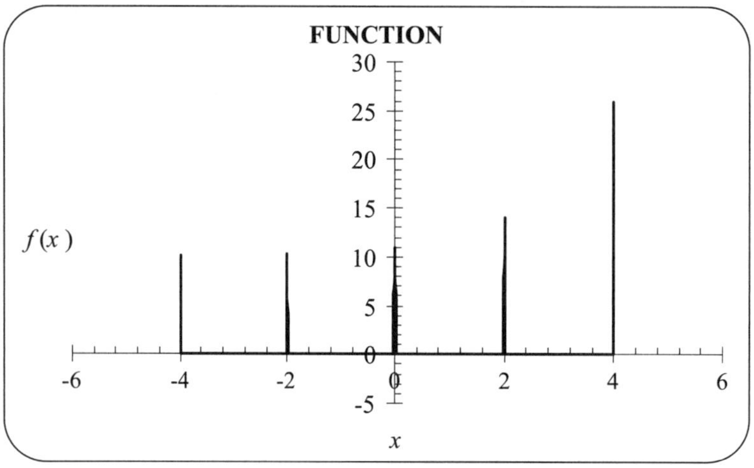


To obtain correct results, enter either of the formulas =-1*u^x+s or =-(u^x)+s in *Cell C18* and the value 2 in *Cell L17*. Then click the **Graph** button to run the macro.

**Definition**

| Formula for $f(x)$ |
|---|
| 9 |

**Computation**

| $x$ | $f(x)$ |
|---|---|
| | 9 |

**Plot Interval**

| $a$ | $b$ |
|---|---|
| -4 | 4 |

**Constants**

| | |
|---|---|
| $s$ | 10 |
| $t$ | |
| $u$ | 2 |
| $v$ | |
| $w$ | |

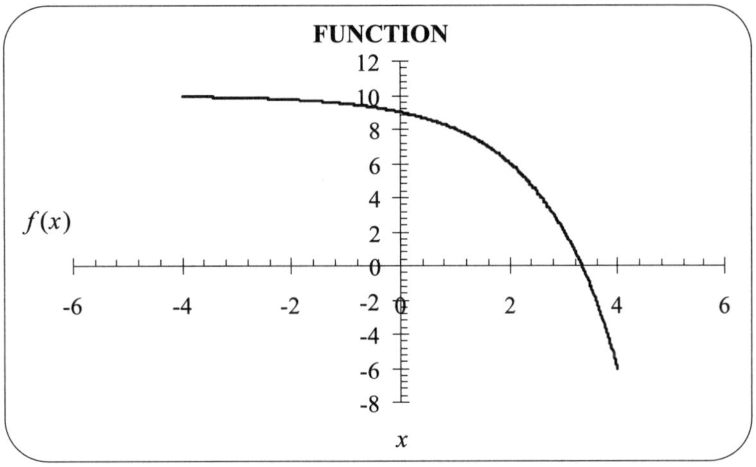


## *Exercises*

1. Use *Excel* to plot the linear function $f(x) = -\frac{3}{5}x + 8$ over the interval $[-7, 7]$. The graph should have the following characteristics:
   a. The chart title is **LINEAR FUNCTION**.
   b. The $x$-axis title is $x$.
   c. The $y$-axis title is $f(x)$.
   d. The legend is not shown.
   e. The line that is plotted in the graph is the maximum weight.

**SOLUTION**

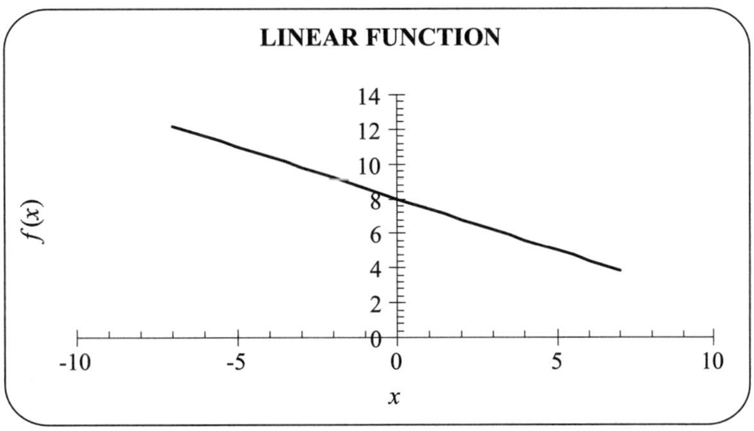


2. Suppose that a function is to be plotted over the interval $[-10, 15]$.
   a. What is the width of each subinterval if the interval is divided into 1000 subintervals of equal width?
   b. How many data points will be plotted?
   c. What is the $x$-coordinate of the second data point?

**SOLUTION**

a. 0.025
b. 1001
c. −9.975

3. Suppose that a third degree polynomial is to be plotted over the interval [−6, 6] using 601 equally spaced points.

   a. What is the value of $\Delta x$?
   b. What is the $x$-coordinate of the 599th data point?

**SOLUTION**

a. 0.02
b. 5.96

4. Use *Excel* to plot the cubic function $g(x) = 3x^3 - 2x^2 + 6x - 25$ over the interval [−5, 5] using 201 equally spaced points. The graph should have the following characteristics:

   a. The chart title is **CUBIC FUNCTION**.
   b. The $x$-axis title is $x$.
   c. The $y$-axis title is $g(x)$.
   d. The legend is not shown.
   e. The line that is plotted in the graph is the maximum weight.

**SOLUTION**

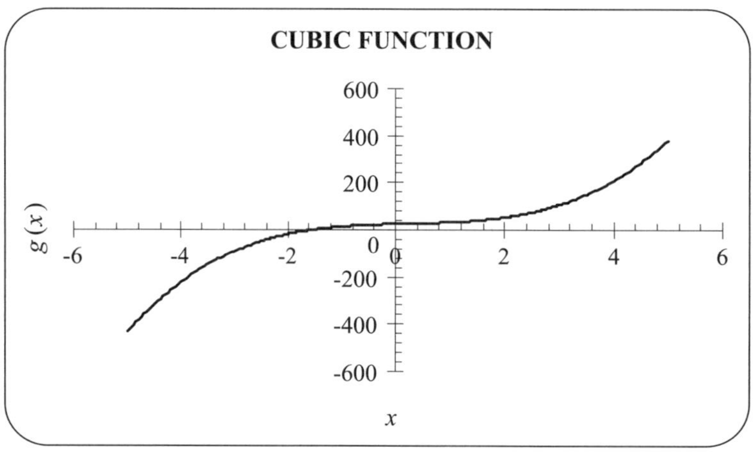


5. Use *Excel* to plot the demand function, $D(q) = -0.000006q^2 - 0.038585q + 303.039287$, over the interval [0, 5000] using 1001 equally spaced points. The graph should have the following characteristics:

   a. The chart title is **DEMAND FUNCTION**.
   b. The title of the horizontal axis is $q$.
   c. The title of the vertical axis is $D(q)$.

d. The legend is not shown.
e. The line that is plotted in the graph is the maximum weight.
f. The number format of the vertical axis tick mark labels is currency with 0 decimal places.

**SOLUTION**

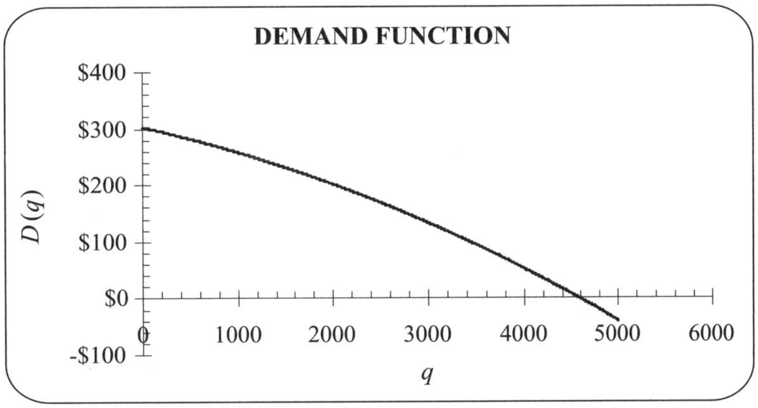


6. Use *Excel* to plot the piecewise function, $f(x)$, over the interval [0, 20] using 401 equally spaced points.

$$f(x) = \begin{cases} 0.1x & \text{if } 0 \le x \le 10 \\ -0.1x + 2 & \text{if } 10 < x \le 20 \end{cases}$$

The graph should have the following characteristics:
a. The chart title is **PIECEWISE FUNCTION**.
b. The $x$-axis title is $x$.
c. The $y$-axis title is $f(x)$.
d. The legend is not shown.
e. The line that is plotted in the graph is the maximum weight.

**SOLUTION**

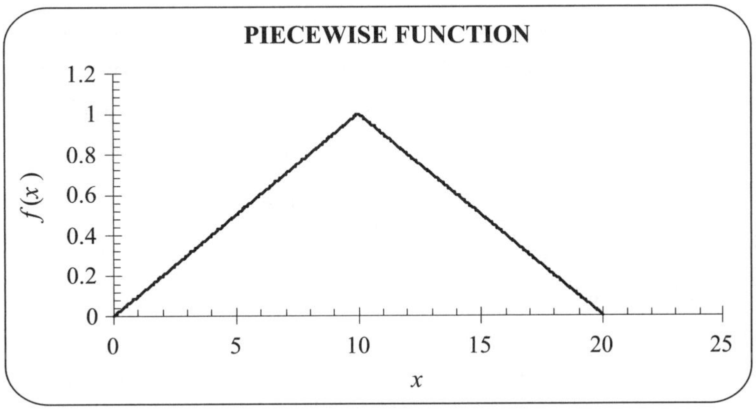


7. Use *Excel* to plot the piecewise function, $g(x)$, over the interval [–10, 15] using 801 equally spaced points.

$$g(x)=\begin{cases} x+1 & \text{if } -5 \le x < 2 \\ 3 & \text{if } 2 \le x < 5 \\ -x+8 & \text{if } 5 \le x \le 12 \\ 0 & \text{elsewhere} \end{cases}$$

The graph should have the following characteristics:

a. The chart title is **PIECEWISE FUNCTION**.

b. The $x$-axis title is $x$.

c. The $y$-axis title is $g(x)$.

d. The legend is not shown.

e. The line that is plotted in the graph is the maximum weight.

f. There are no unwanted vertical lines.

**SOLUTION**

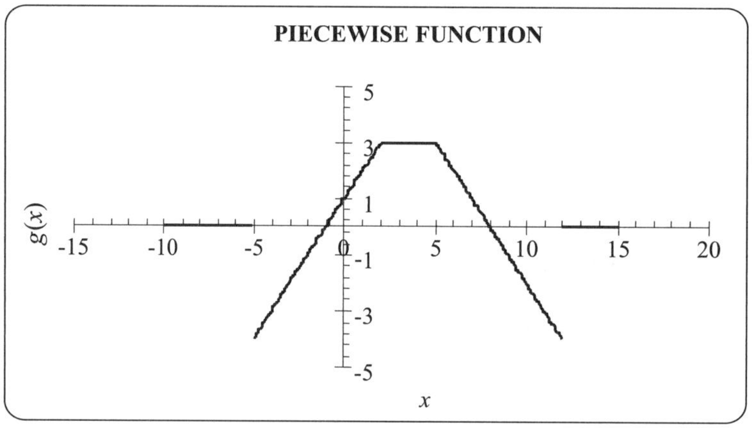


8. A computer company produces desktop and laptop computers. The desktop computers are manufactured in the DHT plant, and the laptops are manufactured in the LHT plant. The cost function for the desktop computers is $C_D(q) = 400q + 500{,}000$, and the cost function for the laptop computers is $C_L(q) = 500q + 200{,}000$. The quantity, $q$, represents the number of computers produced. Plot both cost functions in one graph over the interval [0, 10000]. The graph should have the following characteristics:

   a. The name of the first series is C_D(q), and the name of the second series is C_L(q).

   b. The chart title is **COST FUNCTIONS**.

   c. The title of the horizontal axis is $q$.

   d. The title of the vertical axis is $C(q)$.

   e. The legend is in the default location.

   f. The lines that are plotted in the graph are the maximum weight.

   g. The number format of the vertical axis tick mark labels is currency with 0 decimal places.

**SOLUTION**

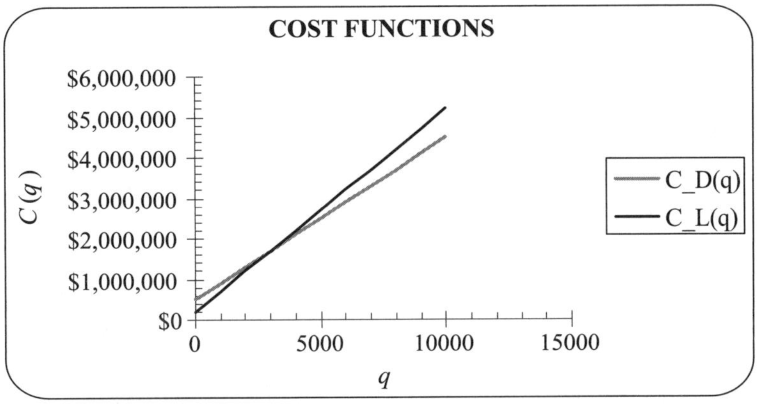


9. Use ***Graphing.xls*** to plot the cumulative distribution function, $F_X(x)$, over the interval [0, 60].

$$F_X(x) = \begin{cases} 0 & \text{if } x < 10 \\ \dfrac{x-10}{40} & \text{if } 10 \le x \le 50 \\ 1 & \text{if } x > 50 \end{cases}$$

**SOLUTION**

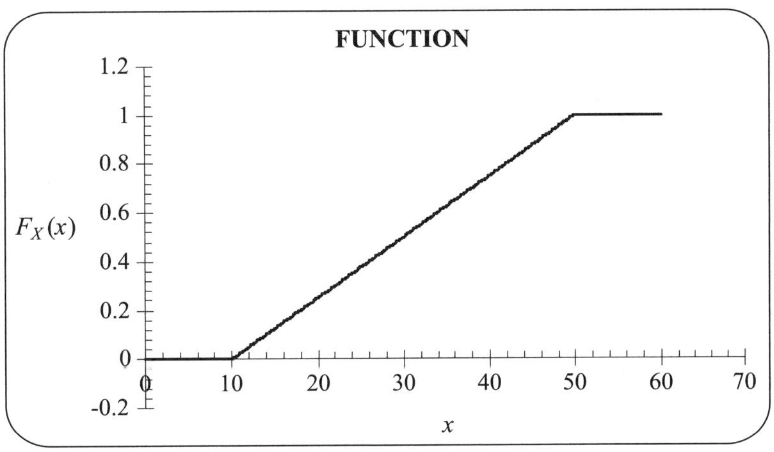


10. Use ***Graphing.xls*** to plot the probability density function, $f_X(x)$, over the interval [0, 60].

$$f_X(x) = \begin{cases} 0 & \text{if } x < 10 \\ \dfrac{1}{40} & \text{if } 10 \le x \le 50 \\ 0 & \text{if } x > 50 \end{cases}$$

Don't forget to remove the unwanted vertical lines.

**SOLUTION**

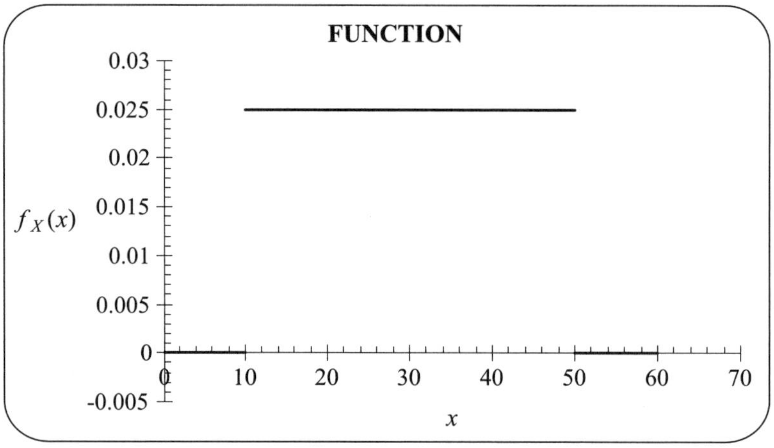


11. Use ***Graphing.xls*** to plot the revenue function,

$$R(q) = -0.000006q^3 - 0.038585q^2 + 303.039287q,$$

over the interval [0, 5000].

**SOLUTION**

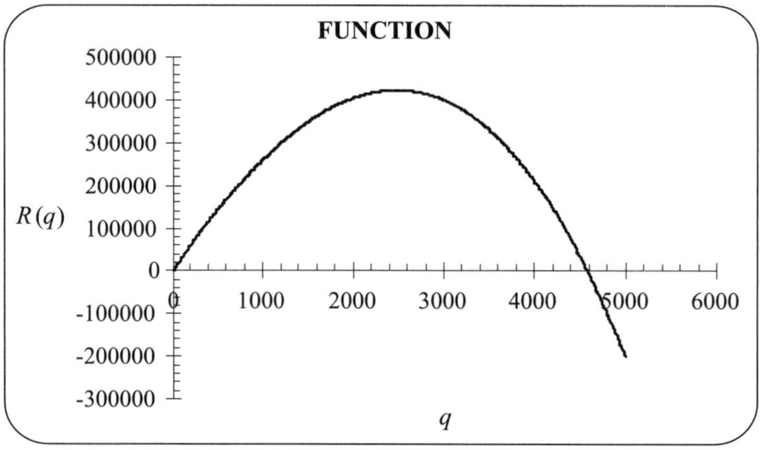


12. Use ***Graphing.xls*** to plot the function $f(x) = sx^2 - tx + 50$ over the interval [–15, 12] with the following conditions:

    a. $s = 2$, $t = 5$; find $f(3)$.

    b. $s = -3$, $t = 9$; find $f(-4)$.

**SOLUTION**

a.

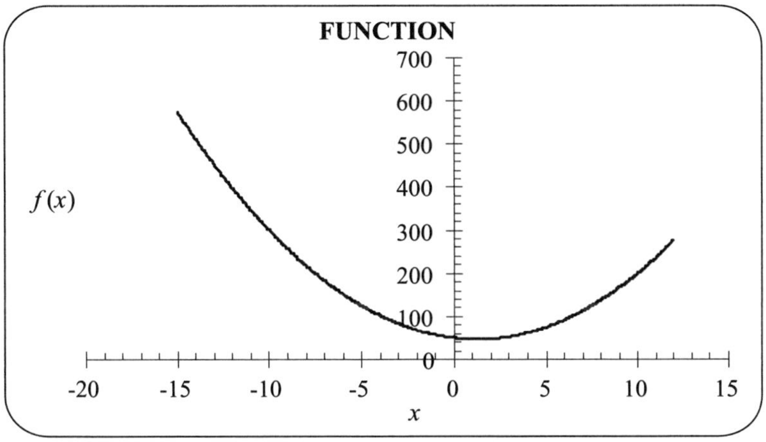


$f(3) = 53$

b.

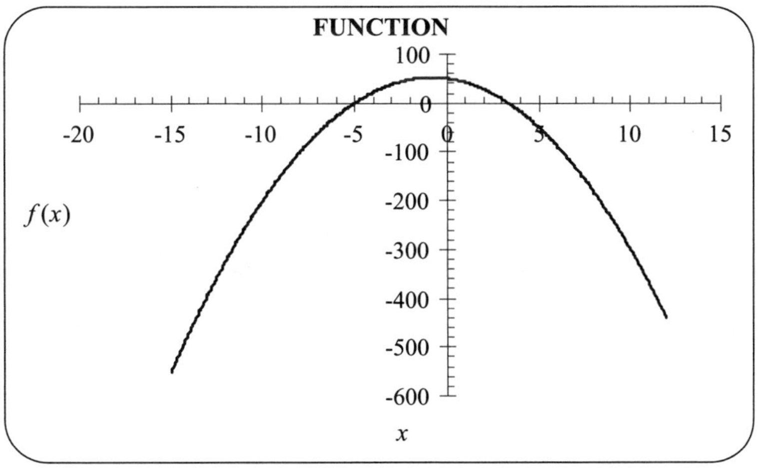


$f(-4) = 38$

# Graphing Functions

(Featuring Microsoft *Excel* 2007)

This section will specifically help you to graph functions using Microsoft *Excel*. In addition you will learn how to graph functions using the utility, ***Graphing.xls***, which is one of your course files.

## *Basic Functions*

Using Microsoft *Excel* to create graphs is fun and easy to do. In the following example you will see how to graph simple functions.

**Example 1:** Plot the quadratic function $f(x) = -0.5x^2 + 1$ over the interval $[-4, 4]$.

## SOLUTION

*Step 1*: Create a table in *Excel* with the same rows and columns as shown below. Column A contains the $x$-values (inputs) and Column B contains the corresponding values of $f(x)$ (outputs).

| | A | B |
|---|---|---|
| 1 | $x$ | $f(x)$ |
| 2 | -4 | -7 |
| 3 | -3 | -3.5 |
| 4 | -2 | -1 |
| 5 | -1 | 0.5 |
| 6 | 0 | 1 |
| 7 | 1 | 0.5 |
| 8 | 2 | -1 |
| 9 | 3 | -3.5 |
| 10 | 4 | -7 |

*Step 2*: Plot the function. Highlight *Cells A2:B10* and then click the **Insert** tab. Click **Scatter** and select **Scatter with Smooth Lines**.

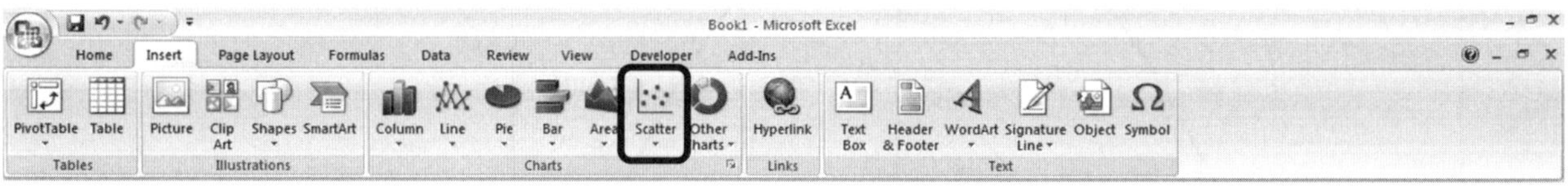


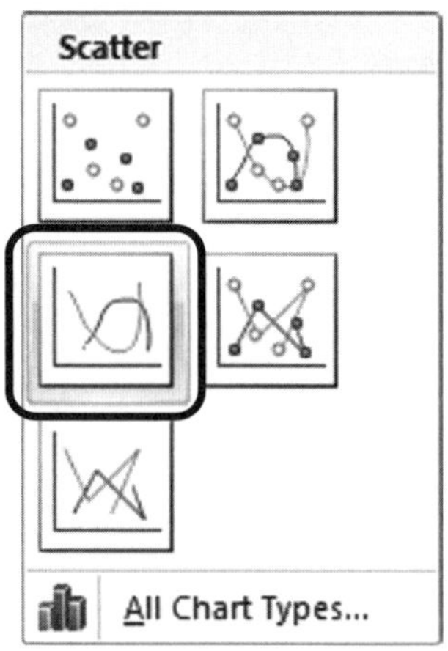


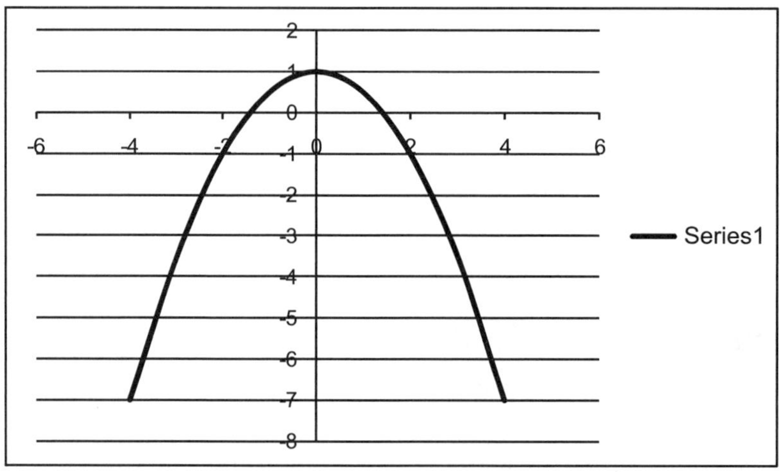


To add chart and axis titles, click the first layout in the **Chart Layouts** group on the **Design** tab.

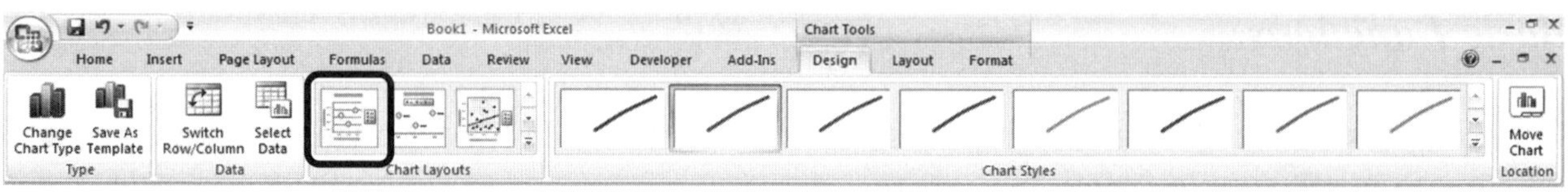

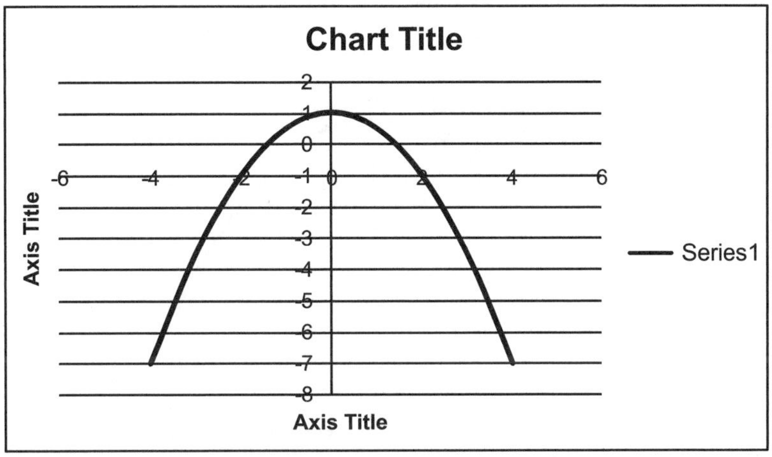


Click the chart title, type QUADRATIC FUNCTION, and press the **Enter** key on your keyboard. Click the $x$-axis title, type $x$, and press the **Enter** key on your keyboard. Click the $y$-axis title, type $y$, and press the **Enter** key on your keyboard.

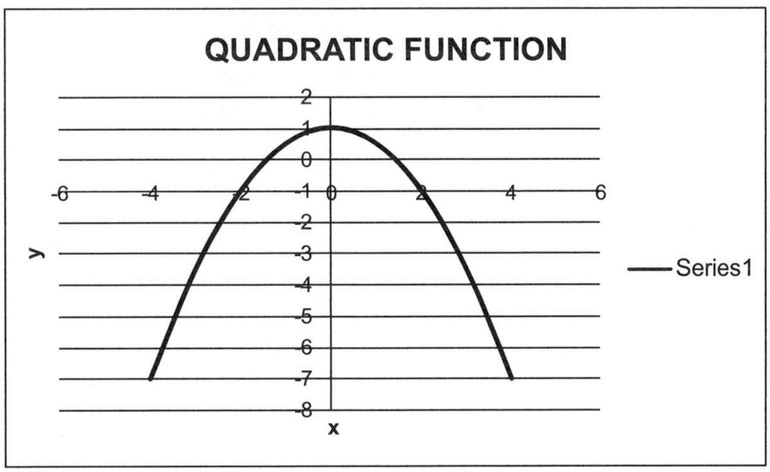


To remove the gridlines, right-click any gridline and select **Delete** from the shortcut menu that appears. To remove the legend, right-click the legend and select **Delete** from the shortcut menu that appears.

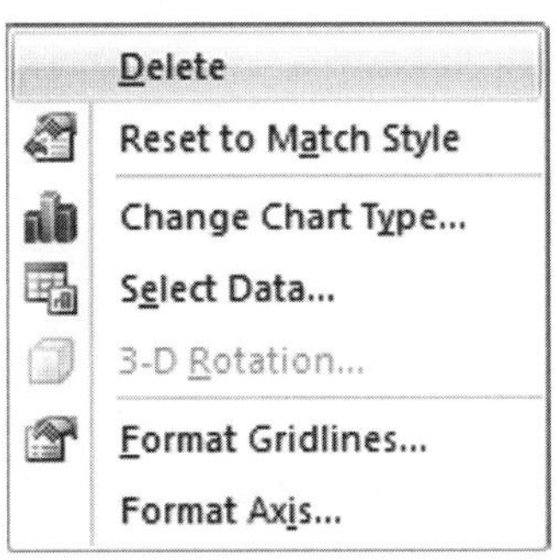


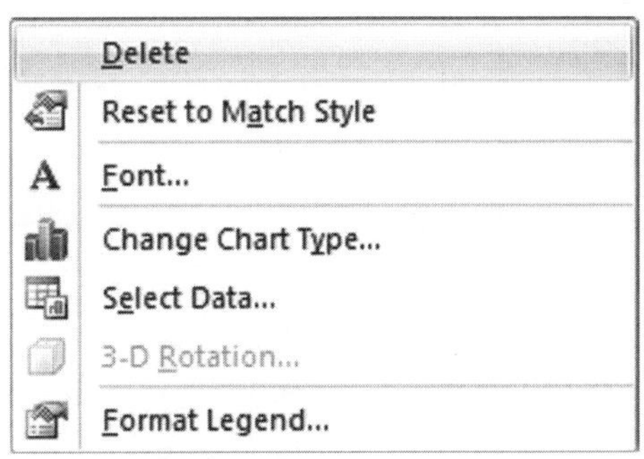


Your graph should look similar to the one shown below.

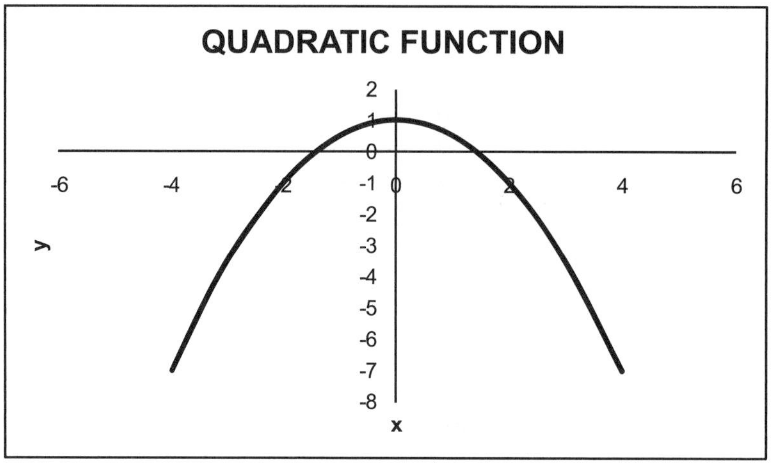


The number of points plotted in **Example 1** was sufficient to show an accurate representation of the function. However, there are many functions that require a large number of data points in order to obtain an accurate representation of the function. The next two graphs are of the same function, $f(x)=\sqrt[3]{x+1}$, and are plotted over the same interval, [–5, 5].

In **GRAPH 1**, 11 equally spaced points are plotted. To compute the difference between the $x$-coordinates of consecutive points, the left endpoint of the interval [–5, 5] is subtracted from the right endpoint, and the result is divided by 10.

$$\frac{5-(-5)}{10}=1$$

| $x$ | $f(x)$ |
|---|---|
| −5.0 | −1.59 |
| −4.0 | −1.44 |
| −3.0 | −1.26 |
| −2.0 | −1.00 |
| −1.0 | 0.00 |
| 0.0 | 1.00 |
| 1.0 | 1.26 |
| 2.0 | 1.44 |
| 3.0 | 1.59 |
| 4.0 | 1.71 |
| 5.0 | 1.82 |

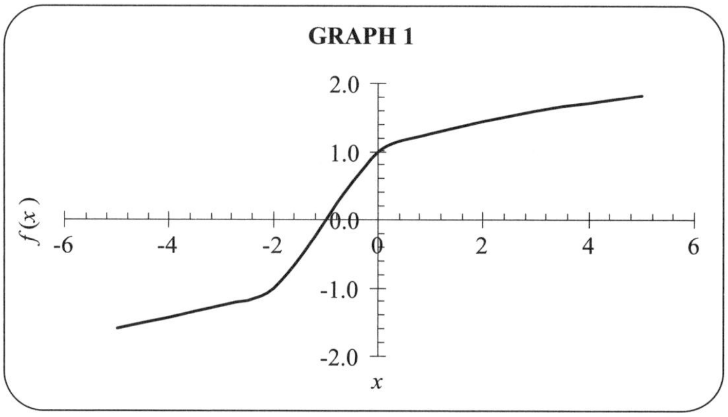


In **GRAPH 2**, 201 equally spaced points are plotted. To compute the difference between the $x$-coordinates of consecutive points, the left endpoint of the interval [–5, 5] is subtracted from the right endpoint, and the result is divided by 200.

$$\frac{5-(-5)}{200}=0.05$$

| $x$ | $f(x)$ |
|---|---|
| −5.00 | −1.59 |
| −4.95 | −1.58 |
| −4.90 | −1.57 |
| −4.85 | −1.57 |
| ⋮ | ⋮ |
| 4.85 | 1.80 |
| 4.90 | 1.81 |
| 4.95 | 1.81 |
| 5.00 | 1.82 |

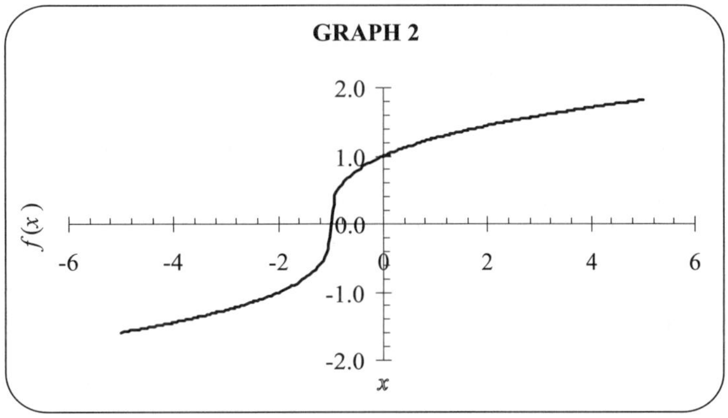


**GRAPH 2** is more accurate than **GRAPH 1**. This is because more points were plotted in **GRAPH 2** than in **GRAPH 1**. As the number of points plotted increases, the more accurate is the graph of the function. However, there are exceptions to this rule. One exception is a linear function (straight line). It doesn't matter whether two points are plotted or one thousand points are plotted; the graph will look the same. Another exception is a quadratic function (parabola). It doesn't matter whether 10 points are plotted or 200 points are plotted; the graph will look the same.

More complicated functions require more data points to obtain an accurate representation of the function. If you are not sure whether or not the function you are plotting would require a large number of data points, it is better to use more data points than fewer; plotting 500 points for any function usually will ensure an accurate graph.

The two previous graphs **GRAPH 1** and **GRAPH 2**, were plotted over the interval [–5, 5], but the interval was divided into a different number of subintervals for each graph. Looking back at these examples, you can see how this was achieved. Any interval $[a, b]$ can be divided into $n$ subintervals of equal width by dividing the width of the interval by $n$. Let $\Delta x$ represent the width of each subinterval. Then

$$\Delta x = \frac{b-a}{n}.$$

This gives a total of $n+1$ data points. The $x$-coordinate of first point will be $a$, and the $x$-coordinate of the last point (the $n+1$ point) will be $b$.

**Example 2**: Suppose that a function is to be plotted over the interval [–7, 7].

a. What is the width of each subinterval if the interval is divided into 400 subintervals of equal width?
b. How many data points will be plotted?
c. What is the $x$-coordinate of the second data point?
d. What is the $x$-coordinate of the second to last data point?

**SOLUTION**

a. $\Delta x = \dfrac{7-(-7)}{400} = \dfrac{14}{400} = 0.035.$

b. The total number of data points is $n+1$ where $n = 400$, so 401 data points will be plotted.

c. The $x$-coordinate of the second data point is equal to the sum of the $x$-coordinate of the first data point and the width of the first subinterval; $-7 + 0.035 = -6.965$.

d. The $x$-coordinate of the second to last data point is equal to the difference between the $x$ coordinate of the last data point and the width of the last subinterval; $7 - 0.035 = 6.965$.

## *Piecewise Functions*

A piecewise function consists of two or more functions, each of which is defined on a different part of the domain. An example is given below.

$$f(x) = \begin{cases} x^2+1 & \text{if } -1 \le x \le 1 \\ 2 & \text{elsewhere} \end{cases}$$

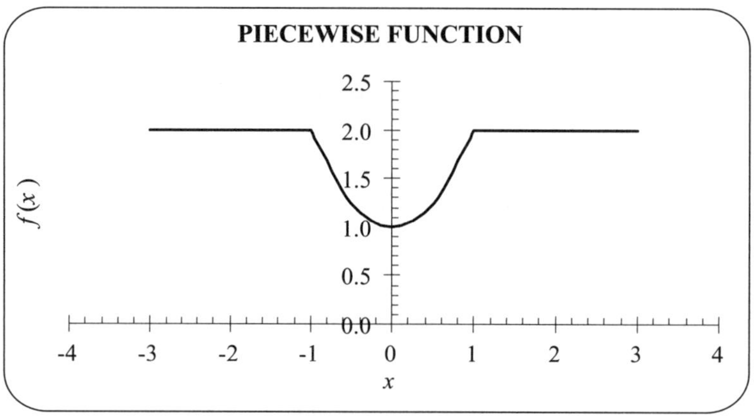


As you can see, when $x$ is between $-1$ and 1 the graph is a parabola, and when $x$ is less than $-1$ or greater than 1 the graph is a horizontal line.

Graphing piecewise functions in *Excel* requires the use of logical operators (*if*, *and*, *or*, *then*). Note that you will not see the logical operator *then* in *Excel* but it is implied with the use of a comma. The *Excel* formula shown below can be read as follows: ***If*** $-1$ is less than or equal to $x$ ($x$ is represented by the cell reference) ***and*** $x$ is less than or equal to 1, ***then*** $f(x) = x^2+1$, ***else*** $f(x) = 2$.

IF(AND(-1<=A2,A2<=1,),A2^2+1,2)

↑ ***and*** ↑ ***then*** ↑ ***else***

**Example 3**: Plot the cumulative distribution function, $F_X(x)$, over the interval [–10, 40] using 501 equally spaced points.

$$F_X(x) = \begin{cases} 0 & \text{if } x < 0 \\ 1 - e^{-0.25x} & \text{if } x \geq 0 \end{cases}$$

**SOLUTION**

$$\Delta x = \frac{40-(-10)}{501-1} = \frac{50}{500} = 0.1$$

| $x$ | $F_X(x)$ |
|---|---|
| −10.0 | 0 |
| −9.9 | 0 |
| −9.8 | 0 |
| −9.7 | 0 |
| −9.6 | 0 |
| −9.5 | 0 |
| ⋮ | ⋮ |
| 39.6 | 0.999950 |
| 39.7 | 0.999951 |
| 39.8 | 0.999952 |
| 39.9 | 0.999953 |
| 40.0 | 0.999955 |

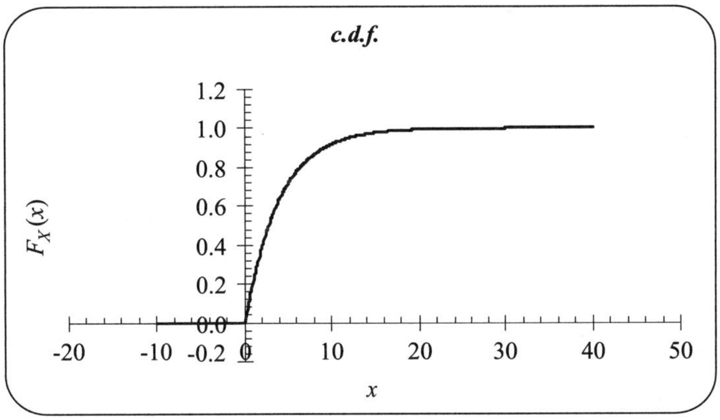


**Example 4**: Plot the probability density function, $f_X(x)$, over the interval [−10, 40] using 501 equally spaced points.

$$f_X(x) = \begin{cases} 0 & \text{if } x < 0 \\ 0.25e^{-0.25x} & \text{if } x \geq 0 \end{cases}$$

**SOLUTION**

$$\Delta x = \frac{40-(-10)}{501-1} = \frac{50}{500} = 0.1.$$

| $x$ | $f_X(x)$ |
|---|---|
| -10.0 | 0 |
| -9.9 | 0 |
| -9.8 | 0 |
| -9.7 | 0 |
| -9.6 | 0 |
| -9.5 | 0 |
| ⋮ | ⋮ |
| 39.6 | 1.25E-05 |
| 39.7 | 1.22E-05 |
| 39.8 | 1.19E-05 |
| 39.9 | 1.16E-05 |
| 40.0 | 1.13E-05 |

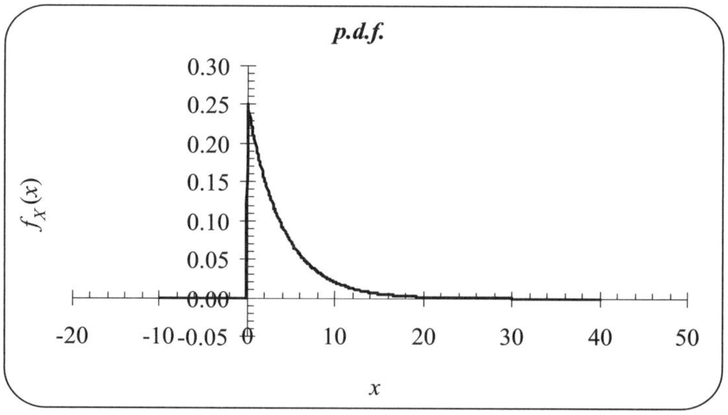


When plotting a piecewise function, unwanted vertical lines will appear when there are vertical gaps between parts of the graph. An unwanted vertical line is shown above in the graph of $f_X(x)$. This vertical line is not part of the graph and it needs to be removed. The removal of the vertical line will be covered in ***Advanced Formatting***.

## *Multiple Functions*

There are times when you may want to plot two functions on one graph. Specifically, in the ***Marketing Project***, you will be asked to plot both the cost function and the revenue function on one graph.

**Example 5:** Plot the functions $f(x) = 10$ and $g(x) = 0.75x + 5$ over the interval [0, 10].

### SOLUTION

*Step 1*: Create a table in *Excel* with the same rows and columns as shown below. Column A contains the $x$-values, and Columns B and C contain the corresponding values of $f(x)$ and $g(x)$, respectively.

| | A | B | C |
|---|---|---|---|
| 1 | $x$ | $f(x)$ | $g(x)$ |
| 2 | 0 | 10 | 5 |
| 3 | 1 | 10 | 5.75 |
| 4 | 2 | 10 | 6.5 |
| 5 | 3 | 10 | 7.25 |
| 6 | 4 | 10 | 8 |
| 7 | 5 | 10 | 8.75 |
| 8 | 6 | 10 | 9.5 |
| 9 | 7 | 10 | 10.25 |
| 10 | 8 | 10 | 11 |
| 11 | 9 | 10 | 11.75 |
| 12 | 10 | 10 | 12.5 |

*Step 2*: Plot the functions. Highlight *Cells A2:C12* and then click the **Insert** tab. Click **Scatter** and select **Scatter with Smooth Lines**. To add chart and axis titles, click the first layout in the **Chart Layouts** group on the **Design** tab. Click the chart title, type MULTIPLE FUNCTIONS, and press the **Enter** key on your keyboard. Click the $x$-axis title, type $x$, and press the **Enter** key on your keyboard. Click the $y$-axis title, type $y$, and press the **Enter** key on your keyboard. To remove the gridlines, right-click any gridline and select **Delete** from the shortcut menu that appears.

To name the series, right-click either of the lines that are plotted in the graph and select **Select Data** from the shortcut menu that appears.

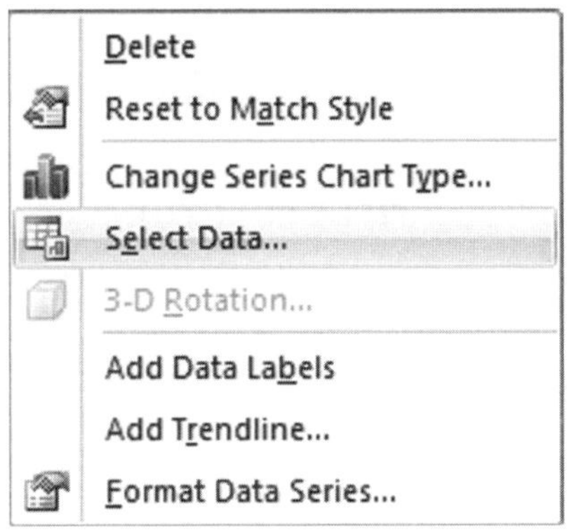


In the **Legend Entries (Series)** section of the **Select Data Source** dialog box, click **Series1** and then click **Edit**.

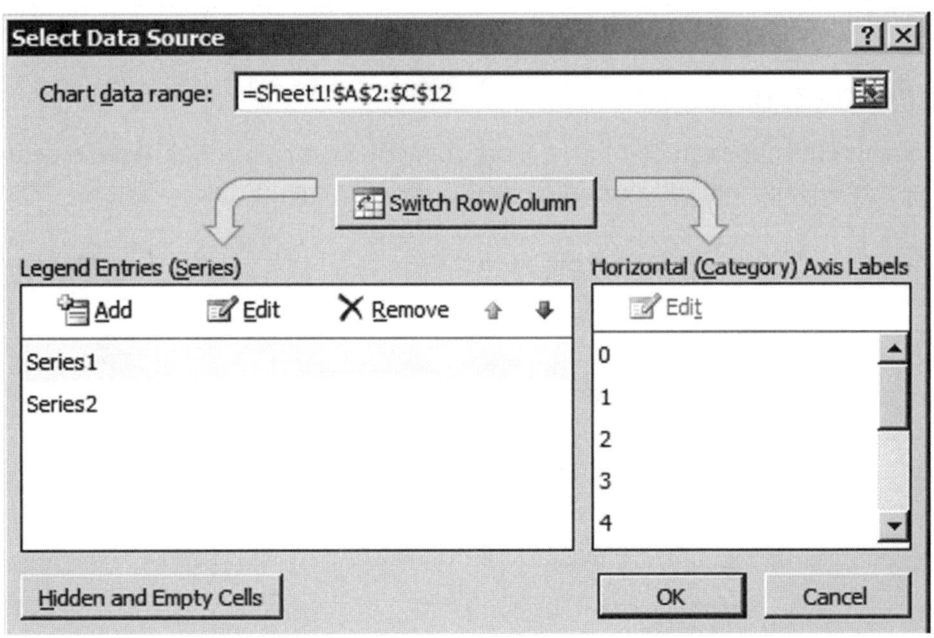


Type the name $f(x)$ in the **Series name** box, and click **OK**. In the **Legend Entries (Series)** section of the **Select Data Source** dialog box, click **Series2** and then click **Edit**. Type the name $g(x)$ in the **Series name** box, and click **OK**.

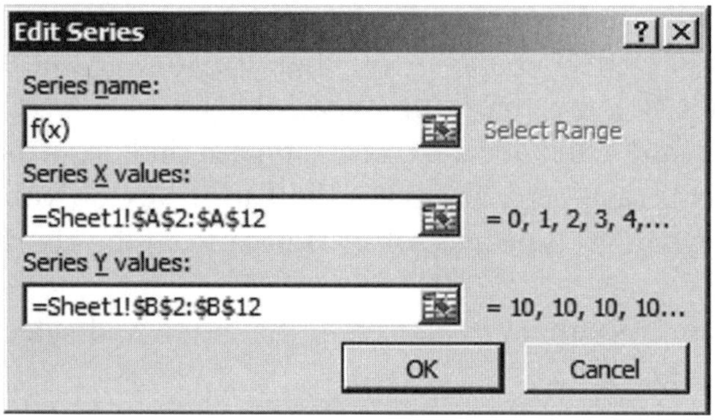


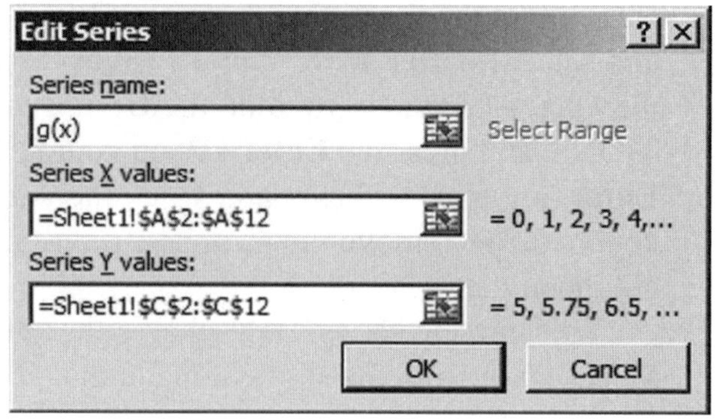


Click **OK** in the **Select Data Source** dialog box. Your graph should look similar to the one shown below.

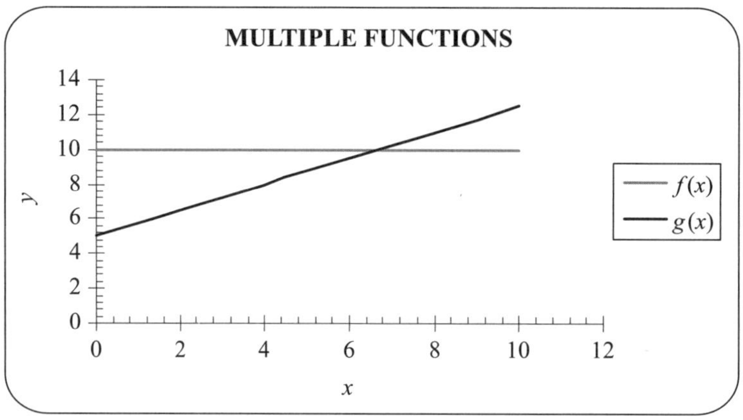


## *Advanced Formatting*

It is easy to change the appearance of any chart element. Just right-click the element you want to modify and select the appropriate option from the shortcut menu that appears.

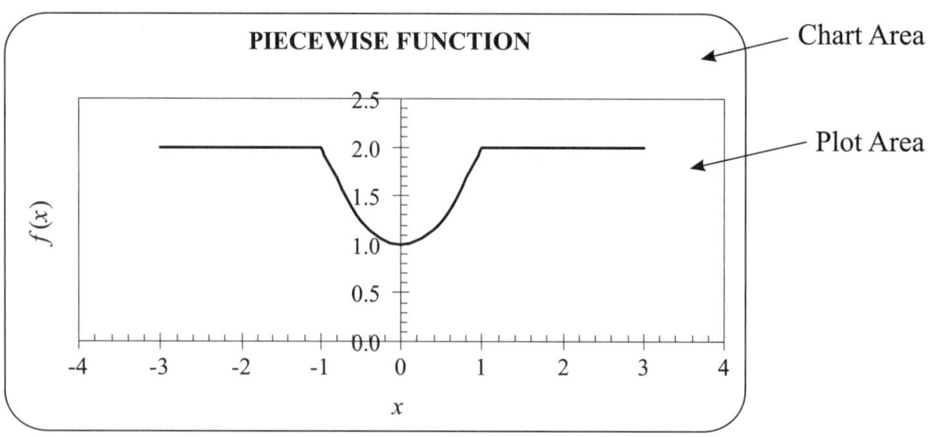


*Formatting the Chart Area*

To change the appearance of the chart area, right-click the chart area and select **Format Chart Area** from the shortcut menu that appears.

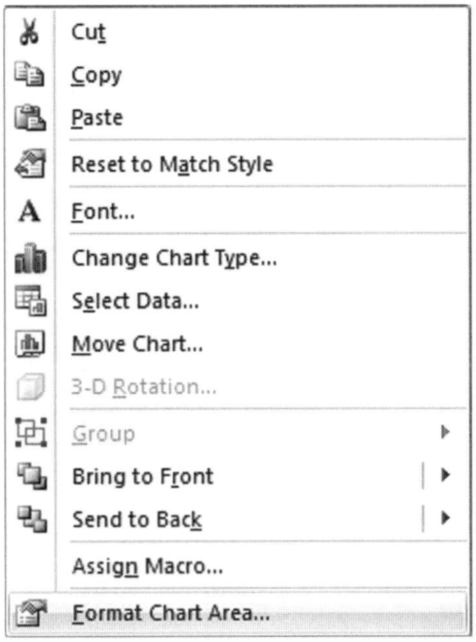


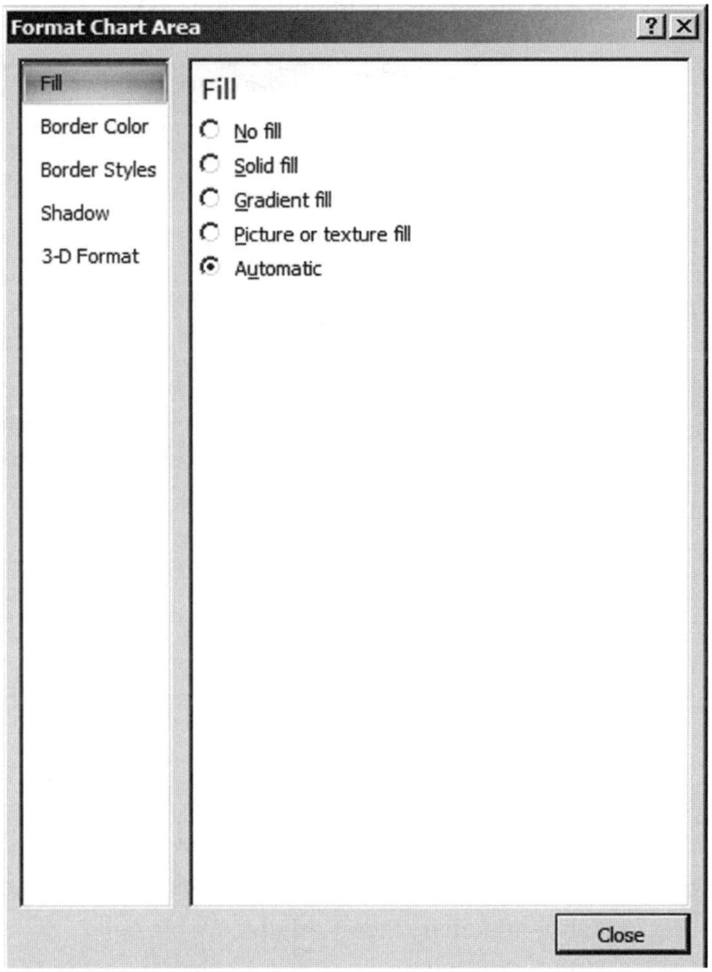


With the **Fill** options, you can change the color of the chart area and add a gradient or texture. With the **Border Color** options, you can change the color of the chart area border. With the **Border Styles** options, you can change the width and type of the chart area border. You can also round the corners of the chart area border.

To change the font, font style, and size of the text and numbers in the chart, right-click the chart area and select the appropriate options from the **Mini toolbar** that appears. Note that changing the attributes of the text and numbers in the chart area will affect all text and numbers in the chart — this includes the chart title, axis titles, and tick mark labels. To change the font, font style, and size of any title individually, right-click the title you want to change and select the appropriate options from the **Mini toolbar** that appears.

You can also change the font, font style, and size of any part of a title. For example, you can format the $X$ in the titles $F_X(x)$ and $f_X(x)$ as a subscript. To do this, click the axis title and highlight the $X$. Click the **Home** tab, and then click the **Font** dialog box launcher next to **Font**.

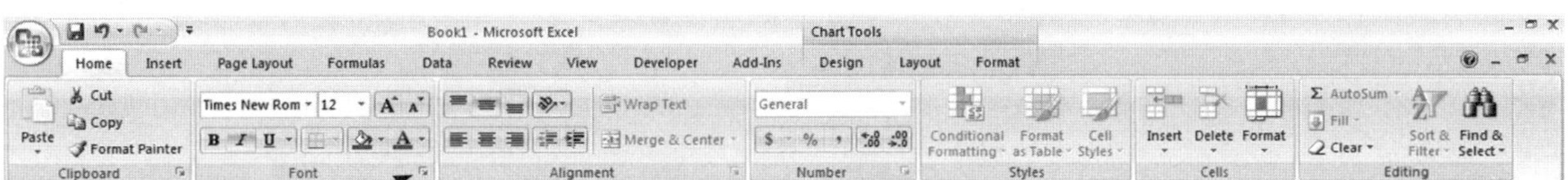


Font dialog box launcher

Check the **Subscript** check box and click **OK**.

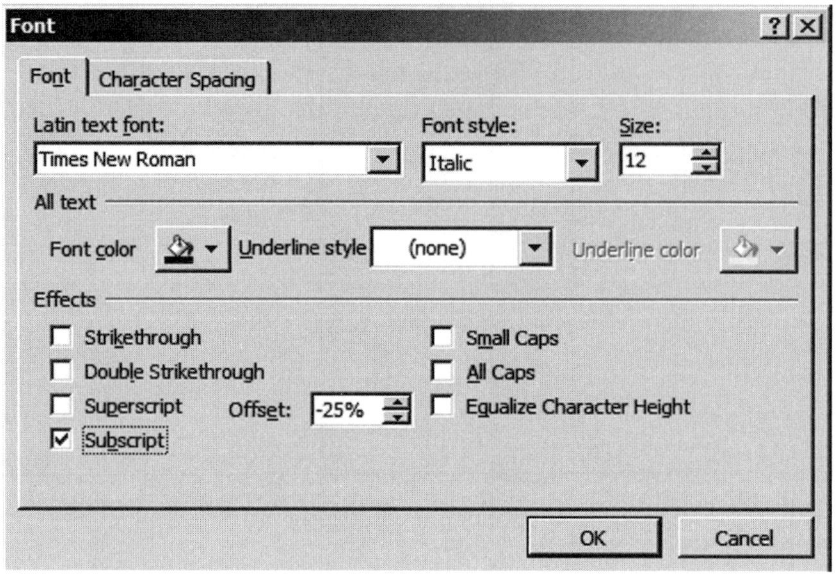


**Example 6**: Format the graph from **Example 1** as follows:

a. Change the chart area border color to black.
b. Change the width of chart area border to 3 pt.
c. Round the corners of the chart area border.
d. Change the chart area font to Times New Roman.
e. Change the chart area font size to 12.
f. Change the font style of the $x$-axis title to italic.
g. Change the font style of the $y$-axis title to italic.

**SOLUTION**

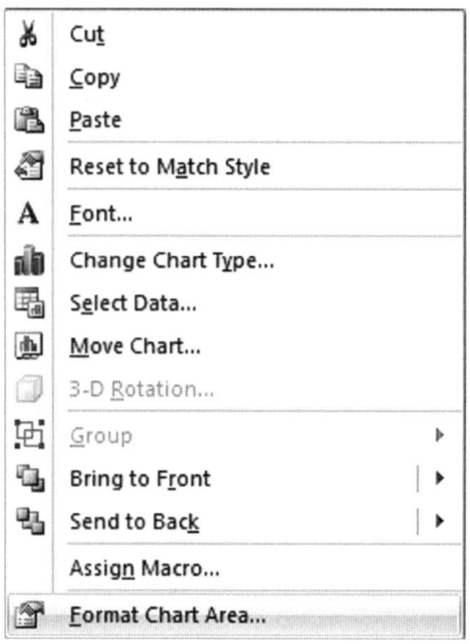

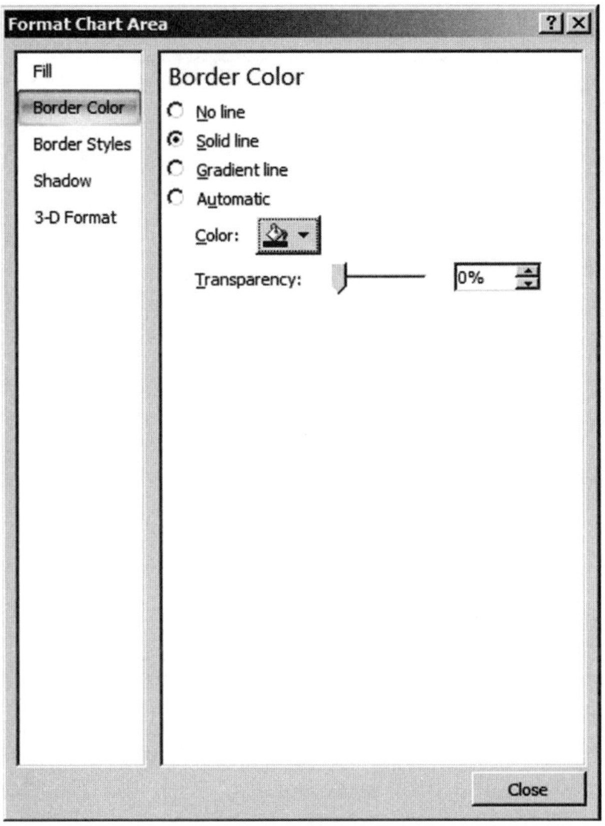


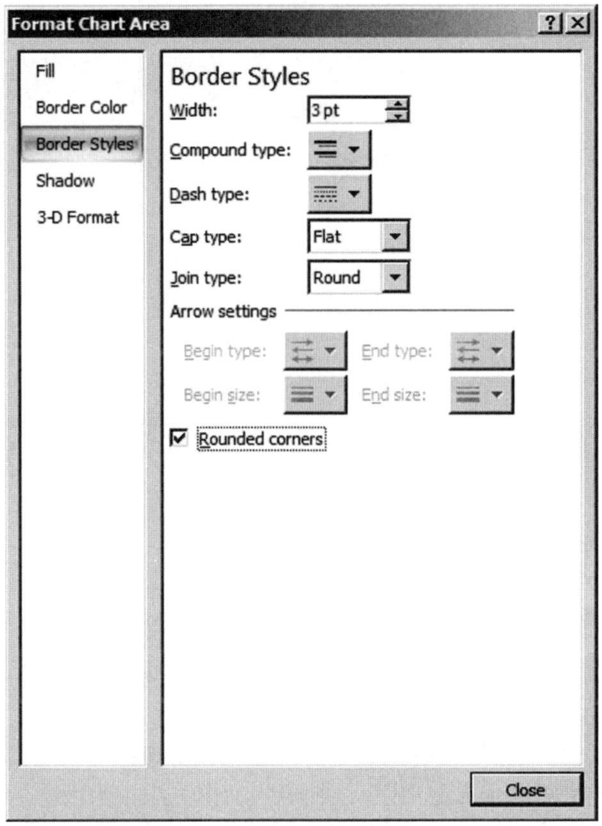


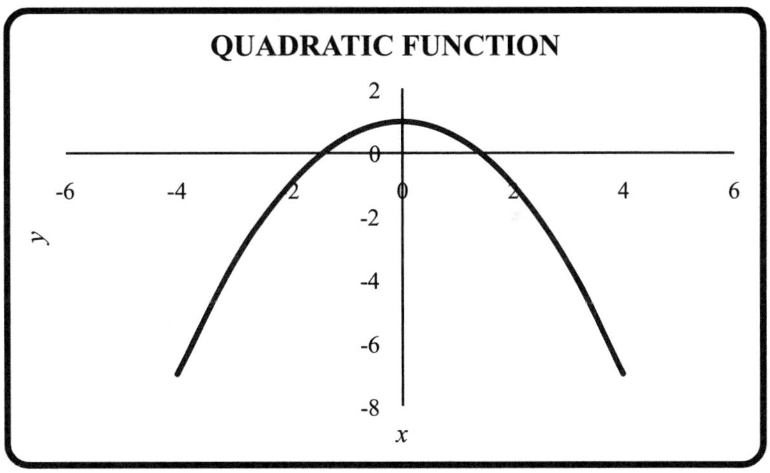


*Formatting the Plot Area*

To change the appearance of the plot area, right-click the plot area and select **Format Plot Area** from the shortcut menu that appears.

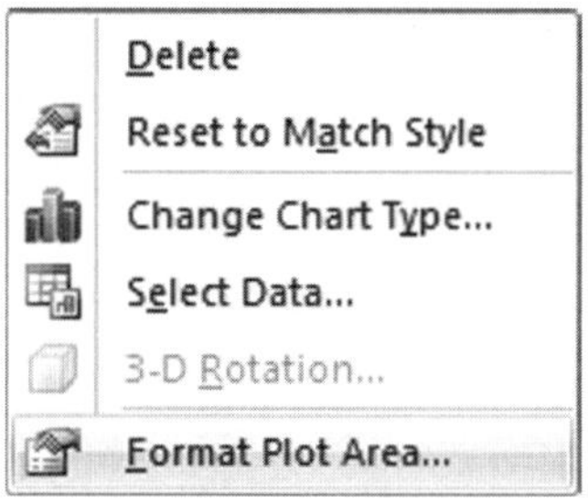


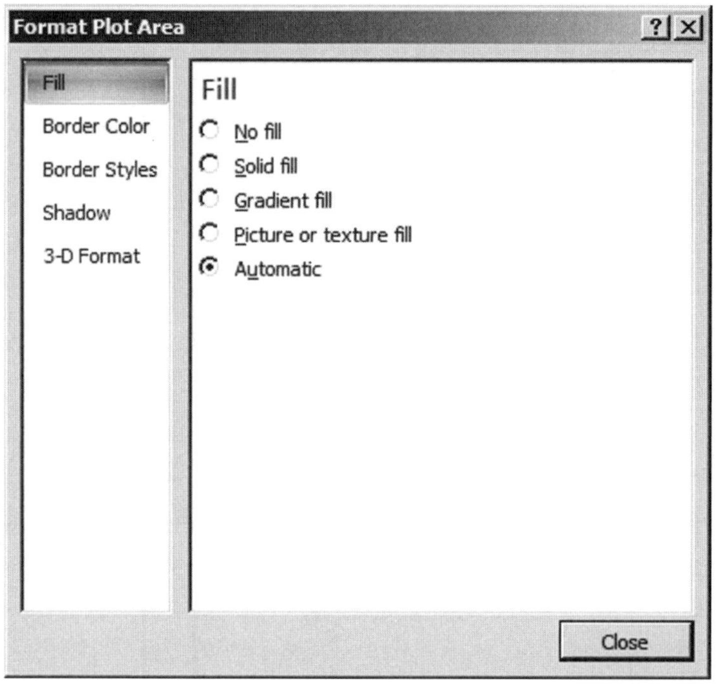


With the **Fill** options, you can change the color of the plot area and add a gradient or texture. With the **Border Color** options, you can change the color of the plot area border. With the **Border Styles** options, you can change the width and type of the plot area border.

*Formatting the Data Series*

To change the appearance of the line that is plotted in the graph, right-click the line and select **Format Data Series** from the shortcut menu that appears.

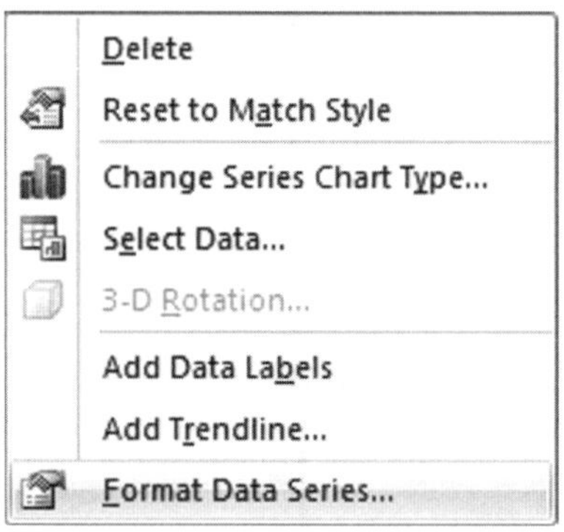


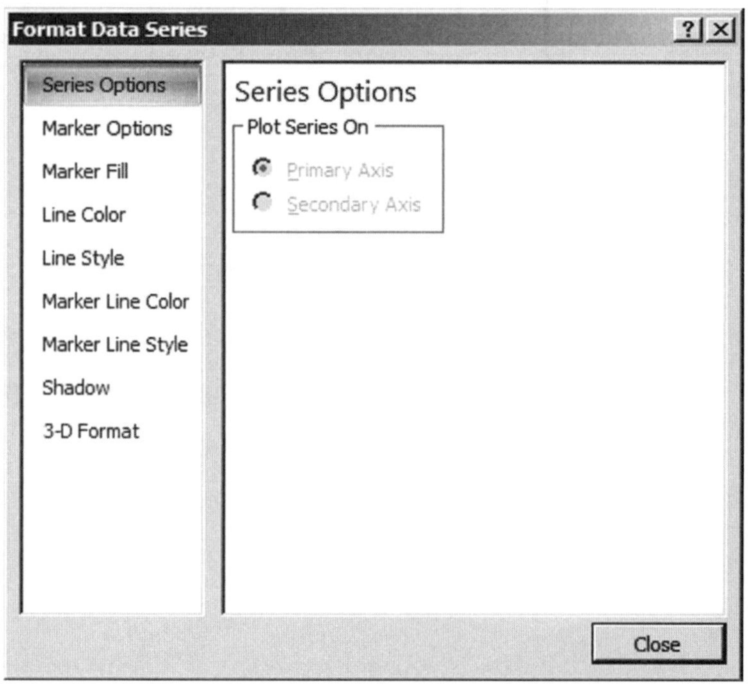

With the **Line Color** options, you can change the color of the line. With the **Line Style** options, you can change the width and type of the line.

**Example 7**: Change the width of the line that is plotted in the graph from **Example 6** to 3 pt.

**SOLUTION**

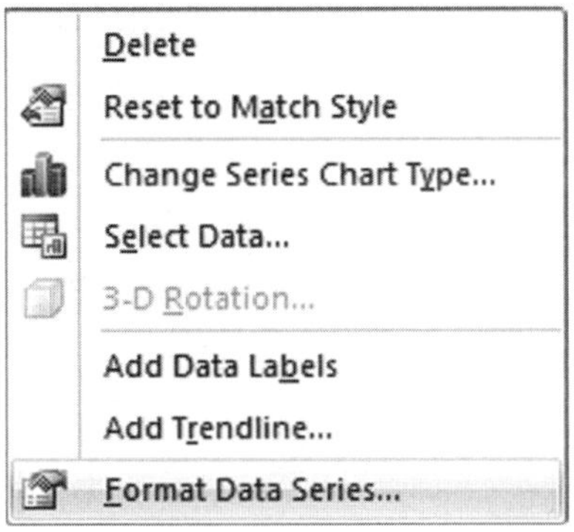


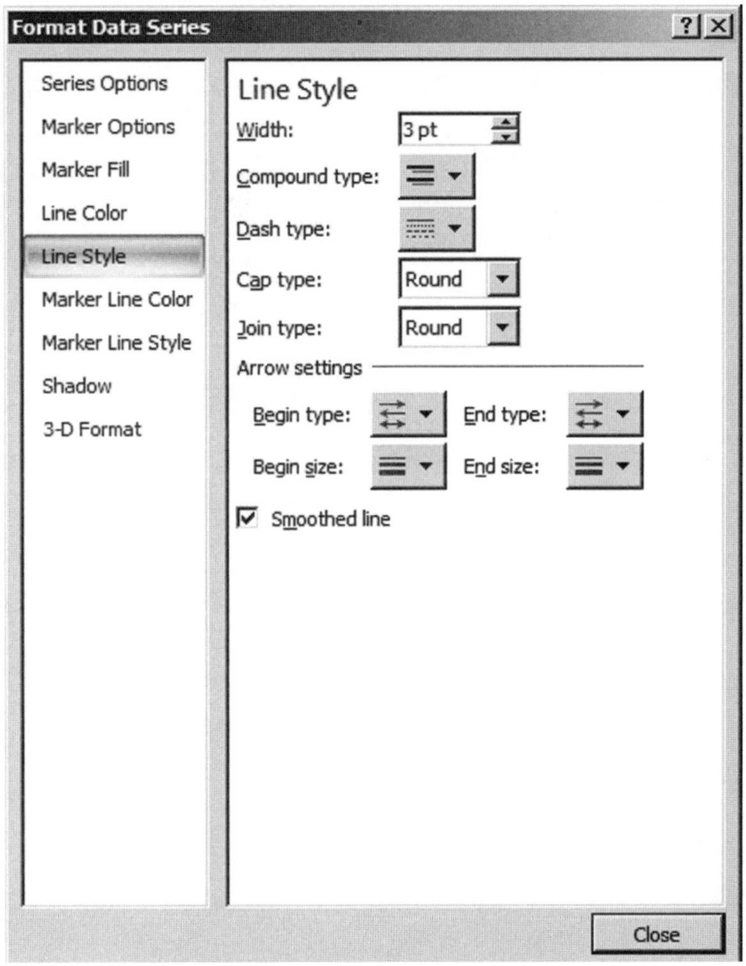


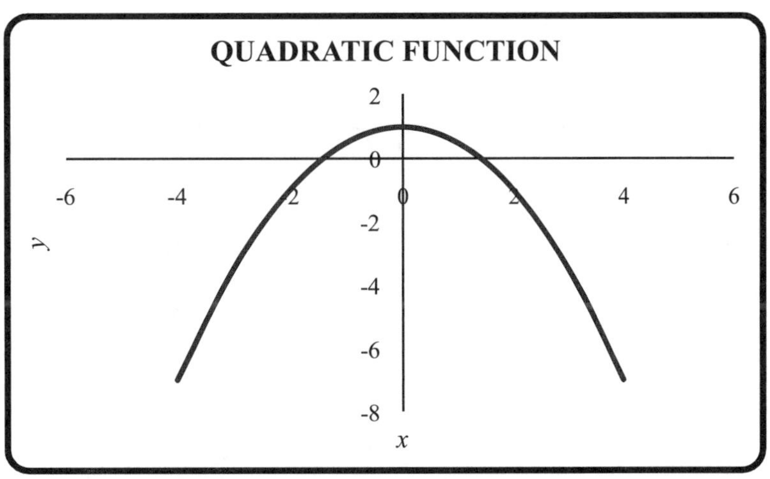

Recall that unwanted vertical lines will appear when there are vertical gaps between parts of the graph of a piecewise function. To remove an unwanted vertical line, click the data series and then click the appropriate point. Right-click the point and select **Format Data Point** from the shortcut menu that appears. Click **Line Color** and select the **No line** option button. Then click **CLOSE**.

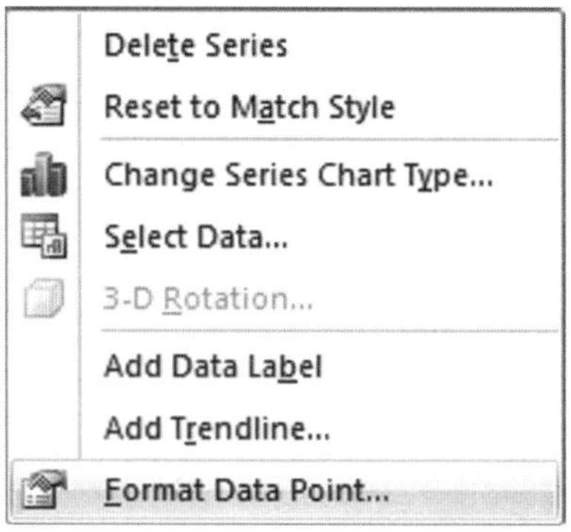


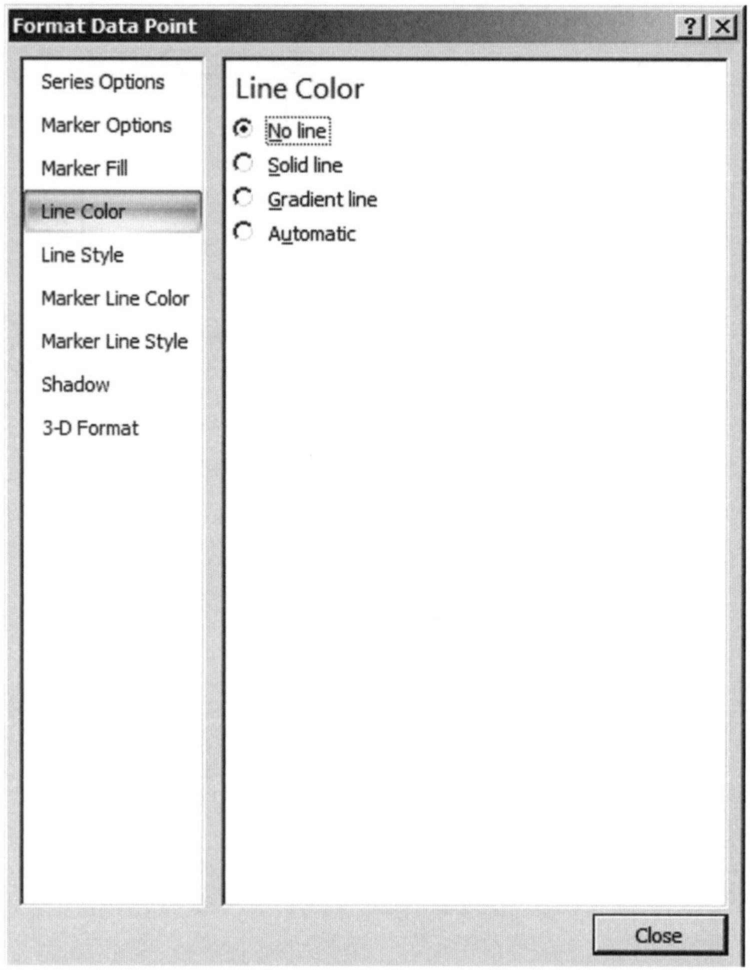


**Example 8:** Remove the unwanted vertical line from the graph from **Example 4**.

## SOLUTION

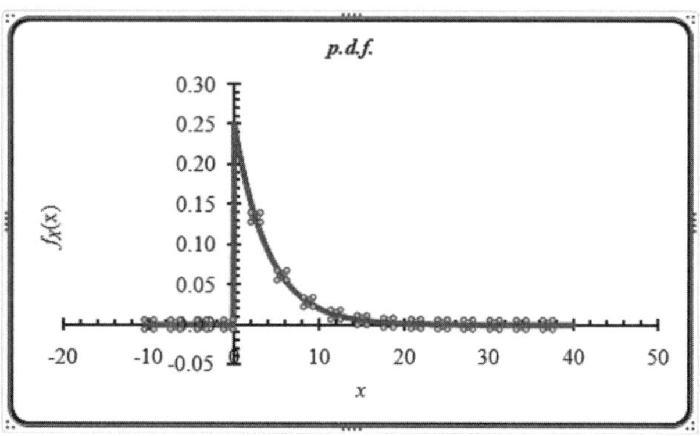


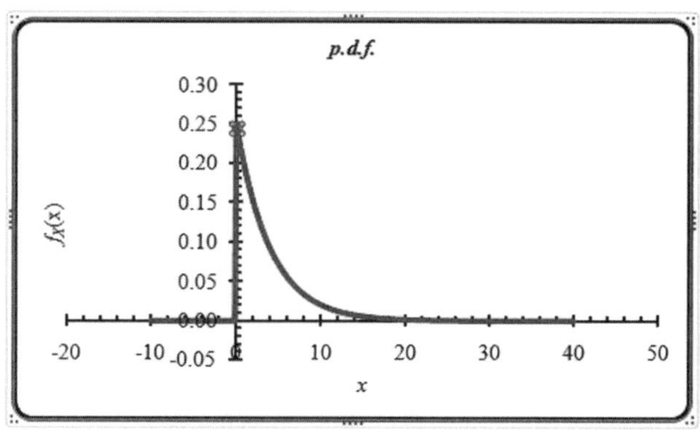

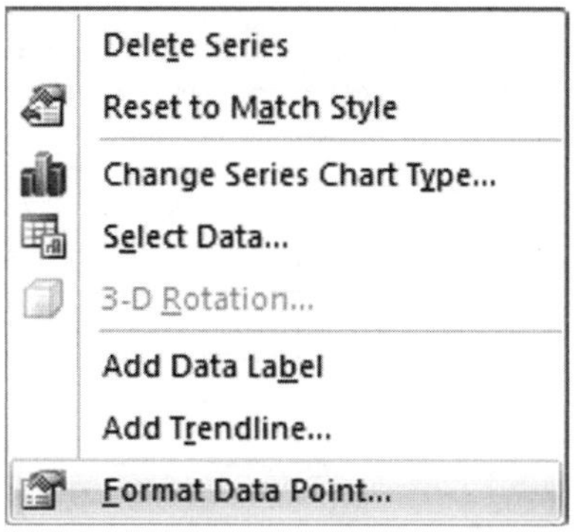


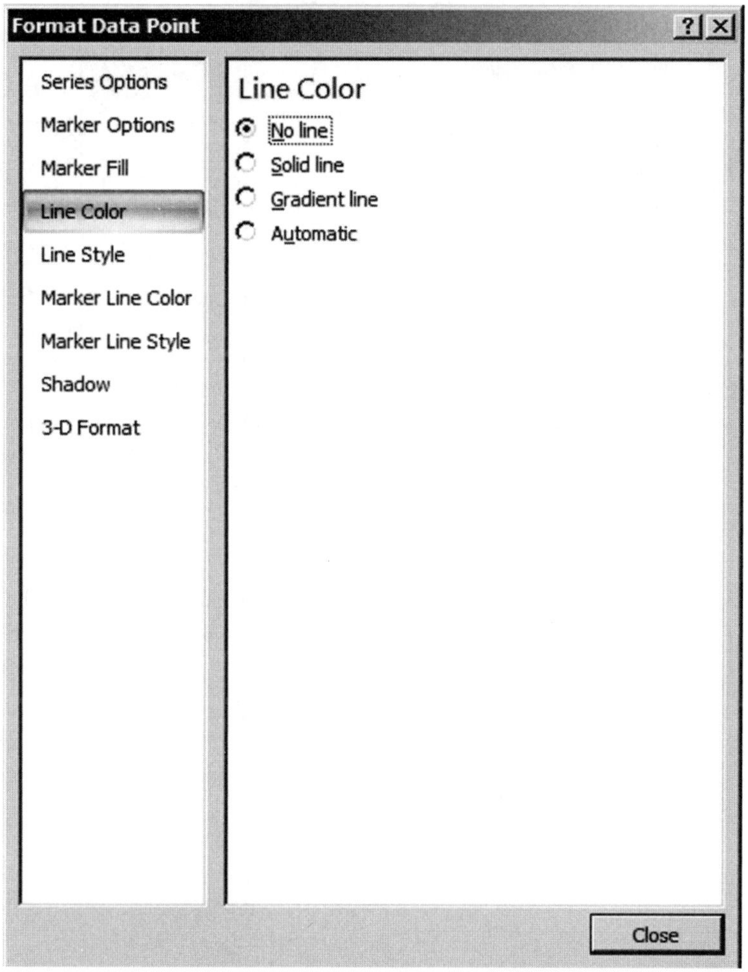


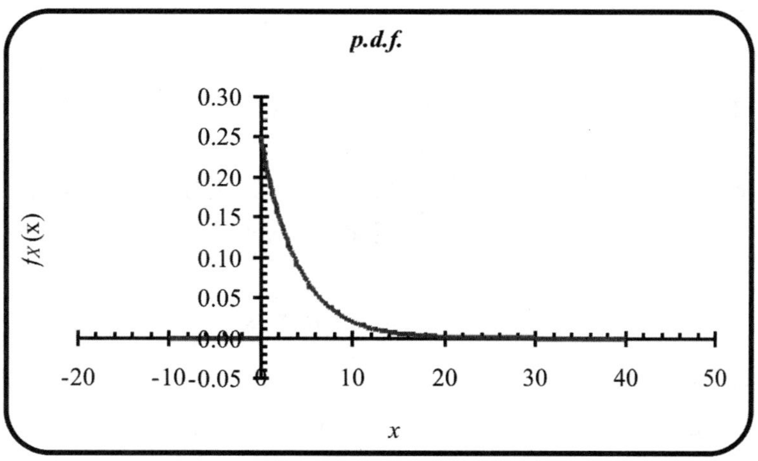


*Formatting the Axes*

To change the appearance of an axis, right-click the axis you want to modify and select **Format Axis** from the shortcut menu that appears.

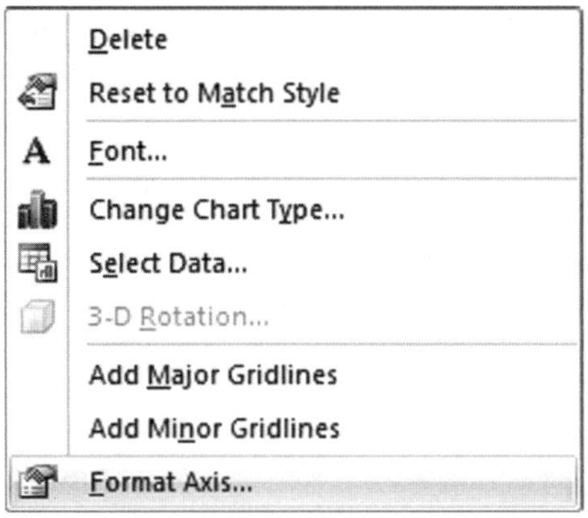


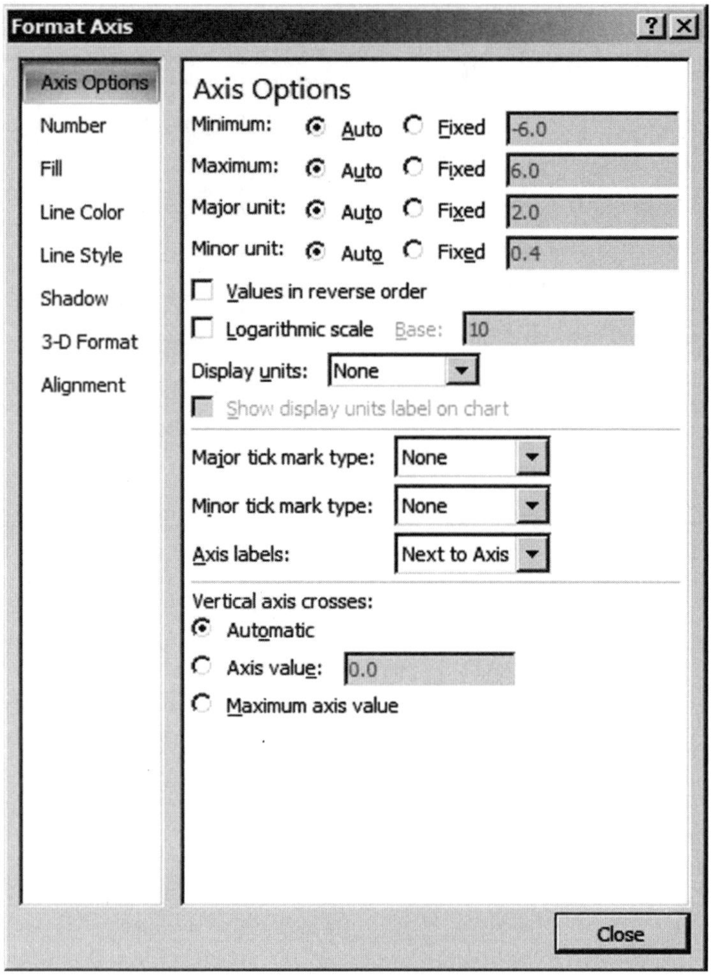


With the **Axis Options**, you can change the minimum value, maximum value, major unit, and minor unit. With the **Number** options, you can change the number format of the tick mark labels. With the **Line Color** options, you can change the color of the axis. With the **Line Style** options, you can change the width and type of the axis.

To change the font, font style, and size of the tick mark labels, right-click the axis and select the appropriate options from the **Mini toolbar** that appears.

**Example 9**: Format the graph from **Example 7** as follows:

a. Change the color of the $x$-axis to black.

b. Change the scale of the $x$-axis so that it has a minimum of –6, a maximum of 6, a major unit of 2, and a minor unit of 1.

c. Change the color of the $y$-axis to black.

## SOLUTION

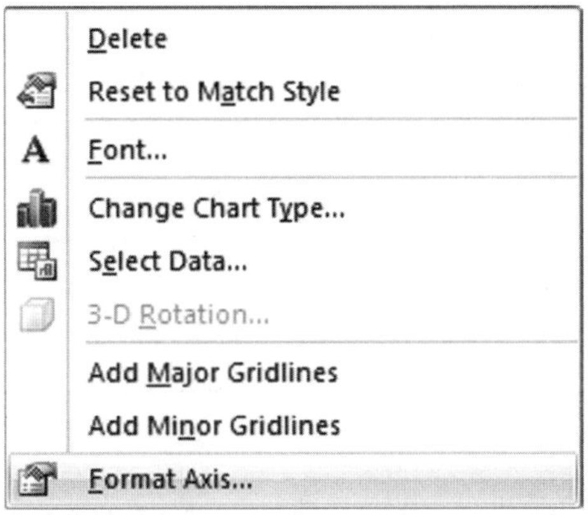
Delete
Reset to Match Style
Font...
Change Chart Type...
Select Data...
3-D Rotation...
Add Major Gridlines
Add Minor Gridlines
Format Axis...

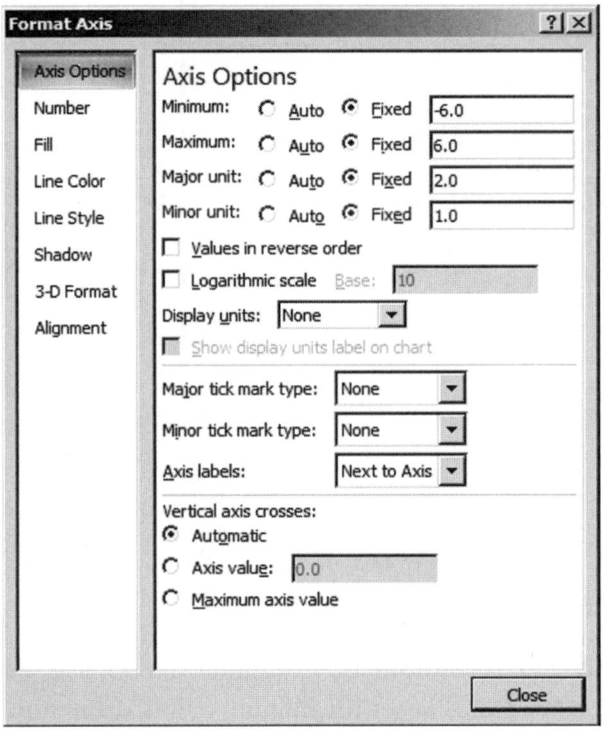
Format Axis
Axis Options
Number
Fill
Line Color
Line Style
Shadow
3-D Format
Alignment
Axis Options
Minimum: Auto Fixed -6.0
Maximum: Auto Fixed 6.0
Major unit: Auto Fixed 2.0
Minor unit: Auto Fixed 1.0
Values in reverse order
Logarithmic scale Base: 10
Display units: None
Show display units label on chart
Major tick mark type: None
Minor tick mark type: None
Axis labels: Next to Axis
Vertical axis crosses:
Automatic
Axis value: 0.0
Maximum axis value
Close

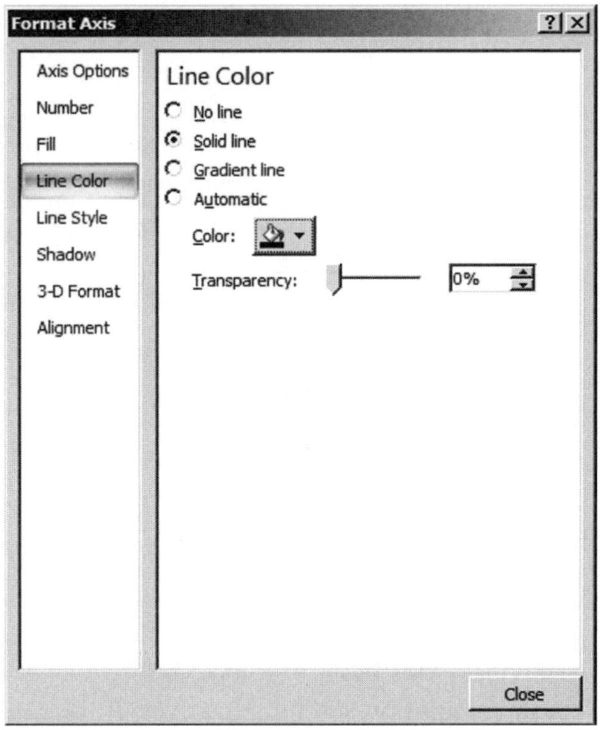
Format Axis
Axis Options
Number
Fill
Line Color
Line Style
Shadow
3-D Format
Alignment
Line Color
No line
Solid line
Gradient line
Automatic
Color:
Transparency: 0%
Close

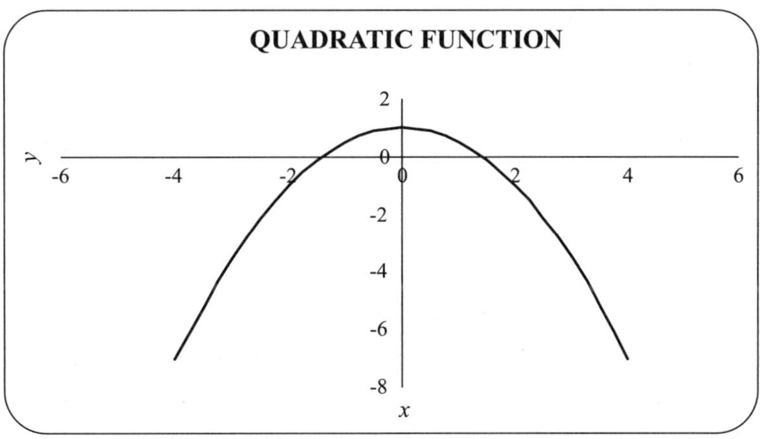
QUADRATIC FUNCTION
y
x
-6
-4
-2
0
2
4
6
-8

## *Using Graphing.xls*

***Graphing.xls*** is a utility found in your course files. Before you actually begin to use ***Graphing.xls***, a few important issues need to be addressed.

First, if the security alert below appears when you open ***Graphing.xls***, you will need to enable the macro.

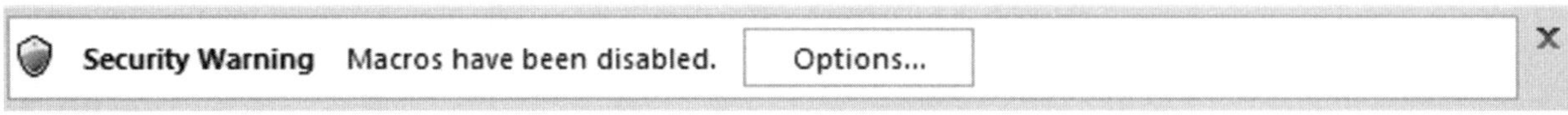


Click **Options** on the Message bar. Then select the **Enable this content** option button and click **OK**.

Second, you need to make certain that *Excel* is set for automatic computation. This is the default calculation setting. However, if for some reason *Excel* is set for manual calculation, click the **Formulas** tab and then click **Calculation Options** and select **Automatic**.

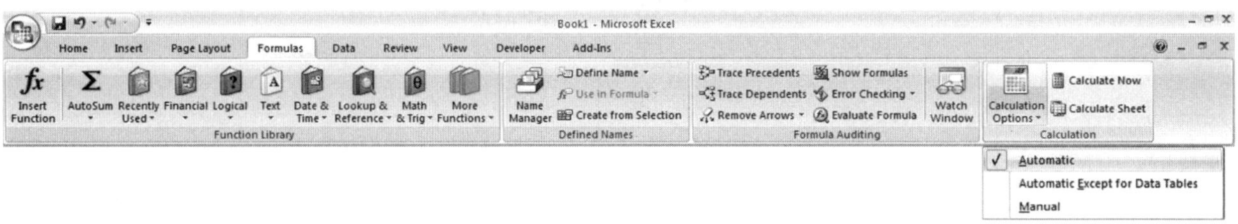


Finally, when you enter formulas in ***Graphing.xls***, always call the variable *x*.

***Graphing.xls*** is easy to use. Simply enter the formula in *Cell C18* and the endpoints of the plot interval in *Cells H18* and *I18*. Then click the **Graph** button to run the macro. Note that if you change the formula in *Cell C18*, you need to rerun the macro.

| | A | B | C | D | E | F | G | H | I | J | K | L |
|---|---|---|---|---|---|---|---|---|---|---|---|---|
| 16 | | **Definition** | | | **Computation** | | | **Plot Interval** | | | **Constants** | |
| 17 | | **Formula for *f*(*x*)** | | | *x* | *f*(*x*) | | *a* | *b* | | *s* | |
| 18 | | | = | | | = | | | | | *t* | |
| 19 | | | | | | | | | | | *u* | |
| 20 | | | | | | | | | | | *v* | |
| 21 | | | | | | | | | | | *w* | |

Initially, there is no need to enter anything in the **Computation** or **Constants** cells.

**Example 10**: Use ***Graphing.xls*** to plot the quadratic function $f(x) = x^2 + 4x + 1$ over the interval $[-10, 10]$.

**SOLUTION**

Enter the formula = *x*^2+4**x*+1 in *Cell C18*, and the values –10 and 10 in *Cells H18* and *I18*, respectively. Then click the **Graph** button to run the macro.

| | A | B | C | D | E | F | G | H | I |
|---|---|---|---|---|---|---|---|---|---|
| 16 | | **Definition** | | | **Computation** | | | **Plot Interval** | |
| 17 | | **Formula for *f*(*x*)** | | | *x* | *f*(*x*) | | *a* | *b* |
| 18 | | | 1 | | | 1 | | -10 | 10 |

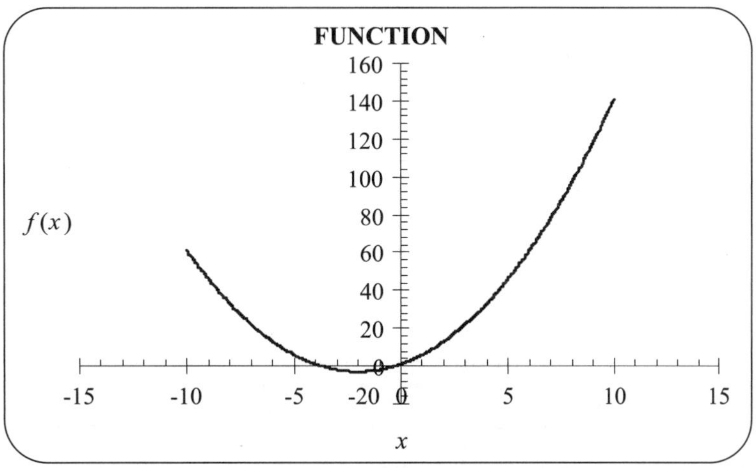


The number 1 in *Cell C18* is the value of the function at $x = 0$. The same is true for the 1 in *Cell F18*. If you wanted to evaluate the function at any other value of $x$, you would enter that value of $x$ in *Cell E18*. That is, if you entered 2 in *Cell E18*, the number 13 would appear in *Cell F18* since $f(2) = 13$.

The formula for the function in **Example 10** contains a constant. If you wanted to see the effects of changing the constant without changing the formula every time, you would use the

**Constants** cells. To do this, you would enter the formula in *Cell C18* using any of the variables *s, t, u, v,* or *w* in place of the constant. If you decided to use *t* as the constant, you would enter the formula =x^2+4*x+t in *Cell C18* and a value for *t* in *Cell L18*. Then you would click the **Graph** button to run the macro.

| | A | B | C | D | E | F | G | H | I | J | K | L |
|---|---|---|---|---|---|---|---|---|---|---|---|---|
| 16 | | **Definition** | | | **Computation** | | | **Plot Interval** | | | **Constants** | |
| 17 | | **Formula for $f(x)$** | | | $x$ | $f(x)$ | | $a$ | $b$ | | $s$ | |
| 18 | | | 50 | | | 50 | | -10 | 10 | | $t$ | 50 |
| 19 | | | | | | | | | | | $u$ | |
| 20 | | | | | | | | | | | $v$ | |
| 21 | | | | | | | | | | | $w$ | |

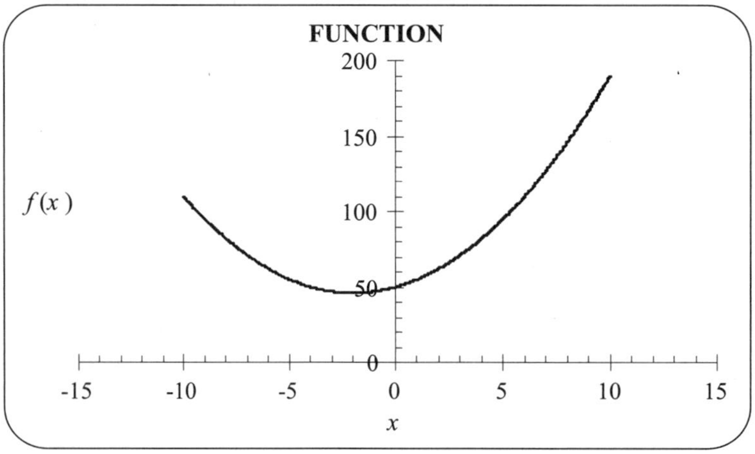


**Example 11**: Use ***Graphing.xls*** to plot the function $f(x) = u^x + s$ over the interval $[-4, 4]$ with the following conditions:

a. $s = 5$, $u = 1$; find $f(2)$.
b. $s = -50$, $u = 2$; find $f(-2)$.
c. $s = 40$, $u = 3$; find $f(3.71)$.

**SOLUTION**

a.

| **Definition** | | **Computation** | | **Plot Interval** | | **Constants** | |
|---|---|---|---|---|---|---|---|
| **Formula for $f(x)$** | | $x$ | $f(x)$ | $a$ | $b$ | $s$ | 5 |
| | 6 | 2 | 6 | -4 | 4 | $t$ | |
| | | | | | | $u$ | 1 |
| | | | | | | $v$ | |
| | | | | | | $w$ | |

Formula for $f(x)$: =u^x+s

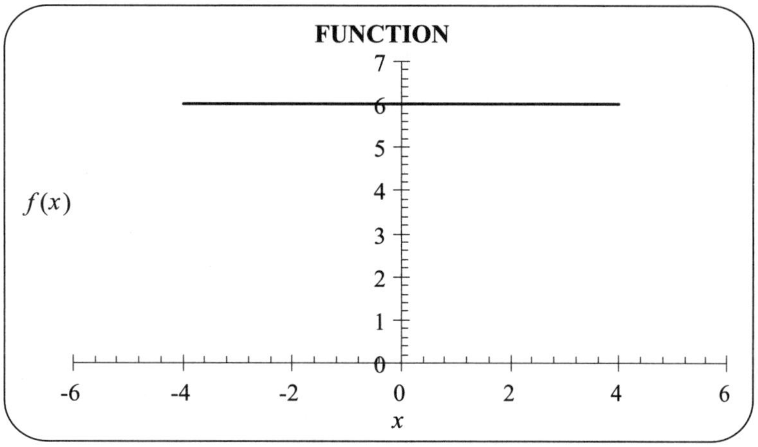


b.

**Definition**

| Formula for $f(x)$ |
|---|
| -49 |

**Computation**

| $x$ | $f(x)$ |
|---|---|
| -2 | -49.75 |

**Plot Interval**

| $a$ | $b$ |
|---|---|
| -4 | 4 |

**Constants**

| | |
|---|---|
| $s$ | -50 |
| $t$ | |
| $u$ | 2 |
| $v$ | |
| $w$ | |

Formula for $f(x)$: =u^x+s

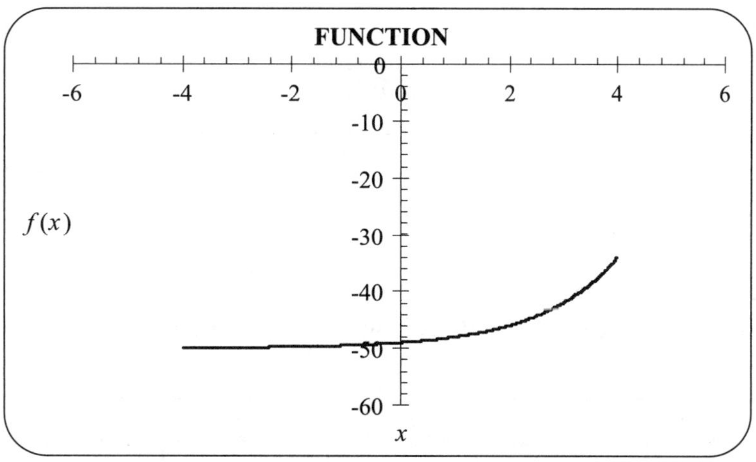


c.

**Definition**

| Formula for $f(x)$ |
|---|
| 41 |

**Computation**

| $x$ | $f(x)$ |
|---|---|
| 3.71 | 98.90062 |

**Plot Interval**

| $a$ | $b$ |
|---|---|
| -4 | 4 |

**Constants**

| | |
|---|---|
| $s$ | 40 |
| $t$ | |
| $u$ | 3 |
| $v$ | |
| $w$ | |

Formula for $f(x)$: =u^x+s

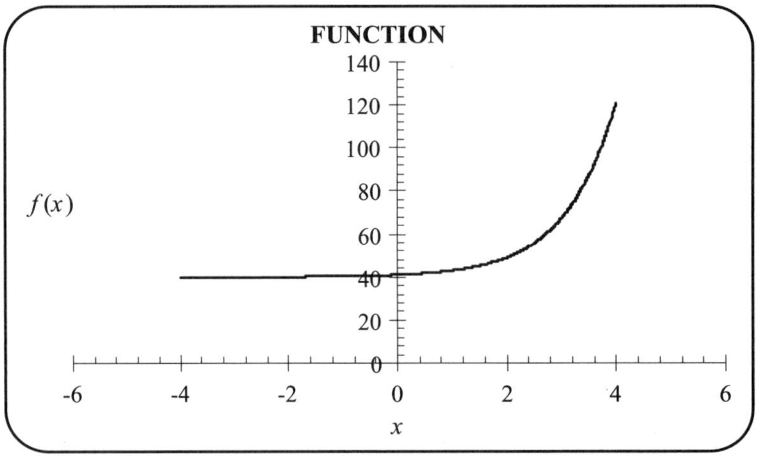


Notice that all of the values used for $u$ in **Example 11** were greater than zero. When $u$ is less than zero, *Excel* will not correctly interpret the formula used in **Example 11**. Look what happens if $s = 10$ and $u = -2$.

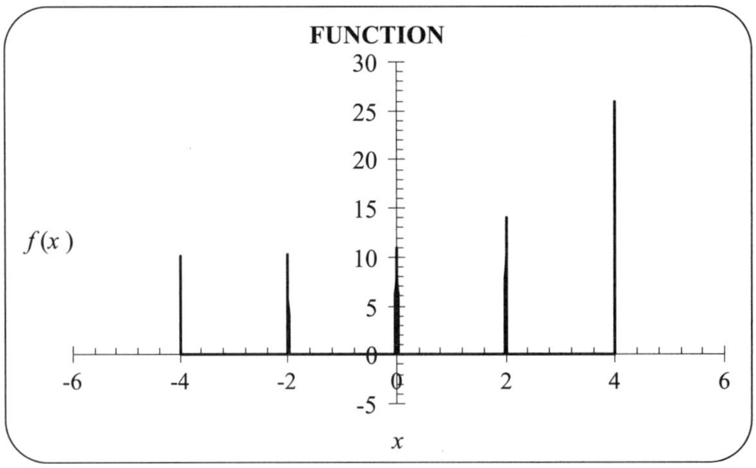


To obtain correct results, enter either of the formulas =-1*u^x+s or =-(u^x)+s in *Cell C18* and the value 2 in *Cell L17*. Then click the **Graph** button to run the macro.

**Definition**

| Formula for $f(x)$ |
|---|
| 9 |

**Computation**

| $x$ | $f(x)$ |
|---|---|
| | 9 |

**Plot Interval**

| $a$ | $b$ |
|---|---|
| -4 | 4 |

**Constants**

| | |
|---|---|
| $s$ | 10 |
| $t$ | |
| $u$ | 2 |
| $v$ | |
| $w$ | |

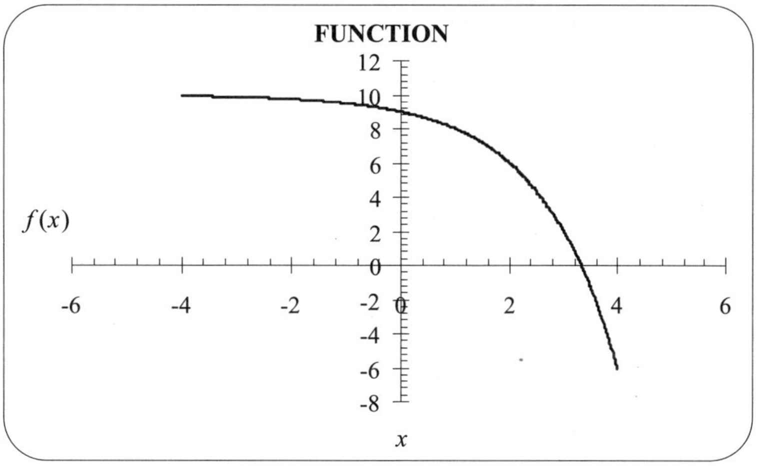


## *Exercises*

1. Use *Excel* to plot the linear function $f(x) = -\frac{3}{5}x + 8$ over the interval [−7, 7]. The graph should have the following characteristics:
   a. The chart title is **LINEAR FUNCTION**.
   b. The $x$-axis title is $x$.
   c. The $y$-axis title is $f(x)$.
   d. The legend is not shown.
   e. The line that is plotted in the graph is the maximum weight.

**SOLUTION**

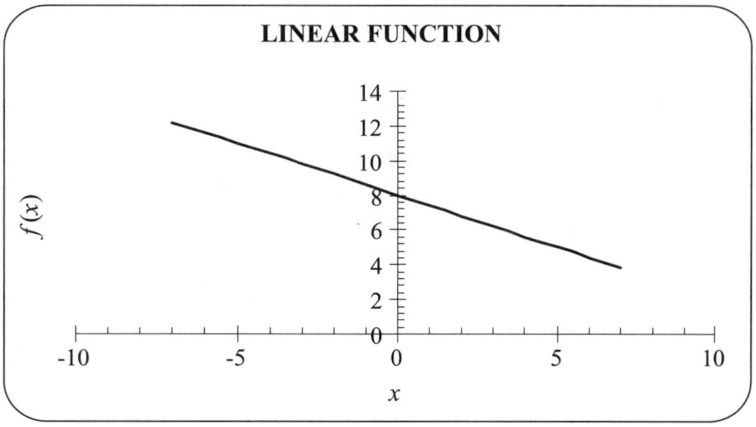


2. Suppose that a function is to be plotted over the interval [−10, 15].
   a. What is the width of each subinterval if the interval is divided into 1000 subintervals of equal width?
   b. How many data points will be plotted?

c. What is the $x$-coordinate of the second data point?

**SOLUTION**

a. 0.025
b. 1001
c. −9.975

3. Suppose that a third degree polynomial is to be plotted over the interval [−6, 6] using 601 equally spaced points.
   a. What is the value of $\Delta x$?
   b. What is the $x$-coordinate of the 599th data point?

**SOLUTION**

a. 0.02
b. 5.96

4. Use *Excel* to plot the cubic function $g(x) = 3x^3 - 2x^2 + 6x - 25$ over the interval [−5, 5] using 201 equally spaced points. The graph should have the following characteristics:
   a. The chart title is **CUBIC FUNCTION**.
   b. The $x$-axis title is $x$.
   c. The $y$-axis title is $g(x)$.
   d. The legend is not shown.
   e. The line that is plotted in the graph is the maximum weight.

**SOLUTION**

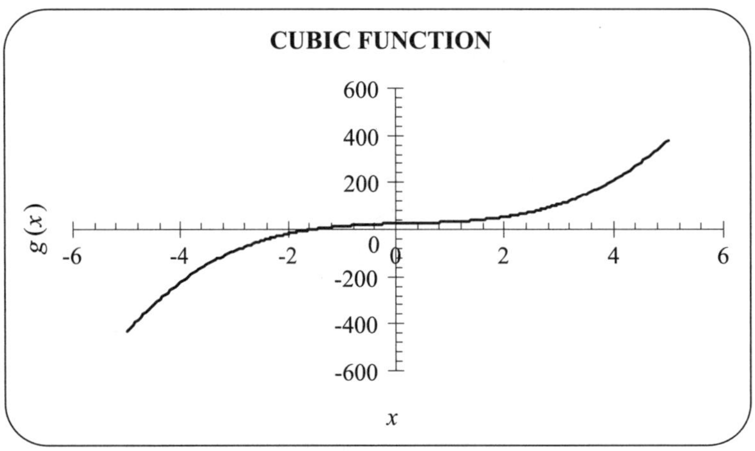


5. Use *Excel* to plot the demand function, $D(q) = -0.000006q^2 - 0.038585q + 303.039287$, over the interval [0, 5000] using 1001 equally spaced points. The graph should have the following characteristics:
   a. The chart title is **DEMAND FUNCTION**.
   b. The title of the horizontal axis is $q$.

c. The title of the vertical axis is $D(q)$.
d. The legend is not shown.
e. The chart area border is black.
f. The line that is plotted in the graph is the maximum weight.
g. The number format of the vertical axis tick mark labels is currency with 0 decimal places.

**SOLUTION**

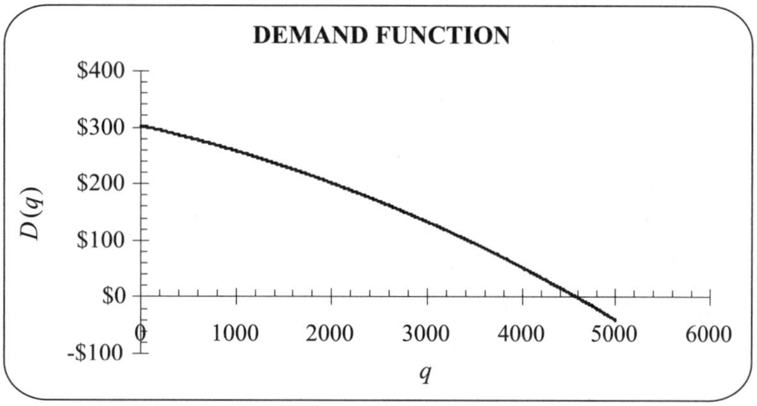


6. Use *Excel* to plot the piecewise function, $f(x)$, over the interval [0, 20] using 401 equally spaced points.

$$f(x) = \begin{cases} 0.1x & \text{if } 0 \leq x \leq 10 \\ -0.1x + 2 & \text{if } 10 < x \leq 20 \end{cases}$$

The graph should have the following characteristics:

a. The chart title is **PIECEWISE FUNCTION**.
b. The $x$-axis title is $x$.
c. The $y$-axis title is $f(x)$.
d. The legend is not shown.
e. The line that is plotted in the graph is the maximum weight.

**SOLUTION**

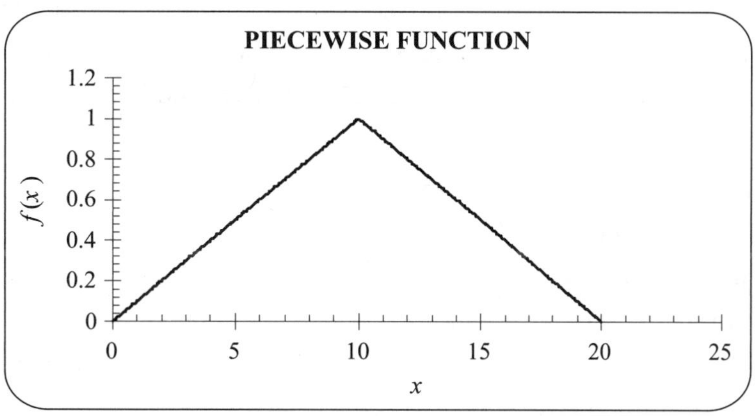

7. Use *Excel* to plot the piecewise function, $g(x)$, over the interval [–10, 15] using 801 equally spaced points.

$$g(x)=\begin{cases} x+1 & \text{if } -5\le x<2 \\ 3 & \text{if } 2\le x<5 \\ -x+8 & \text{if } 5\le x\le 12 \\ 0 & \text{elsewhere} \end{cases}$$

The graph should have the following characteristics:

a. The chart title is **PIECEWISE FUNCTION**.

b. The $x$-axis title is $x$.

c. The $y$-axis title is $g(x)$.

d. The legend is not shown.

e. The line that is plotted in the graph is the maximum weight.

f. There are no unwanted vertical lines.

**SOLUTION**

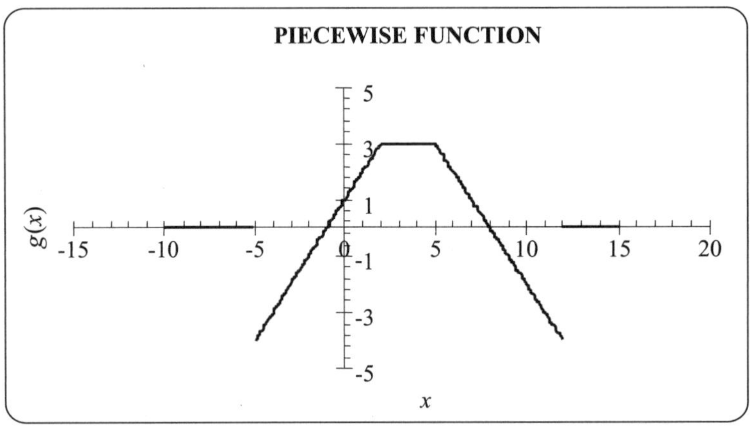


8. A computer company produces desktop and laptop computers. The desktop computers are manufactured in the DHT plant, and the laptops are manufactured in the LHT plant. The cost function for the desktop computers is $C_D(q) = 400q + 500{,}000$, and the cost function for the laptop computers is $C_L(q) = 500q + 200{,}000$. The quantity, $q$, represents the number of computers produced. Plot both cost functions in one graph over the interval [0, 10000]. The graph should have the following characteristics:

a. The chart title is **COST FUNCTIONS**.

b. The title of the horizontal axis is $q$.

c. The title of the vertical axis is $C(q)$.

d. The legend is in the default location.

e. The name of the first series is C_D(q), and the name of the second series is C_L(q).

f. The chart area border is black.

g. The axes are black.
h. The number format of the vertical axis tick mark labels is currency with 0 decimal places.

**SOLUTION**

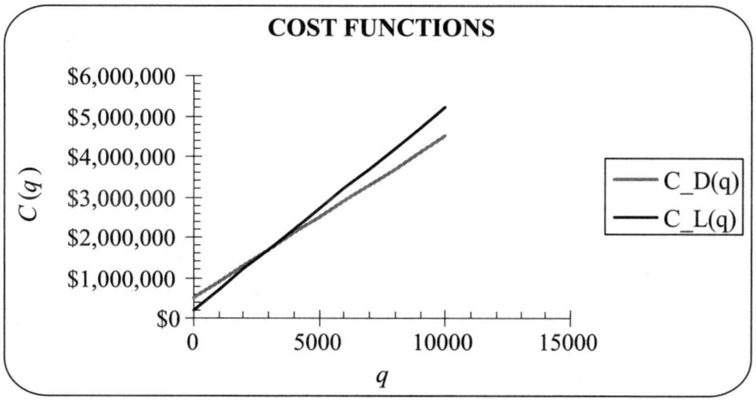


9. Use ***Graphing.xls*** to plot the cumulative distribution function, $F_X(x)$, over the interval [0, 60].

$$F_X(x) = \begin{cases} 0 & \text{if } x < 10 \\ \dfrac{x-10}{40} & \text{if } 10 \le x \le 50 \\ 1 & \text{if } x > 50 \end{cases}$$

**SOLUTION**

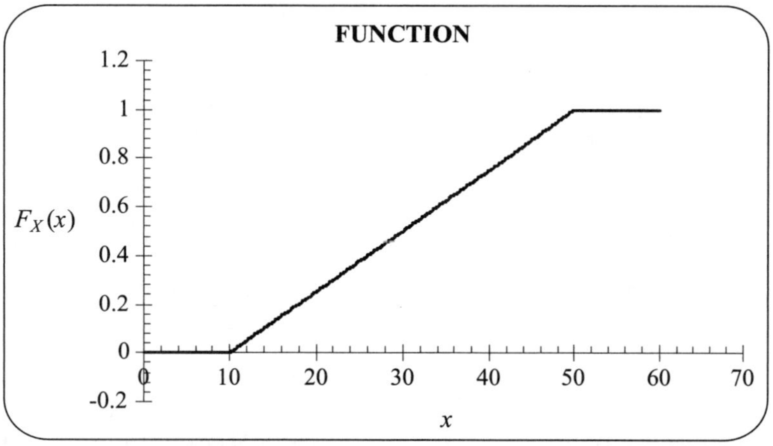


10. Use ***Graphing.xls*** to plot the probability density function, $f_X(x)$, over the interval [0, 60].

$$f_X(x) = \begin{cases} 0 & \text{if } x < 10 \\ \dfrac{1}{40} & \text{if } 10 \le x \le 50 \\ 0 & \text{if } x > 50 \end{cases}$$

Don't forget to remove the unwanted vertical lines.

**SOLUTION**

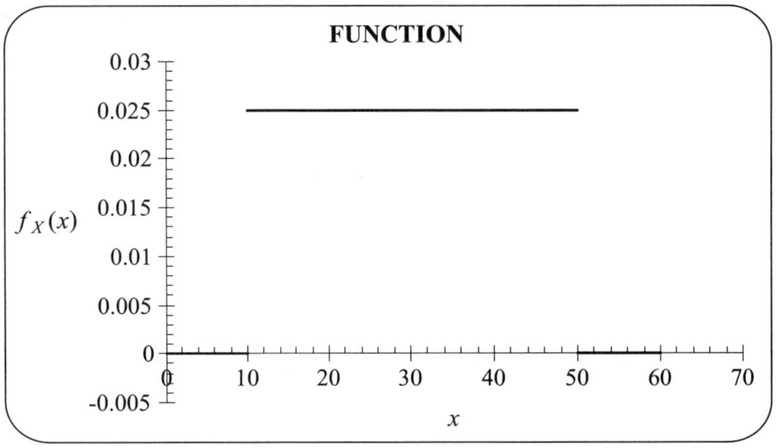


11. Use ***Graphing.xls*** to plot the revenue function,

$$R(q) = -0.000006q^3 - 0.038585q^2 + 303.039287q,$$

over the interval [0, 5000].

**SOLUTION**

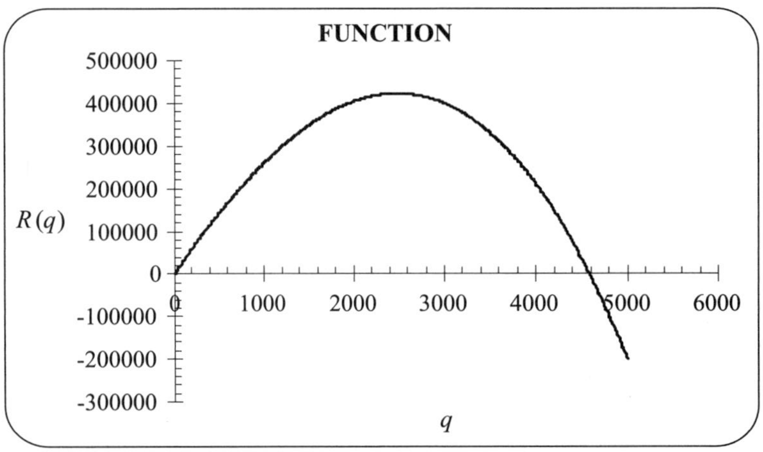


12. Use ***Graphing.xls*** to plot the function $f(x) = sx^2 - tx + 50$ over the interval [–15, 12] with the following conditions:

    a. $s = 2$, $t = 5$; find $f(3)$.

    b. $s = -3$, $t = 9$; find $f(-4)$.

## SOLUTION

a.

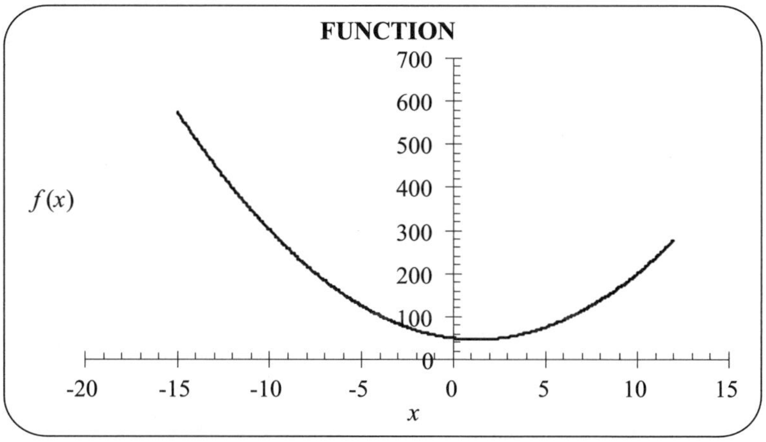


$f(3) = 53$

b.

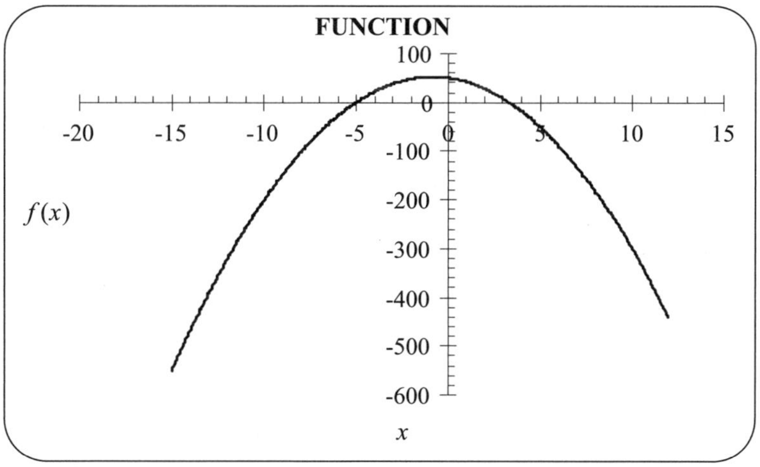


$f(-4) = 38$

# Trendlines

## *Fitting Models*

A picture may be worth a thousand words, but how many mathematics symbols is it worth? Quantitative information is often presented visually– via graphs, charts, or tables. In this section, we will work with paired data that can be represented graphically through mathematical/statistical constructs called **trendlines**.

### What is a trend/regression line/curve?

Collections of paired quantitative data, such as (year, population size), (quantity sold, price per unit), (height, weight), (time elapsed, distance traveled), (time of study, % grade), (quantity produced, cost of production), etc., can be graphed as a collection of points in the plane. This data may have no particular pattern or shape, but often the points fall in a configuration that resembles a common graph such as a line, parabola, exponential curve, or others. Choosing a curve and analyzing how well the curve represents the data is a large subfield of statistics called regression.

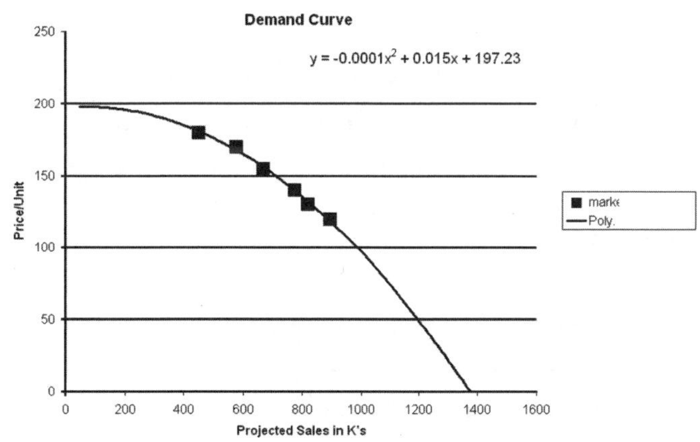


Regression models or trend lines are used to describe possible relationships between two measurable quantitative variables. While one can create, for instance, an exponential trendline for a great variety of paired data, if the data falls in a linear pattern, then the exponential trendline will be useless for predictive purposes. Choosing the right type of trendline to represent your data is crucial.

One statistic that is used to assess how well a given trend line fits a data set is $R^2$, the coefficient of determination. The closer $R^2$ is to 1, the better is the fit of your trendline to your data. Roughly speaking, the coefficient of determination is the percentage of the total variation (total squared deviation from the mean) in the observed values of the response variable (output) that is explained by the trendline. If all the points lie exactly on the chosen trendline, the value of $R^2$ will be 1; that is, 100% of the total variation in the observed values of the response variable is reflected in the relationship between the response variable and the predictor variable (input) as represented by the chosen trendline. This perfect fit is rarely seen in empirically generated data sets.

Before we practice with trendlines, we will briefly review a few of the basic functions that are frequently used as trend "lines."

**Linear functions**

General form: $y = mx + b$, where $m$ is the slope and $(0, b)$ is the $y$-intercept.

**Example 1:** The linear relationship between Fahrenheit and Celsius temperature measurements is given by the equation: $C = \frac{5}{9}(F - 32)$.

a. Fill in the table (round to 1 decimal place when necessary).

| Fahrenheit | Celsius |
|---|---|
| 0° | |
| 32° | |
| 72° | |
| 98.6° | |
| 212° | |

b. Use the above data to sketch a graph of $C = \frac{5}{9}(F - 32)$.

c. What is the $C$-intercept of this line? What does it represent?

d. What is the slope of this line? Interpret the slope in terms of the change in the Celsius temperature compared to the change in Fahrenheit temperature.

**SOLUTION**

a.

| Fahrenheit | Celsius |
|---|---|
| 0° | –17.8° |
| 32° | 0° |
| 72° | 22.2° |
| 98.6° | 37° |
| 212° | 100° |

b.

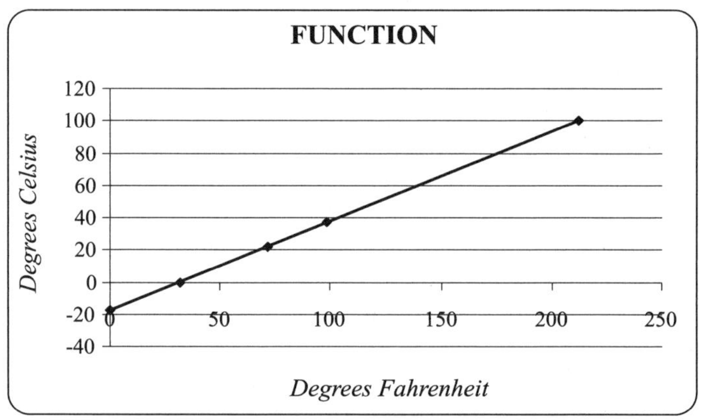


c. The Celsius temperature that corresponds to 0° Fahrenheit is –17.8°.

d. The slope is 5/9. In practical terms, this indicates that the Celsius temperature increases by 5 degrees for every nine-degree increase in Fahrenheit temperature.

**Quadratic functions**

General form: $y = ax^2 + bx + c$.

**Example 2:** If Johnny Rocket, Jr. blasts off his toy rocket straight up into the air with an initial velocity of 64 feet per second, then the height of the rocket in feet $t$ seconds after liftoff (neglecting air resistance) can be approximated by $h = -16t^2 + 64t$.

a. Find the height at $t = 0$ seconds and plot this point on a time-height axes system.

b. Do the same for $t = 1, 2, 3$, and 4 seconds.

c. Connect these points with a smooth curve.

d. What is the significance of $t = 2$ seconds?

e. What is the significance of $t = 4$ seconds?

**SOLUTION**

a. At $t = 0$ seconds, $h = -16(0)^2 + 64(0) = 0$ feet.

b. At $t = 1$ second, $h = -16(1)^2 + 64(1) = 48$ feet.

At $t = 2$ seconds, $h = -16(2)^2 + 64(2) = 64$ feet.

At $t = 3$ seconds, $h = -16(3)^2 + 64(3) = 48$ feet.

At $t = 4$ seconds, $h = -16(4)^2 + 64(4) = 0$ feet.

c.

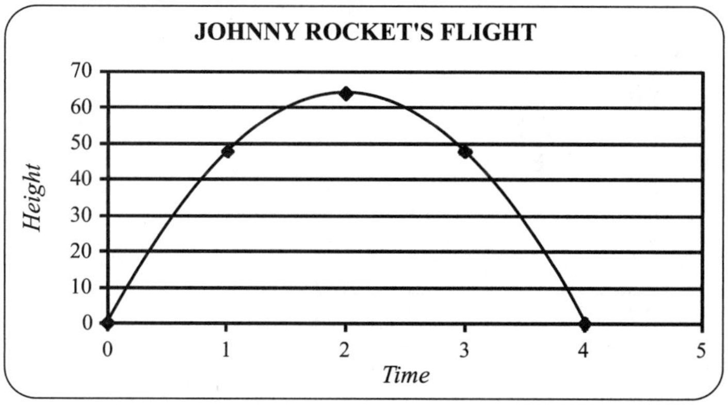


d. At $t = 2$ seconds, the rocket reaches maximum height.

e. At $t = 4$ seconds, the rocket crashes back down to the ground.

**Exponential functions**

General form: $y = Ce^{kx}$.

**Example 3:** Suppose that the population of a certain town, Whereami, Montana, can be represented by an exponential function of the form $P(t) = P_0e^{kt}$, where $t = 0$ corresponds to the year 1990.

a. If $P_0 = P(0)$, what does $P_0$ represent?

b. Suppose that Whereami's population was 1000 in 1990 and 1200 in 1998. Find $k$ and write the specific exponential function that represents this data.

c. Use this model to approximate the population of Whereami, MT in 1995 (this is called *interpolation*).

d. Use this model to approximately predict the population of Whereami, MT in 2002 (this is called *extrapolation*).

**SOLUTION**

a. $P_0$ represents the population in 1990.

b. We find $k$ by solving the equation $P(8) = P_0 e^{k(8)}$ for $k$.

The exponential function that represents this data is $P(t) = 1000e^{\frac{\ln 1.2}{8}t}$.

c. Using $P(t) = P_0 e^{kt}$ and $t = 5$, we obtain $P(5) = 1000e^{(\ln 1.2)(5)/8} \cong 1121$. The population of Whereami, MT in 1995 is approximately 1121 people.

d. Using $P(t) = P_0 e^{kt}$ and $t = 12$, we obtain $P(12) = 1000e^{(\ln 1.2)(12)/8} \cong 1315$. The population of Whereami, MT in 2002 is approximately 1315 people.

**Example 4:** Below is a chart downloaded from `finance.yahoo.com` giving the closing prices of the Coca-Cola Co. prices from April 25, 2007 to May 7, 2007 in reverse order. Saturdays and Sundays are not counted.

| | | |
|---|---|---|
| 7-May-07 | 8 | 53.61 |
| 4-May-07 | 7 | 53.21 |
| 3-May-07 | 6 | 52.91 |
| 2-May-07 | 5 | 52.71 |
| 1-May-07 | 4 | 52.63 |
| 30-Apr-07 | 3 | 52.19 |
| 27-Apr-07 | 2 | 52.06 |
| 26-Apr-07 | 1 | 52.08 |
| 25-Apr-07 | 0 | 51.74 |

a. Create a scatter plot and linear trendline for the given data using April 25, 2007 as $t = 0$ and May 7, 2007 as $t = 8$.

b. Use the trendline from Part a to predict the closing price of the stock on May 11, 2007.

c. Look up the actual historical closing price of the Coca-Cola stock on May 11, 2007 and compare this with your result.

d. Do you think your prediction method was reasonable? Why or why not?

**SOLUTION**

a. The linear equation for the trendline is $y = 0.2182x + 51.698$ and the graph is given below.

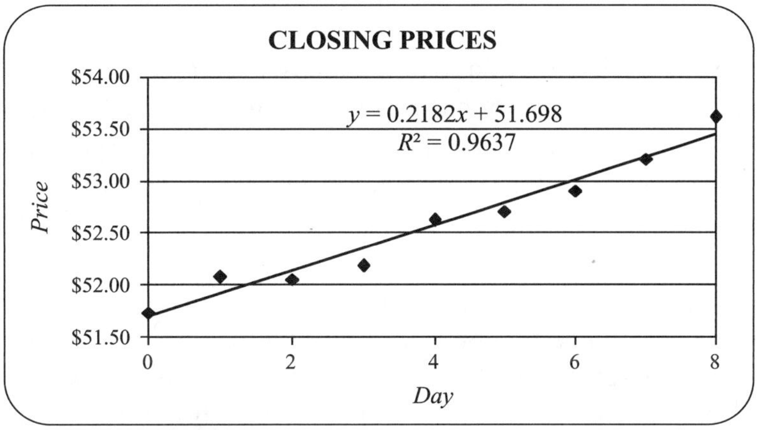


b. Since May 11, 2007 would correspond to $t = 12$, using our linear model, $y = 0.2182x + 51.698$, we obtain an estimated closing price of $y = 0.2182(12) + 51.698 = 54.31$.

c. The actual closing price on May 11, 2007 was $52.69. Since the total change from April 25, 2007 to May 7, 2007 is $1.87, this error of $1.63 is considerable.

d. The prediction method was not reasonable for at least three serious reasons. First, you may recall from Mathematics for Business I that the past history of a stock cannot be used to predict the future value of that stock. Second, using trendlines to predict values for input outside the range of original values, called extrapolation, often results in predictions that are way off the mark. Third, the stock market often shows upswings in value followed by downswings. The small amount of data used to construct the trendline showed only a period of time in which the stock value rose.

**Example 5:** The scatter plot given below gives the monthly sales figures for a card, gift and party supply store for the year 2006 as a function of the month, with January 2006 corresponding to $t = 1$ and December 2006 corresponding to $t = 12$. The trendline graphed is a 6th-degree polynomial. Explain why such a high degree polynomial is needed to fit this curve and why using this polynomial and $t = 15$ to correspond to March 2007 would be a bad idea.

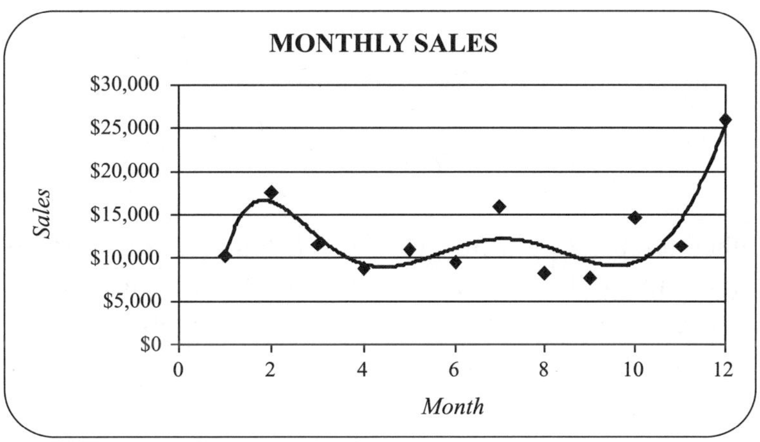

## SOLUTION

A high degree polynomial is needed due to the many ups and downs of the data. As the store is a card, gift and party store, months having national holidays or special events such as Valentine's Day have much higher sales than months without such special occasions. It would be a bad idea to use this polynomial to predict future sales because, again, this would be an example of extrapolation.

**Example 6:** Suppose that a shop owner finds, after many months of experimentation, that he can sell 100 units per month of a certain product at \$60/item, 350 units at \$45/item, and 600 units at \$20/item. Find a quadratic (2nd-degree polynomial) trendline that models this data. Use your model to estimate the price per unit the shop owner should charge if he hopes to sell 500 units per month.

## SOLUTION

First input the points (100, 60), (350, 45), and (600, 20) in two columns in Microsoft *Excel* with quantity as the input and price as the output. Then plot the data and add a 2nd-degree polynomial trendline. Do not forget to use the options to include the equation on the graph. The shop owner should charge $-0.00008(500)^2 - 0.024(500) + 63.2 = 31.20$ or \$31.20/item if he wishes to sell 500 units/month.

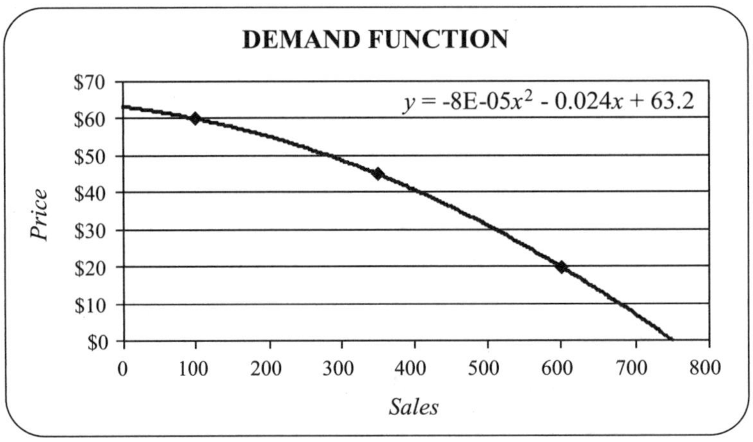


## *Exercises*

**1.** When the Fahrenheit temperature is 32°, the Celsius temperature is 0°. When the Fahrenheit temperature is 212°, the Celsius temperature is 100°. Use this information to write the Fahrenheit temperature as a function of the Celsius temperature.

## SOLUTION

$$F = \frac{9}{5}C + 32$$

**2.** Use the data given below to construct a linear trendline giving price as a function of sales.

| National Yearly Sales | Price |
|---|---|
| 1550 | \$245.00 |
| 1423 | \$252.00 |
| 1217 | \$260.00 |
| 1104 | \$275.00 |
| 924 | \$287.00 |
| 789 | \$298.00 |
| 526 | \$304.00 |
| 472 | \$310.00 |

**SOLUTION**

The graph of the scatter plot with trendline is shown below.

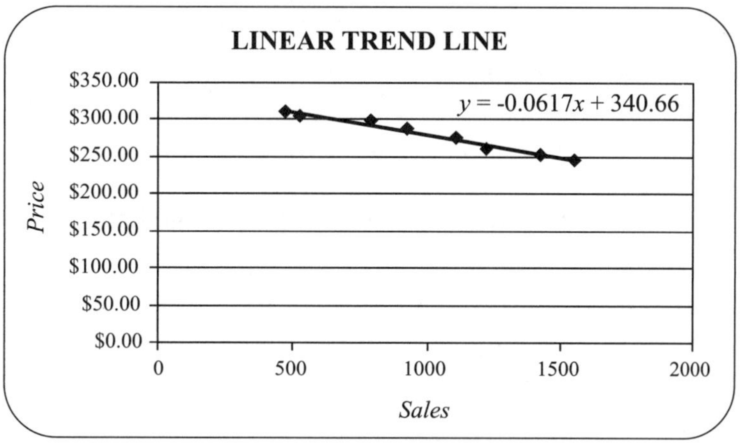


**3.** Use the equation from Exercise 2 above to predict the price for a sales level of $q = 1000$. Is this an example of interpolation or extrapolation?

**SOLUTION**

The price would be approximately \$278.96. This is an example of interpolation.

**4.** Power-lifter Johnny Large read somewhere that eating a lot of carbohydrates shortly before a physical competition will improve his performance. He keeps a record of the grams of carbohydrates he consumes one hour before a bench-pressing competition and the amount he is able to press on four separate competitions. His results were as follows: {(100 g, 300 lbs), (150 g, 320 lbs), (200 g, 345 lbs), (250 g, 350 lbs)}.

a. What is the equation of this linear trendline?

b. What is the coefficient of determination, $R^2$?

c. Do you think a linear model makes sense for this situation?

## SOLUTION

The graph of the scatter plot with trendline is shown below.

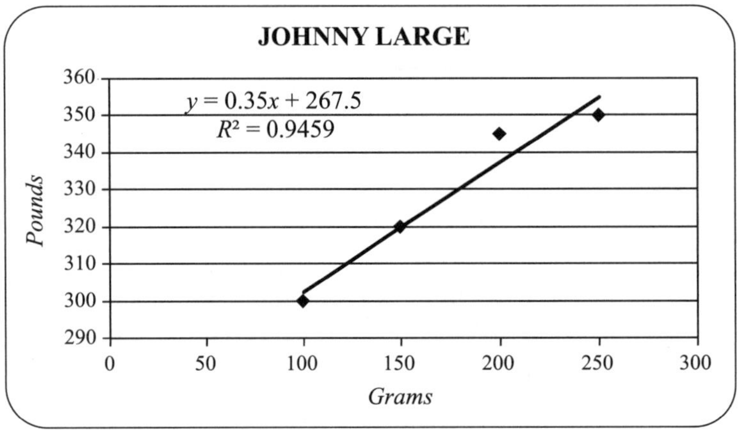


a. $y = 0.35x + 267.5$

b. $R^2 = 0.9459$

c. Probably not, because the rate of increase is larger for smaller levels than for higher levels. Linear functions have a constant rate of change. As Johnny approaches his maximum potential bench-press, increasing the amount of carbohydrates he consumes before a competition will probably not help him.

**5.** Use in Part a of **Example 1** the *Excel* trend line feature to generate a linear trend line for the paired data.

a. What is the equation of this linear trendline?

b. What is the coefficient of determination, $R^2$?

c. Why is $R^2$ so close to 1?

## SOLUTION

The graph of the scatter plot with trend line is shown below.

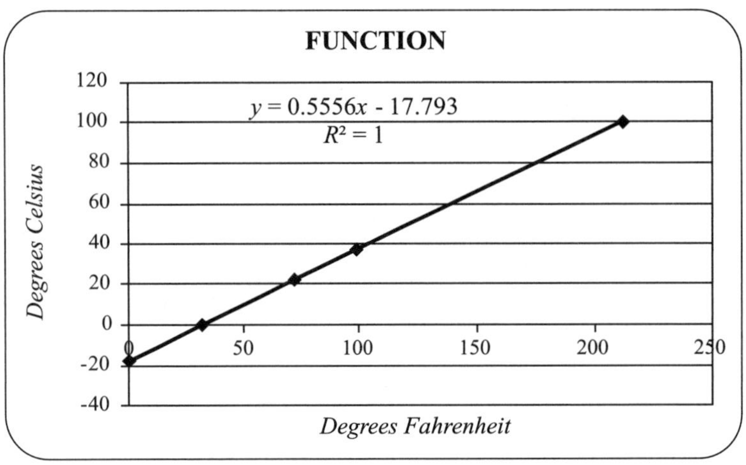

a. $y = 0.5556x - 17.793$

b. $R^2 = 1$

c. Because the two quantities are in fact linearly related.

**6.** Sally Smart is just finishing her sophomore year. She noticed that during her freshman year, if she did not study enough for an exam the night before, she did not do well. She also noticed that if she studied too much, she also did poorly due to lack of sleep. During her sophomore year, Sally kept track of how many hours she studied the night before an exam and the percent grade she earned on the exam. Sally's results, in the form (hours studied the night before, % score earned), are: {(2,72), (3,81), (2,68), (4,95), (8,56), (6,84), (3,75), (5,89), (10,48), (1,52), (5,91), (3,78), (2,78)}.

a. Generate a linear trendline for this data; display the equation as well as $R^2$.

b. Generate a quadratic trendline for this data (polynomial, order 2); display the equation as well as $R^2$.

c. Generate an exponential trendline for this data; display the equation as well as $R^2$.

d. Which of the above trendlines is the best fit for Sally's data?

e. Using the quadratic model, estimate Sally's optimal study time and predict her score on an exam if she studies this number of hours the night before an exam.

f. Explain why it might be better for Sally to study five hours instead of four hours the night before an exam even though she scored a 95% on an exam for which she had studied four hours.

g. Use the quadratic model to predict Sally's grade on a test if she were to study seven hours the night before.

h. Use the quadratic model to predict Sally's grade on a test if she were to study 15 hours the night before.

## SOLUTION

Graphs of the three trend lines are given below with their equations and the values of $R^2$.

a. $y = -1.4049x + 80.22$

$R^2 = 0.0606$

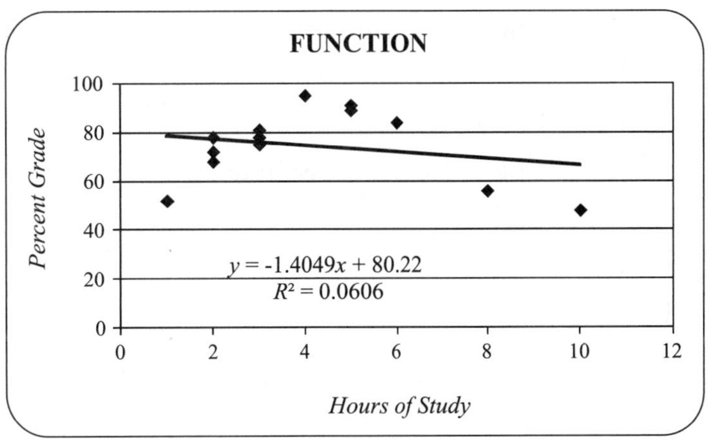

b. $y = -1.8673x^2 + 18.64x + 40.912$
$R^2 = 0.8132$

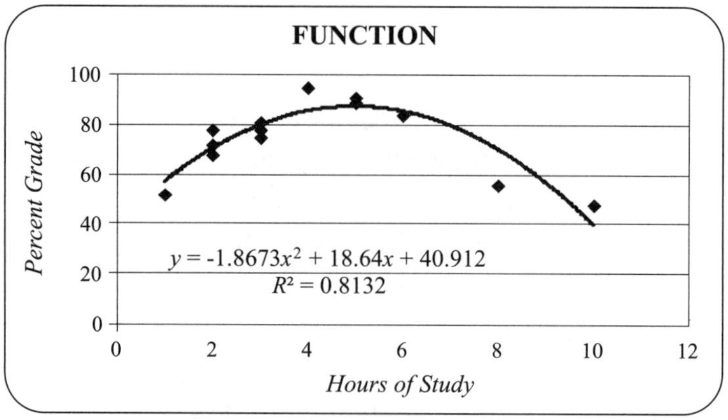


c. $y = 80.74e^{-0.0247x}$
$R^2 = 0.0883$

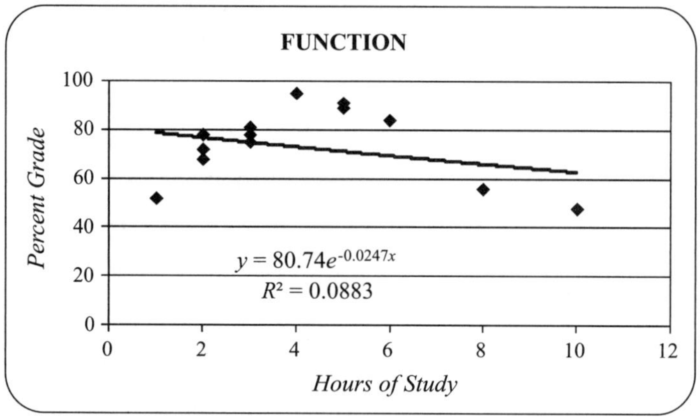


d. None of these fit very well, but the best of the three is the quadratic Part b.

e. According to the quadratic model, Sally's optimal study time appears to be approximately five hours.

f. According to the model, five hours appears to be the optimal number of hours for Sally to study. The high score for four hours might just have been because it was an unusually easy test.

g. Approximately 80%.

h. According to the model, Sally would score a negative 99%. Clearly, the model breaks down by this point.

# Demand, Revenue, Cost, and Profit

## *Income*

So your company has a shiny new product and you are the director of the Marketing Department. Your boss expects you to come up with the optimal selling price — the price that will yield the highest profit. Optimal selling prices, even for shiny new products, seldom reveal themselves as a result of meditation alone, and 1-800-BEST-PRICE would correspond to too many digits to be a good hotline number. You know, deep down, what you need — don't you? Mathematics!

Let's start out with something relatively simple. You want to find a price, $p$, in dollars per unit at which to sell your new product. Let the variable $q$ represent the quantity, the number of units, to be sold. If you want to sell 10,000 units, then you had better charge *less* per unit than if you want to sell only 5,000 units. As the number of units you want to sell increases, the price at which you can hope to sell that many units will, generally speaking, decrease. Thus, you should expect that the price per unit, $p$, is a *decreasing* function of the quantity, $q$, of the product to be sold. The equation that gives the relationship between the price per unit, $p$, and the quantity to be sold, $q$, is called the *demand function*, and is represented by $p = D(q)$. What is the simplest decreasing function? How about a straight line with a negative slope?

**Example 1**: Suppose that Don is planning to introduce a new menu item, tofu tarts, at his breakfast nook, Don's Delicious Delicacies, at Pima Community College's Downtown Campus. On a few trial runs, Don discovered that he was able to sell 100 tarts at \$1.25 per tart, but he was able to sell only 80 tarts at \$1.50 per tart. Assuming a linear relationship between the price per unit, $p$, and the number of tarts, $q$, to be sold, find the formula for the demand function for Don's tofu tarts.

**SOLUTION**

We are trying to write $p$ as a linear function of $q$; that is $p = D(q) = mq + b$, for some constants $m$ and $b$. We already have two points on our line: (80, 1.50) and (100, 1.25). Two distinct points determine a line, so we can use these two points to find the formula for our demand function.

**Step 1:** Find the slope:

$$m = \frac{rise}{run} = \frac{\text{change in price per unit}}{\text{change in sales}} = \frac{1.25-1.50}{100-80}$$
$$= \frac{-0.25}{20} = -0.0125.$$

This tells Don that for every additional tofu tart he wants to sell he should reduce the price by \$0.0125/tart.

**Step 2:** Find the equation relating $p$ and $q$. You have a point (actually you have two points) and a slope, so you can use the point-slope formula to find the formula for the demand function.

Recall the point-slope form of the equation of a line: $y - y_1 = m(x - x_1)$, where $m$ is the slope of the line and $(x_1, y_1)$ is any point on the line. Instead of $y$ and $x$, we are using $p$ and $q$; so the equation of the line becomes $p - p_1 = m(q - q_1)$ or $p = m(q - q_1) + p_1$. Substituting – 0.0125 for $m$, 100 for $q_1$, and 1.25 for $p_1$, we obtain the equation:

$$\begin{aligned} p &= -0.0125(q-100)+1.25 \\ &= -0.0125q+1.25+1.25 \\ &= -0.0125q+2.50. \end{aligned}$$

Thus we have the formula for the demand function: $p = D(q) = -0.0125q + 2.50$.

**Step 3:** Graph $p = D(q)$.

Note that when $q = 0$, $p = -0.0125(0) + 2.50 = 2.50$; and when $p = 0$, we solve $0 = -0.0125q + 2.50$ for $q$ to obtain $q = -2.50/-0.0125 = 200$. Recall from algebra that this is similar to finding the $x$ and $y$-intercepts. In this case, however, they are the $q$ and $p$-intercepts.

Also note that it does not make sense to graph this function for negative values of either $q$ or $p$, as this would correspond to selling a negative number of units (not possible) or charging negative dollars per unit (not a very good business practice).

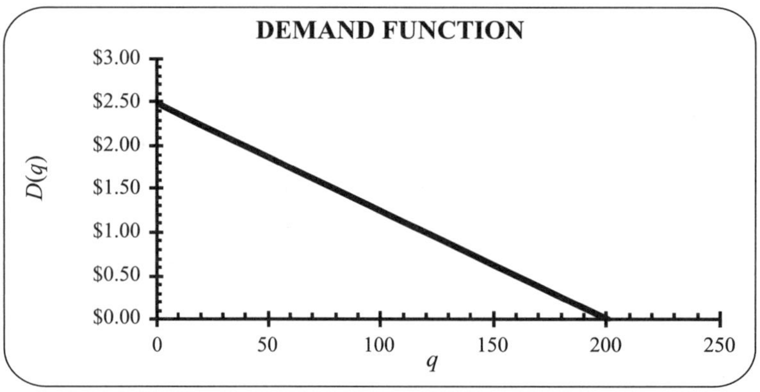


**Example 2**: Use the demand function from **Example 1** to answer the following questions:

a. How many tofu tarts can Don sell at $2.25/tart?

b. At what price could 175 tarts be sold?

**SOLUTION**

a. In this case, we are given that $p = 2.25$ and asked to find $q$. Thus we must solve the equation given below.

$$\begin{aligned} 2.25 &= -0.0125q+2.50 \\ -0.25 &= -0.0125q \\ 20 &= q \end{aligned}$$

If Don charged $2.25 per tart, he would sell only 20 tarts.

b. To achieve a sales level of $q$ = 175 tarts, Don should charge

$$p = D(175) = -0.0125(175) + 2.50 = 0.3125.$$

Thus Don should charge approximately 31¢ per tart if he wants to sell 175 tarts.

**Example 3:** Find and interpret the intercepts of the graph of $p = D(q)$.

**SOLUTION**

The $q$-intercept is obtained by solving the equation $D(q) = 0$ as shown below.

$$-0.0125q + 2.50 = 0$$
$$-0.0125q = -2.50$$
$$q = 200$$

If Don wanted to sell 200 or more tarts, he would have to give them away — i.e., charge \$0 per tart.

The $p$-intercept is obtained by evaluating $p = D(q)$ when $q = 0$. Doing this, we obtain $D(0) = 2.50$. Don would be able to sell no tarts if he charged \$2.50 or more per tart.

When Don sold 100 tarts at \$1.25 each, the revenue was 100 tarts · \$1.25/tart = \$125. When he sold 80 tarts at \$1.50 each, his revenue was 80 tarts · \$1.50/tart = \$120. Even though Don was charging less at \$1.25 per tart than at \$1.50 per tart, his revenue for the sale of the 100 tarts at \$1.25/tart was greater than the revenue for the sale of 80 tarts at \$1.50/tart. Does this mean his profit was greater too? Try to think of a scenario where it would be better for Don to charge the \$1.50/tart than the \$1.25/tart even though he would take in less money.

**POSSIBLE SOLUTON**

Suppose each tofu tart cost Don 50 cents to make. The additional 20 tarts (100 – 80 = 20) would cost him an additional variable cost of 20 tarts · \$0.50/tart = \$10 to produce. Thus, although he made an extra \$5 in revenue (\$125 – \$120 = \$5), it cost him \$10 more to produce the tarts, so his net gain was –\$5. That is, he lost \$5 by producing and selling 20 more tarts.

**Let's bring it all together**:

1) You have a decreasing demand function, $p = D(q)$, which gives the price that you should charge per unit to sell $q$ units of your product.
2) The revenue brought in from selling $q$ units at \$$p$/unit is simply the price per unit multiplied by the number of units sold at that price; thus $R(q) = p \cdot q = D(q) \cdot q = q \cdot D(q)$.
3) Note that graphically $p \cdot q$ is the area of the rectangle under the curve, $p = D(q)$, with upper right-hand corner at $(q, p)$ and lower left-hand corner at (0, 0). See the graph on the right.

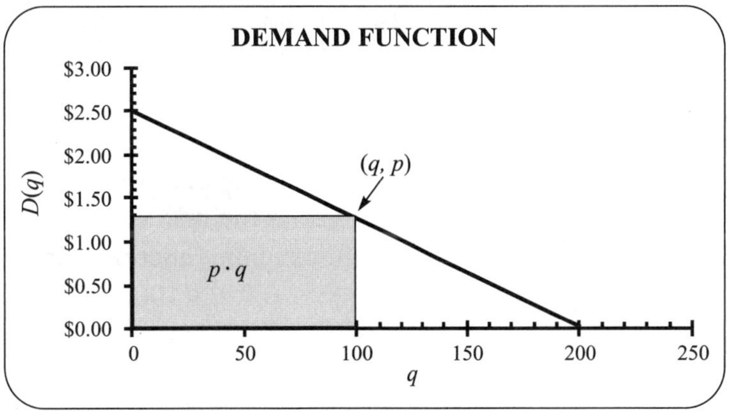

As the area of the rectangle numerically gives the revenue generated by the sale of $q$ units at $p$ dollars per unit, if you wish to find the sales level that gives the greatest revenue, then you should find the inscribed rectangle with the greatest area. The tall skinny areas (high price per unit with low sales level) do not provide much area (revenue); neither do the wide short rectangles (high sales at a low price per unit).

For the tofu tarts at Don's, the highest revenue that we have noted so far was \$125, which resulted from selling 100 tarts at \$1.25/tart. Is that the maximum revenue that could be generated? How can you tell? To this end, let's take a closer look at the revenue function:

$$\begin{aligned} R(q) &= q \cdot (-0.0125q + 2.50) \\ &= -0.0125q^2 + 2.50q\ . \end{aligned}$$

Thus revenue is a *quadratic* function of $q$. You may recall from algebra that the graph of a quadratic function is a parabola. Furthermore, this parabola will open downwards as its leading coefficient, $-0.0125$, is negative.

**Example 4:** Find and interpret the intercepts and vertex of this parabola and sketch its graph.

**SOLUTION**

The $q$-intercepts are found by solving $R(q) = 0$ for $q$.

$$\begin{aligned} q(-0.0125q + 2.50) &= 0 \\ -0.0125q + 2.50 &= 0 \text{ or } q = 0 \\ -0.0125q &= -2.50 \\ q &= 200 \text{ or } q = 0 \end{aligned}$$

The $q$-intercepts are the quantities that correspond to revenue of \$0. When Don sells no tarts, he receives \$0 in revenue. If Don wanted to "sell" 200 tarts, the demand function dictates that he should "sell" them at \$0/tart, so he would again receive \$0 in revenue. Note that in this case, one of the $q$-intercepts, $(0, 0)$ is also the $R$-intercept.

**Finding the vertex ($h$, $k$)**

Recall from algebra that $h$, the $x$-coordinate of the vertex of the graph of $y = ax^2 + bx + c$, is $h = -b/2a$. To find $k$, the corresponding $y$-coordinate, simply plug the $x$-coordinate into the equation for $y$. In this case, $q$ replaces $x$ and $R(q)$ replaces $y$. In $R(q) = -0.0125q^2 + 2.50q$, $a = -0.0125$, $b = 2.50$, and $c = 0$. So the $q$-coordinate of the vertex is $h = -2.50/2(-0.0125) = 100$ and the $R$-coordinate is $k = -0.0125(100)^2 + 2.50(100) = 125$. Thus the vertex is $(100, 125)$.

**Sketch the graph**

See the graph below. The practical interpretation of the vertex of the revenue function is that, in the case of Don's tofu tarts, \$125 is the maximum revenue (based on the linear demand function used and the resulting quadratic revenue function) that Don can realize in a morning from the sale of 100 tofu tarts at $p = D(100) = -0.0125(100) + 2.50 = \$1.25$ per tart.

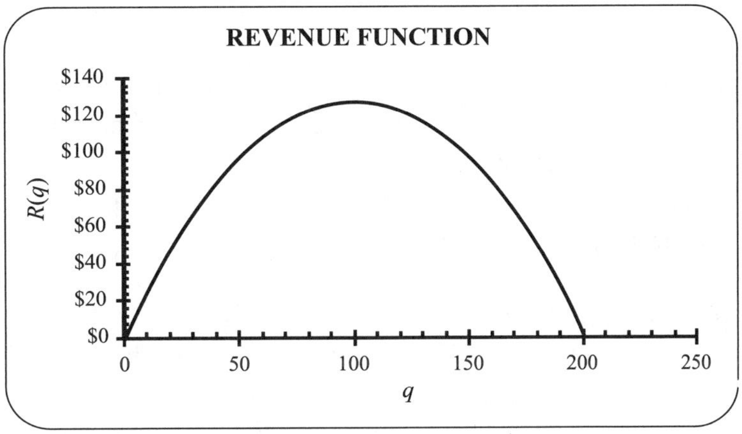


**Example 5:** Angelita owns a small Native American gift shop in Winslow, Arizona just off Route 66. Recently, she acquired a large collection of hand-woven Navajo rugs. The woman who sold Angelita the rugs advised her that she would be able to sell approximately 40 rugs a month at \$150/rug, but only 20 rugs a month at \$220/rug. Assuming a linear relationship between the price per rug and the number of rugs to be sold, find the formula for the demand function.

**SOLUTION**

This example is very similar to Don's tofu tarts. The rugs may be more expensive, but the basics are the same. First, we find the slope; then we use one of the points with the point-slope formula to find the formula for the demand function.

**Step 1:** Find the slope:

$$m = \frac{\text{rise}}{\text{run}} = \frac{\text{change in price per unit}}{\text{change in sales}}$$
$$= \frac{150-220}{40-20} = \frac{-70}{20} = -3.5$$

This tells Angelita that for every additional rug she wants to sell, she should reduce the price by \$3.50/rug.

**Step 2:** The point-slope form of the equation of a line in $p$ and $q$ is $p = m(q - q_1) + p_1$. Using the point (40 rugs, \$150),

$$p = -3.5(q-40)+150$$
$$= -3.5q + 290.$$

Thus we have the formula for the demand function: $p = D(q) = -3.5q + 290$.

## *Expense and Profit*

The question for Don still remains, however, as to whether or not $q = 100$ tarts will also yield the maximum profit. Since *profit* is the difference between revenue and cost, we must construct a cost function before we can go any further. As discussed in the **Business Background**, total cost is usually comprised of two major components: fixed costs and variable costs. The variable costs vary with the number of units produced, but the fixed costs do not.

**Example 6:** Don splits the fixed rental costs for his breakfast nook space equally among his items. Including tofu tarts, Don sells 30 different products on his breakfast menu. The rental for his space is \$450 per morning, so he assigns a fixed cost of \$450/30 = \$15 to his tofu tarts production. Based on the cost of the ingredients and the time and energy required to make the tarts, Don calculates that the variable production cost per tart is 50 cents or \$0.50/tart. Find the formula for and graph the cost function for Don's tofu tarts.

**SOLUTION**

Total cost is equal to the sum of the fixed costs and the variable costs, so we have

$$\text{Cost} = \text{variable cost} + \text{fixed cost}$$
$$C(q) = 0.50q + 15$$

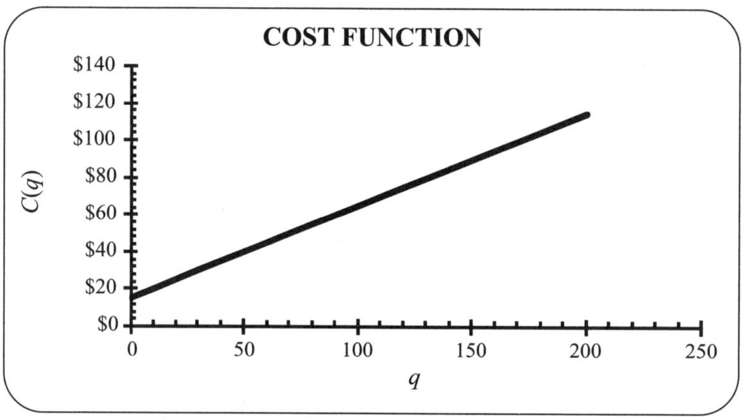


**Example 7:** Use the cost function from **Example 6** to answer the following questions.

a. What is the cost of producing 100 tofu tarts?

b. How many tofu tarts could be produced at a cost of \$75?

**SOLUTION**

a. $C(100) = 15 + 0.5(100) = 65$. It costs Don \$65 to produce 100 tofu tarts.

b. The number of tofu tarts that could be produced at a cost of \$75 is obtained by solving the equation $C(q) = 75$ as shown below.

$$\begin{aligned} 75 &= 15 + 0.5q \\ 0.05q &= 60 \\ q &= 120 \end{aligned}$$

Don can produce 120 tofu tarts at a cost of \$75.

Profit is simply the difference between revenue and cost, so we have

$$\text{Profit} = \text{Revenue} - \text{Cost}$$
$$P(q) = -0.0125q^2 + 2.5q - (0.5q + 15)$$
$$= -0.0125q^2 + 2q - 15$$

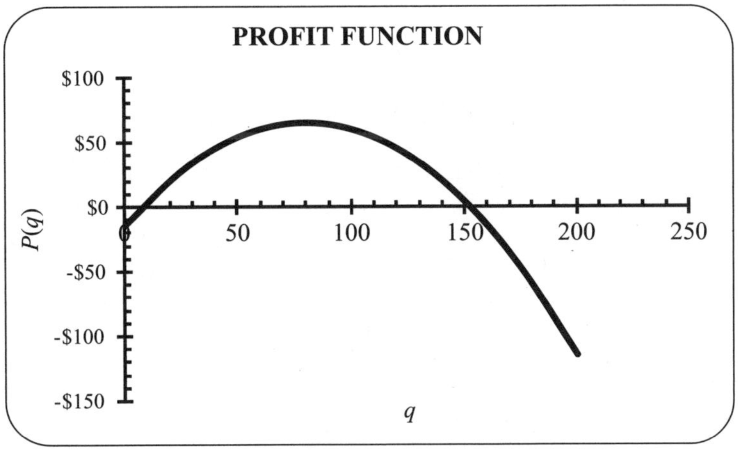


**Example 8:** Compute the profit from the production and sale of 100 tofu tarts.

**SOLUTION**

$p(100) = -0.0125(100)^2 + 2(100) - 15 = \$60$. Don's profit from the sale of 100 tarts at \$1.25 per tart is \$60. Notice that $\$60 = 100(1.25) - 65 = 125 - 65$. That is, *profit = revenue – cost.*

**Example 9:** Compute the profit from the sale of 90 tofu tarts.

**SOLUTION**

We can compute the profit from the sale of 90 tofu tarts in two ways. One, we can compute the revenue and cost individually using $R(q)$ and $C(q)$, respectively, and then take the difference. Two, we can use $P(q)$, which we find by taking the difference between $R(q)$ and $C(q)$. If we compute the revenue and cost individually and then take the difference, we have

$$R(90) = -0.0125(90)^2 + 2.50(90)$$
$$= \$123.75$$
$$C(90) = 15 + 0.5(90)$$
$$= \$60.00$$
$$P(90) = R(90) - C(90)$$
$$= 123.75 - 60$$
$$= \$63.75$$

Using $P(q)$ directly,

$$P(90) = -0.0125(90)^2 + 2(90) - 15$$
$$= \$63.75$$

If Don sells 90 tarts instead of 100 tarts, his revenue decreases from \$125 (maximum revenue) to \$123.75, but his profit increases from \$60 to \$63.75. This is because the cost increases by more than the revenue. You will learn more about this in **Differentiation: Marginal Analysis**.

**Example 10:** Find and interpret the intercepts of the graph of the profit function.

**SOLUTION**

The $q$-intercepts are found by solving $P(q)$ for $q$.

$$-0.0125q^2 + 2q - 15 = 0$$

$$q = \frac{-2 \pm \sqrt{(2)^2 - 4(-0.0125)(-15)}}{2(-0.0125)}$$

$$q \cong 8 \text{ or } q \cong 152$$

The $q$-intercepts are the quantities that correspond to a profit of \$0. Since profit is equal to *revenue – cost*, a profit of \$0 implies that revenue is equal to cost. If Don sells approximately 8 tarts or approximately 152 tarts, his revenue is equal to the cost of producing the tarts, so he just breaks even.

The $P$-intercept is found by evaluating $P(q)$ when $q = 0$. Doing this, we obtain $P(0) = -15$. When Don sells no tarts, he loses \$15, the fixed cost of his tofu tarts production.

Note that, in this case, profit is also a quadratic function of $q$, so you can find where the maximum profit occurs (for what $q$-value) and what that maximum profit is using the same methods that were used to find maximum revenue. Try this as an exercise before you read the solution below.

**Finding the maximum profit**

First, find the $q$-coordinate of the vertex of the downward opening parabola.

$$h = \frac{-2}{2(-0.0125)} = 80 \text{ units}$$

Second, plug this into the formula for the profit function to find the profit at this production/sales level.

$$k = P(80) = -0.0125(80)^2 + 2(80) - 15 = \$65 .$$

Thus, based on the model used, Don can expect a maximum profit of \$65 from the sale of 80 tofu tarts at $p = D(80) = -0.0125(80) + 2.50 = \$1.50$ per tart. Notice that, in this case, the maximum profit occurred at a production/sales level that is less than the level that yielded the maximum revenue.

**Example 11:** A plot of the revenue and cost functions for Don's tofu tarts is shown below.

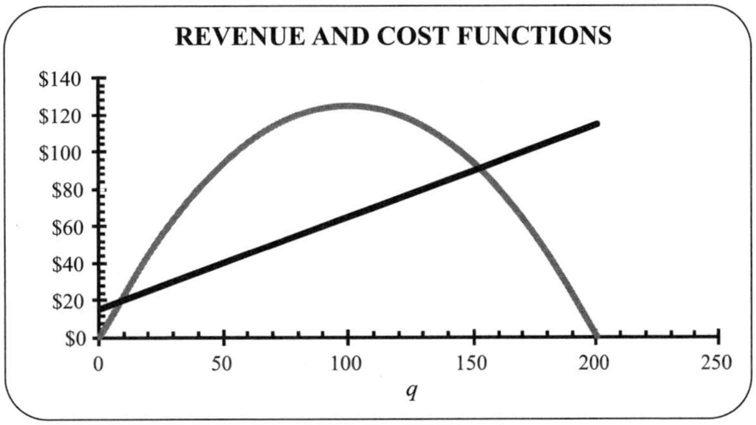


a. For what values of $q$ values will there be a ***positive profit***?

b. Illustrate graphically the optimal profit of \$65 at a quantity of 80 tarts.

**SOLUTION**

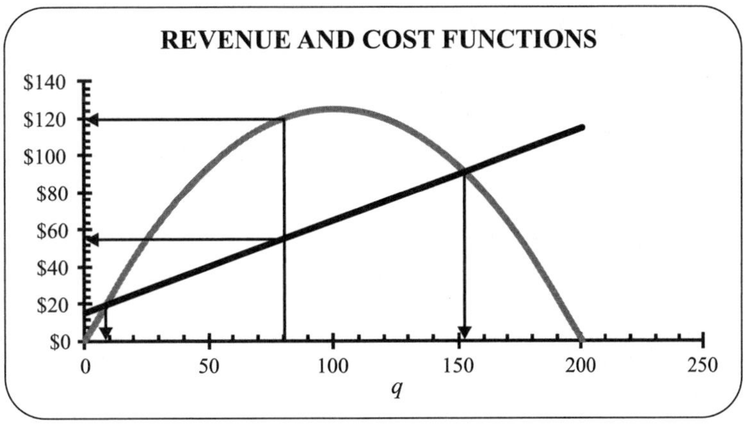


a. There will be a positive profit whenever the revenue exceeds the cost, so from $q \cong 18$ to $q \cong 190$ tarts.

b. The optimal profit of \$65 at a quantity of 80 tarts is represented by the difference between $R(80)$ and $C(80)$.

**Example 12:** The demand function for table saws is given by $D(q) = 200 - \frac{q}{30}$ where $q$ is the quantity and $D(q)$ is the price. The fixed cost for producing the saws is \$75,000, and the variable cost is \$60 per saw.

a. Find the formula for the revenue function, $R(q)$.

b. Find the formula for the cost function, $C(q)$.

c. Find the formula for the profit function, $P(q)$.

d. What profit is generated from the sale of 300 table saws? 1000 table saws?

e. Graph the profit function.

f. For what values of $q$ is there a loss? Profit?

g. How many table saws should be sold in order to maximize profit?

h. What is the maximum profit?

**SOLUTION**

a. $R(q) = q \cdot D(q) = q\left(200 - \frac{q}{30}\right) = 200q - \frac{q^2}{30}$

b. $C(q)$ = Fixed Costs + Variable Costs

$= 75{,}000 + 60q$

c. $P(q) = R(q) - C(q)$

$$= \left(200q - \frac{q^2}{30}\right) - (75{,}000 + 60q)$$

$$= -\frac{q^2}{30} + 140q - 75{,}000$$

d. $P(300) = -\frac{300^2}{30} + 140(300) - 75{,}000$

$= -36{,}000$

A loss of \$36,000 is generated from the sale of 300 table saws.

$P(1000) = -\frac{1000^2}{30} + 140(1000) - 75{,}000$

$= \$31{,}667$

A profit of \$31,667 is generated from the sale of 1000 table saws.

e.

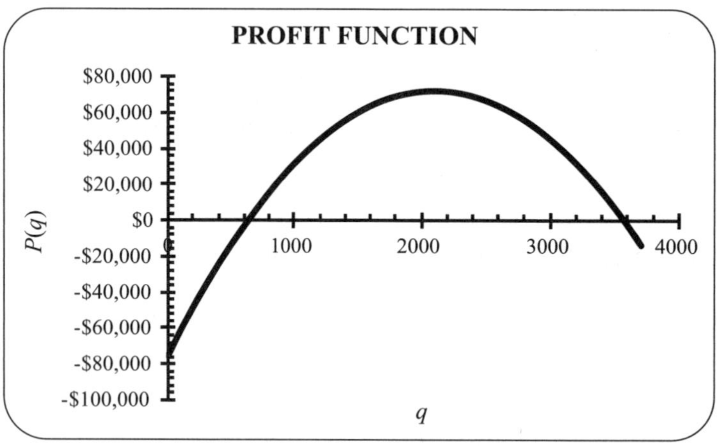


f. We need to find the $q$ intercepts (the values of $q$ for which $P(q) = 0$).

$P(q) = a \cdot q^2 + b \cdot q + c$ where $a = -\frac{1}{30}$, $b = 140$, and $c = -75{,}000$

$$q = \frac{-b \pm \sqrt{b^2 - 4ac}}{2a}$$
$$= \frac{-140 \pm \sqrt{(140)^2 - 4(-1/30)(-75,000)}}{2(-1/30)}$$
$$\approx 630.31 \text{ or } 3569.69$$

After rounding to the nearest whole number, we have $q = 630$ and 3570. For $q < 630$ and for $q > 3570$, the curve is below the horizontal axis. Thus a loss is generated on the intervals [0, 630] and [3570, ∞). A profit is generated on the interval [630, 3570].

g. We need to find the $q$-coordinate of the vertex of the profit function.

$$\text{vertex} = \left(-\frac{b}{2a}, P\left(-\frac{b}{2a}\right)\right)$$

$$q = -\frac{b}{2a} = -\frac{140}{2(-1/30)} = 2100$$

Thus 2100 table saws should be sold in order to maximize profit.

h. $P(2100) = \frac{-2100^2}{30} + 140(2100) - 75,000 = \$72,000.$ The maximum profit is \$72,000.

## *Exercises*

**1.** Suppose the demand function for a product is $D(q) = -0.2q + 300$. Also suppose that the fixed cost of production is \$20,000 and the variable cost of production is \$25 per unit.

a. Graph the demand function.

b. Find the formula for the revenue function.

c. Find the formula for the cost function.

d. Find the formula for the profit function.

e. Find the price per unit that yields the maximum profit.

**SOLUTION**

a. See the graph to the right.

b. $R(q) = -0.2q^2 + 300q$

c. $C(q) = 20,000 + 25q$

d. $P(q) = -0.2q^2 + 275q - 20,000$

e. Approximately \$162.50/unit.

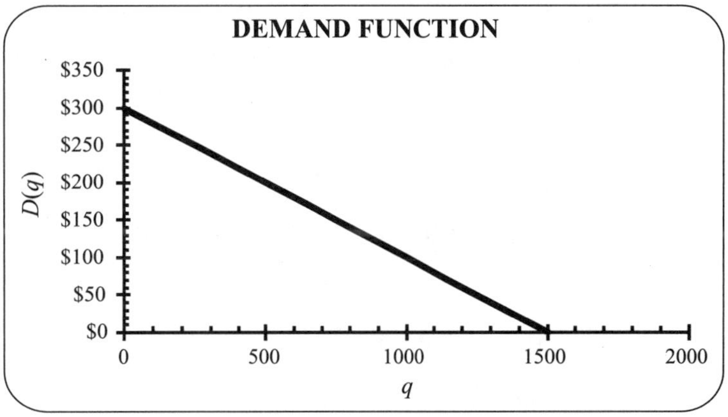

**2.** A company would like to use the data given to the right to find the demand function for their product.

a. Fit a linear trend line to the data.

b. Use the equation of the trendline to estimate the price at which 1000 units could be sold.

| National Yearly Sales | Price |
|---|---|
| 1550 | $245.00 |
| 1423 | $252.00 |
| 1217 | $260.00 |
| 1104 | $275.00 |
| 924 | $287.00 |
| 789 | $298.00 |
| 526 | $304.00 |
| 472 | $310.00 |

**SOLUTION**

a.

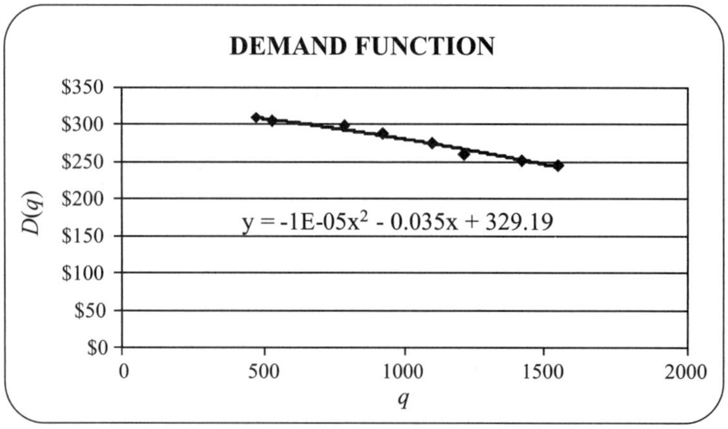


b. $D(1000) \approx \$280$

**3.** The X-Tech Car Company sells many products for all makes and models of vehicles. Sales have been excellent for their X-Car Alarm due to the rising number of car thefts. The owner of the company is considering adjusting the price of the alarm due to an increase in demand. Currently the company is basing the price of the X-Car Alarm on the demand function that has been used since the product was introduced: $D(q) = -0.00002169x^2 - 0.02069234x + 367.03453809$. Recent data has been collected to determine if an adjustment in price for the X-Car Alarm needs to be made based upon the new test market data.

| National Yearly Sales | Price |
|---|---|
| 1271 | $350.00 |
| 1128 | $354.00 |
| 1023 | $360.00 |
| 938 | $364.00 |
| 754 | $370.00 |
| 678 | $375.00 |
| 414 | $377.00 |
| 379 | $380.00 |

a. Using the old demand function, what is the price of the X-Car Alarm at 2,000 alarms? At 3,000 alarms?

b. Using the new test market data find the formula for a quadratic demand function.

c. Using the new demand function, what should the price of the X-Car Alarm be at 2,000 alarms? At 3,000 alarms?

d. Use your answers from Parts a and c to determine whether the owner should increase or decrease the price of the X-Car Alarm.

**SOLUTION**

a. $D(2000) = \$238.89$

$D(3000) = \$109.75$

b.

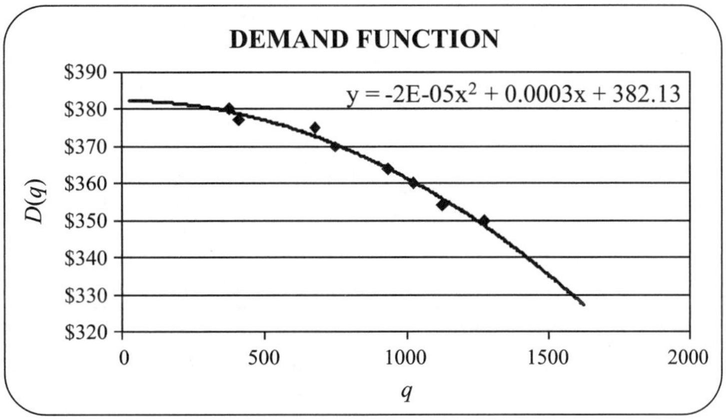


c. $D(2000) = \$302.73$
   $D(3000) = \$203.03$

d. The owner should increase the price of X-Car Alarm.

**4.** Graphs of revenue and cost functions are given below.

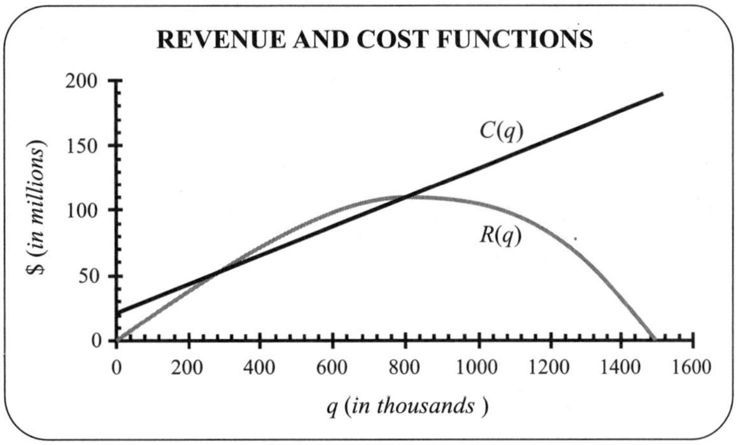


Revenue, $R(q)$, and cost, $C(q)$, are in millions of dollars and the quantity, $q$, is in thousands of units. Use the graph to estimate your answers to the questions below.

a. On what interval(s) does profit occur?

b. On what interval(s) does loss occur?

c. For what values of $q$ does the company break even?

**SOLUTION**

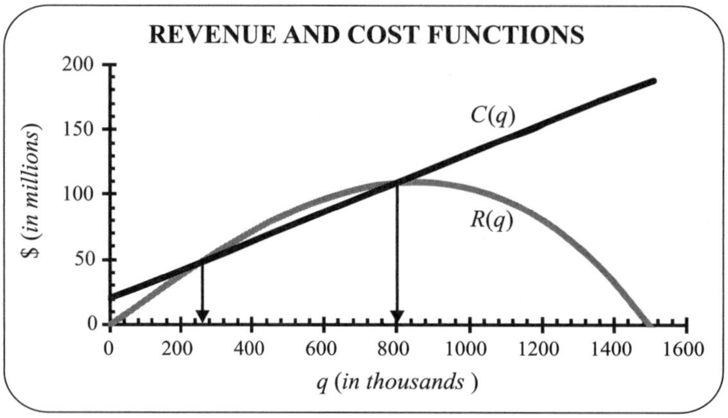


a. On the interval [260, 800] or for values of $q$ between 260,000 to 800,000.

b. On the intervals [0, 260] and [800, ∞) or for values of $q$ between 0 and 260,000 as well as for values of $q$ exceeding 800,000.

c. $q \cong 260{,}000$ and $q \cong 800{,}000$

**5.** At Big Art's Tool Company, fixed costs are three million dollars per year. The total cost of producing 1,000 MIG Welders is \$4,500,000 and the total cost of producing 3,000 MIG Welders is \$7,500,000. The demand function for MIG Welders is $D(q) = -\frac{4}{3}q + 8$ where $q$ is in thousands of MIG Welders and $D(q)$ is in millions of dollars.

a. Find the formula for the cost function, $C(q)$.

b. Find the formula the revenue function, $R(q)$.

c. Use ***Graphing.xls*** to graph $C(q)$.

d. Use ***Graphing.xls*** to graph $R(q)$.

e. At what quantity is $R(q) = 0$?

f. Use ***Excel*** to graph $C(q)$ and $R(q)$ on the same set of axes.

g. For what values of $q$ does the company break even?

h. On what interval(s) does loss occur?

i. On what interval(s) does profit occur?

**SOLUTION**

a. $C(q) = 1.5q + 3$

b. $R(q) = -\frac{4}{3}q^2 + 8q$

c.

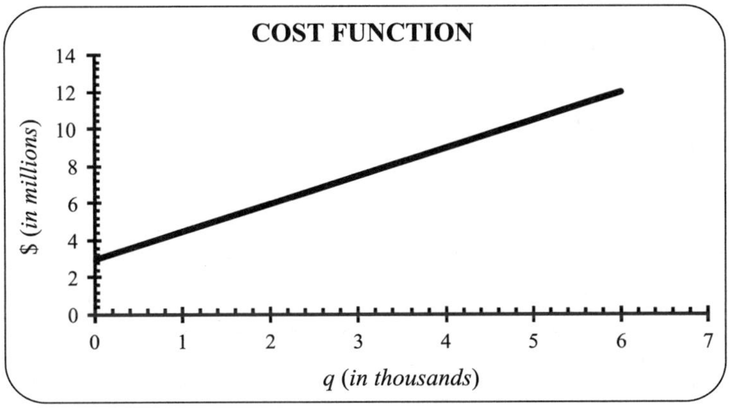


d.

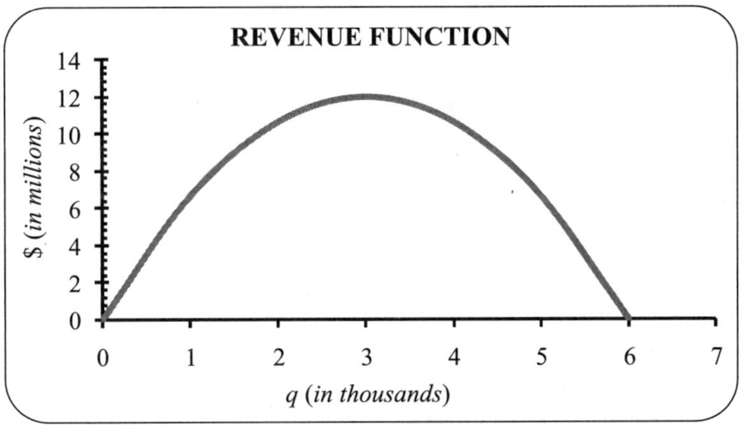


e. $q = 6$

f.

| $q$ | $C(q)$ | $R(q)$ |
|---|---|---|
| 0 | 3 | 0 |
| 1 | 4.5 | 6.67 |
| 2 | 6 | 10.67 |
| 3 | 7.5 | 12 |
| 4 | 9 | 10.67 |
| 5 | 10.5 | 6.67 |
| 6 | 12 | 0 |

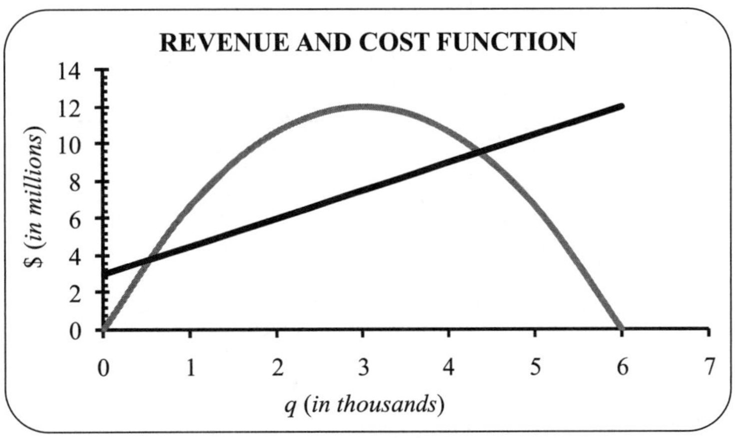


g. $q \cong 4.359$ and $q \cong 0.516$

h. A loss is incurred on the intervals [0, 0.516] and [4.359, $\infty$).

i. A profit is generated on the interval [0.516, 4.359].

**6.** The graph of a demand function is given below.

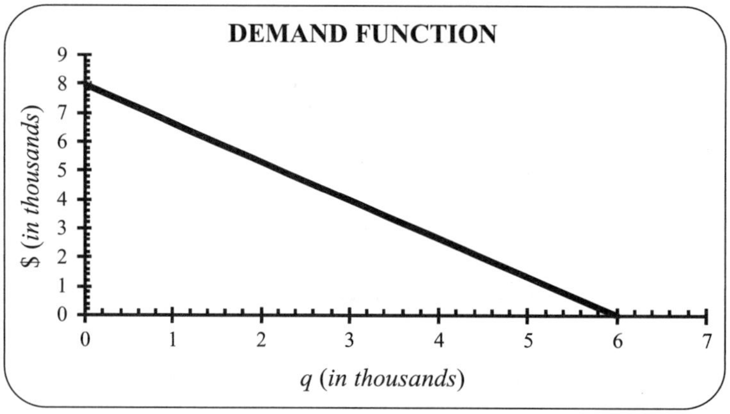


The price, $D(q)$, is in thousands of dollars and the quantity, $q$, is in thousands of units.

a. Use the graph to estimate the price per unit at 2000 units.

b. Find the formula for the demand function, $D(q)$.

c. Use the formula for the demand function to find the price per unit at 2000 units.

d. What is the revenue generated at 2000 units? Illustrate your answer by shading the region on the graph that represents the revenue generated at 2000 units.

**SOLUTION**

a. Remember that $q$ is in thousands, so for 2000 units, we use $q = 2$. $D(2) \cong 5$ or \$5,000/unit.

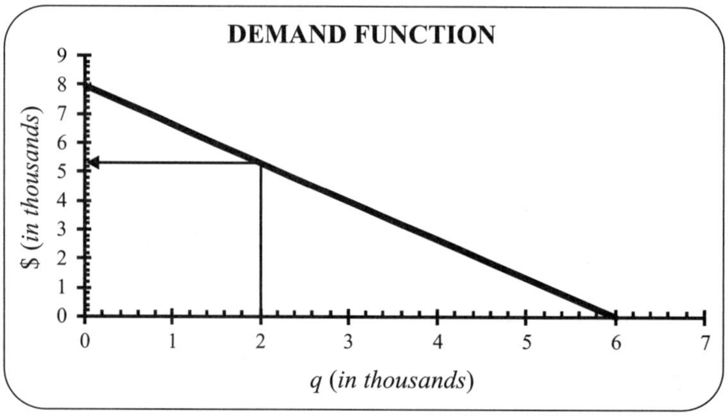


b. $D(q) = -\frac{4}{3}q + 8$

c. Approximately \$5,333/unit

d. The revenue generated at 2000 units is approximately $2000 \cdot 5{,}333.33 = \$10{,}666{,}660$.

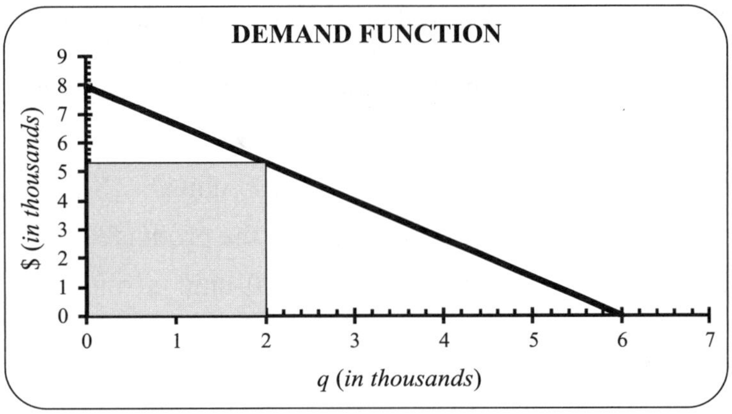


**7.** Use the demand function computed in Part b of **Exercise 6** to find the formula for the revenue function, $R(q)$, and use ***Graphing.xls*** to plot it.

**SOLUTION**

$$R(q) = -\frac{4}{3}q^2 + 8q$$

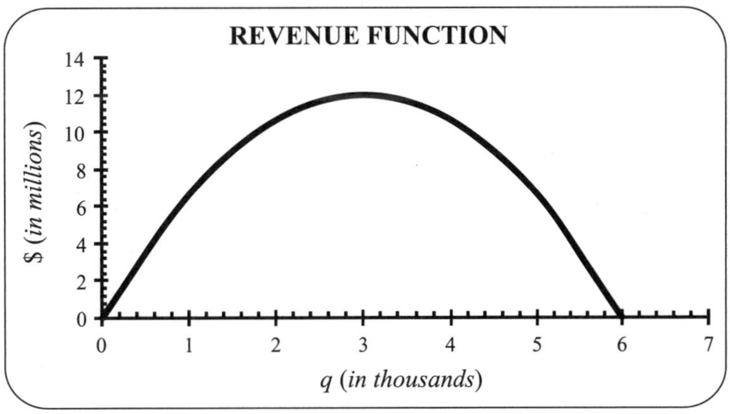


**8.** Refer to **Exercise 7** to answer the questions below.

a. Use the graph of the revenue function to estimate the maximum revenue and the quantity at which revenue is maximized.

b. Compute the maximum revenue and the quantity at which revenue is maximized using algebraic techniques.

**SOLUTION**

a. At $q \cong 3{,}000$,
$R(q) \cong 12$ million dollars

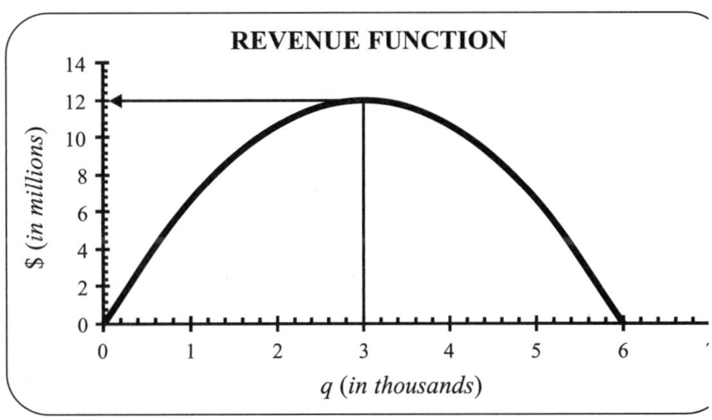

b. The maximum occurs at $q = 3000$ and is equal to 12 million dollars.

**9.** A company estimates that the demand function for its product is given by

$$D(q) = -0.00075q^2 + 900.$$

a. What price per unit would result in the sale of 500 units?

b. What is the equation of the revenue function for the product?

c. What revenue would result from the sale of 500 units at the price that would produce exactly 500 sales?

**SOLUTION**

a. $D(500) = \$712.50$

b. $R(q) = -0.00075q^3 + 900q$

c. $R(500) = \$356{,}250$

**10.** Angelita's rugs, from **Example 5**, cost her \$80 per rug. She calculates fixed cost by taking the same percentage of her average monthly costs to run the store as the percentage of display space needed by the product in question. Her average monthly cost is \$3000 and her total amount of display space is 1200 square feet.

a. If the rug display uses 120 square feet of display space, find the formula for the cost function for Angelita's rugs.

b. Use the demand function from **Example 5** to find the formula for the revenue function for Angelita's rugs.

c. Find the formula for the profit function for Angelita's rugs.

d. What is the maximum monthly profit from the rugs?

**SOLUTION**

a. $C(q) = 300 + 80q$

b. $R(q) = -3.5q^2 + 290q$

c. $P(q) = -3.5q^2 + 210q - 300$

d. \$2,850.00

# Differentiation

## *Rates of Change*

It is often said that the only thing that we can count on to remain the same is the constant presence of change. Understanding change and how changes in one quantity can affect the change in another is often crucial in predictive forms of analysis. The study of change and rates of change can be quite complicated and is the motivation behind differential calculus. In an effort to get a bit of a grip, we will start with something for which the rate of change is very simple. Can you guess what? Yes, it is again our friend the straight line.

**Example 1:** Julie and Jeff live along a trail in the foothills of the Santa Rita Mountains to the east of Tucson, Arizona. Jeff lives one mile south of Julie along the trail. They leave their respective homes at the same time for a hike northbound along the trail. Julie walks at a constant rate of 2 miles per hour while Jeff walks at a constant of rate of 2.5 miles per hour.

a. Write linear equations representing their distance *from Jeff's house* as a function of time, $t$, *in hours*.

b. Find Julie's distance from Jeff's house when $t = 2$ hours.

c. Graph both of these lines on the same set of axes (assume that they travel northbound for three hours).

d. Describe what is happening to the distance between Jeff and Julie over the course of the three hours.

e. How many hours does it take Jeff to catch up with Julie? What do you notice about the graphs at this time?

**SOLUTION**

a. Julie: At time $t = 0$, she is one mile from Jeff's house. If we let $y$ represent the distance from Jeff's house in miles, then we have a $y$-intercept of (0, 1) and a slope of 2 (miles per hour). So the equation giving Julie's distance from Jeff's house as a function of time (at least for the three hours that she is walking away) is given by $y = 2t + 1$.

Jeff: Jeff starts out 0 miles from his own home and his rate is 2.5 mph; so his equation is given by $y = 2.5t$.

b. $y = 2(2) + 1 = 5$ miles from Jeff's house.

c.

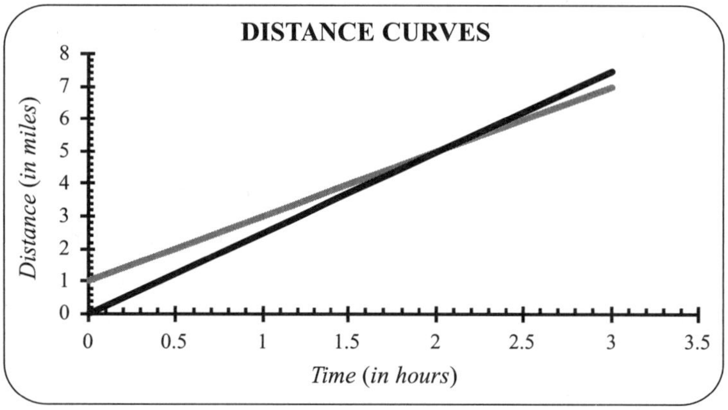


d. They start off one mile apart, but the distance between them shrinks at a constant rate of 0.5 miles per hour until Jeff catches up with Julie. At that point they are 0 miles apart, but as they both keep walking at their respective constant rates, the distance between them starts growing at a constant rate of 0.5 mile per hour.

e. When Jeff catches up with Julie, they will momentarily be the same distance from Jeff's house, so:

$$\begin{aligned} 2t+1 &= 2.5t \\ 1 &= 0.5t \\ 2 &= t. \end{aligned}$$

It takes Jeff two hours to catch up with Julie. This is the point in time at which the two graphs intersect, thus the distance between them is zero.

**Example 2:** On another day of hiking, Julie and Jeff start out from their respective homes again at the same time. Julie walks for two hours northbound at her steady pace of 2 mph. Jeff however starts out fast, begins to slow down, and after one hour turns around. His distance from his house as a function of time during those two hours is given by $y=-5t^2+10t$. Julie's distance from Jeff's house is still $y=2t+1$.

a. Sketch the graphs of Julie and Jeff's distances from Jeff's house as a function of time $t$ in hours for $0 \le t \le 2$.
b. Determine the period over which Jeff is north of Julie.
c. Determine Jeff's maximum distance from his house. After how many hours does he reach this distance?
d. At the time found in Part c, how far is Jeff from Julie?
e. Consider only the time period found in Part b in which Jeff is ahead of Julie. What is the maximum distance that Jeff is ahead of Julie? Hint: It is not the answer from Part d. Why not?

## SOLUTION

a. Note that the straight line represents Julie's distance from Jeff's house as a function of time and the parabolic curve represents Jeff's.

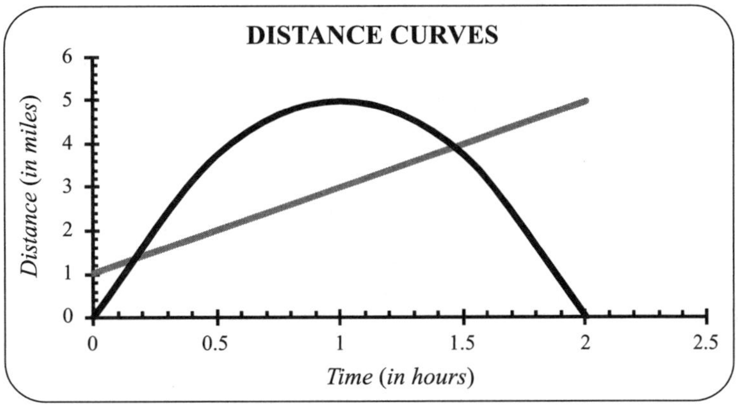


b. First find the points in time at which they meet.

$$-5t^2 + 10t = 2t + 1$$
$$5t^2 - 8t + 1 = 0$$
$$t = \frac{-(-8) \pm \sqrt{(-8)^2 - 4(5)(1)}}{2(5)}$$
$$t \cong 1.46 \text{ or } t \cong 0.14.$$

As the parabolic curve is above the straight line between the approximate points of 0.14 and 1.46, Jeff is north of Julie from approximately $t = 0.14$ hours to $t = 1.46$ hours.

c. The graph of Jeff's distance from his house as a function of time is quadratic with a negative leading coefficient; the maximum distance, therefore, will occur at its vertex. Jeff's equation is $y = -5t^2 + 10t$.

$t$-coordinate of vertex: $h = \frac{-b}{2a} = \frac{-10}{2(-5)} = 1$

$y$-coordinate of vertex: $k = -5(1)^2 + 10(1) = -5 + 10 = 5$
Therefore Jeff reaches a maximum distance from his house of five miles after one hour.

d. At $t = 1$, the distance between Jeff and Julie is $-5(1)^2 + 10(1) - (2(1) + 1) = 5 - 3 = 2$; Jeff is two miles ahead of Julie after one hour.

e. Note that if Jeff is ahead of Julie *and* walking faster than Julie, then his distance from her will continue to increase. Certainly once Jeff turns around and starts heading back, the distance between them will be shrinking, but the distance started shrinking even before Jeff turned around. The instant Jeff started walking slower than Julie, Julie began to close the gap. Therefore, the maximum distance between them occurred at the instant in which Jeff changed his *rate* from a rate faster than Julie's rate to a rate slower than Julie's (the instant that their walking rates were the same).

Julie's walking rate is a constant 2 mph — the slope of the line representing Julie's distance in time. Jeff's distance function, however, is nonlinear — his walking rate is itself changing with time. The function that will give his walking rate is called a **derivative**.

The derivative is a mathematical tool from calculus. It is commonly used to help find maximums or minimums, especially if one of the functions involved is nonlinear. In this problem, however, we can get by without using derivatives, as our difference function is again a quadratic with a negative leading coefficient. Therefore, the maximum will occur at the vertex.

Let $d(t) = \text{Jeff's distance} - \text{Julie's distance}$. Then

$$\begin{aligned} d(t) &= -5t^2 + 10t - (2t+1) \\ &= -5t^2 + 8t - 1. \end{aligned}$$

The $t$-coordinate of the vertex is given by

$$h = \frac{-b}{2a} = \frac{-8}{2(-5)} = 0.8,$$

and the $y$-coordinate of the vertex is given by $k = -5(0.8)^2 + 8(0.8) - 1 = 2.2$.

The maximum distance ahead of Julie that Jeff gets is thus 2.2 miles after 0.8 hours.

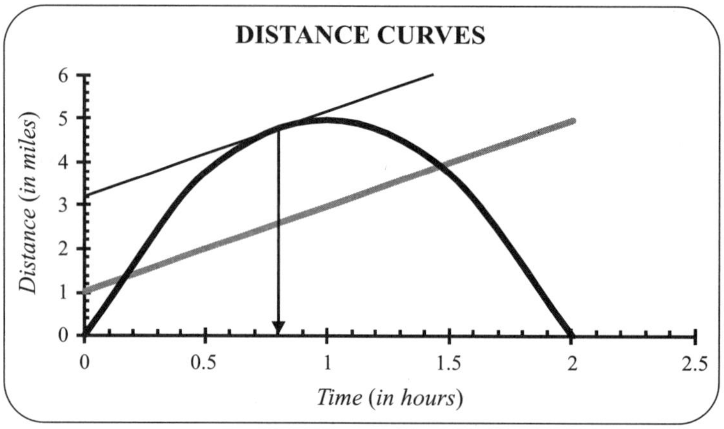


"As its campfires glow against the dark, every culture tells stories to itself about how the gods lit up the morning sky and set the wheel of being into motion. The great scientific culture of the West—our culture—is no exception. The calculus is the story this world first told itself as it became the modern world."

— David Berlinski
A Tour of the Calculus

Before we look too closely at our first calculus concept, the **derivative**, we will briefly review the slope of the line, the concept on which the derivative is built. Recall from algebra the slope-intercept form of the line is $y = mx + b$, where $\boldsymbol{m}$ is the slope of the line and $\mathbf{(0,}\ \boldsymbol{b}\mathbf{)}$ is the $y$-intercept.

**Example 3:** If the point $(x_1, y_1)$ is on the line $y = mx + b$, then $y_1 = mx_1 + b$. What happens if you replace $x_1$ with $x_1 + 1$. How does the new value of $y$ compare to the value $y_1$? Explain this in terms of rate of change. What if you replace $x_1$ with $x_1 + 10$?

**SOLUTION**

When $x = x_1$, $y_1 = mx_1 + b$. At $x = x_1 + 1$,

$$\begin{aligned} y &= m(x_1+1)+b \\ &= mx_1 + m + b \\ &= (mx_1 + b) + m \\ &= y_1 + m. \end{aligned}$$

Each time $x$ increases by 1, $y$ increases by $m$.

When $x$ is replaced by $x_1 + 10$,

$$\begin{aligned} y &= m(x_1+10)+b \\ &= mx_1 + 10m + b \\ &= (mx_1 + b) + 10m \\ &= y_1 + 10m. \end{aligned}$$

When $x$ increased by 10, $y$ increases by $10m$.

**Example 4:** We now consider a nonlinear function $f(x) = x^2$. Recall that the average rate of change for a given function over an interval $[a,b]$ is defined by the ratio (or the quotient) of the change in the output to the change in the input; i.e., $\Delta y / \Delta x$. In function notation, this becomes

$$\frac{\Delta y}{\Delta x} = \frac{f(b) - f(a)}{b - a}.$$

a. Use this ratio to find the average rate of change of $f(x) = x^2$ for $x$ in the intervals [0, 1], [1, 2], [10, 11], and [100, 101]. What conclusion can be made about the average rates of change on these intervals?

b. Let's zoom in on the two intervals [0, 1] and [1, 2]. Compute the average rates of change on the intervals [0.5, 1] and [1, 1.5]. Do the same for [0.9, 1] and [1, 1.1]. Describe any patterns that you note.

**SOLUTION**

a. On [0, 1], the average rate of change is $\dfrac{f(1) - f(0)}{1 - 0} = \dfrac{1^2 - 0^2}{1} = 1.$

On [1, 2], the average rate of change is $\dfrac{f(2) - f(1)}{2 - 1} = \dfrac{2^2 - 1^2}{1} = 4 - 1 = 3.$

On [10, 11], the average rate of change is $\dfrac{f(11) - f(10)}{11 - 10} = \dfrac{11^2 - 10^2}{1} = 121 - 100 = 21.$

On [100, 101], the average rate of change is

$$\frac{f(101) - f(100)}{101 - 100} = \frac{101^2 - 100^2}{1} = 10201 - 10000 = 201.$$

Although the value of $\Delta x$ is always 1, $\Delta y$ keeps increasing. So the rate of change, $\Delta y/\Delta x$, is not constant.

b. On [0.5, 1], the average rate of change is $\dfrac{f(1)-f(0.5)}{1-0.5}=\dfrac{1^2-(0.5)^2}{0.5}=\dfrac{1-0.25}{0.5}=1.5$.

On [1, 1.5], the average rate of change is $\dfrac{f(1.5)-f(1)}{1.5-1}=\dfrac{(1.5)^2-1^2}{0.5}=\dfrac{2.25-1}{0.5}=2.5$.

On [0.9, 1], the average rate of change is $\dfrac{f(1)-f(0.9)}{1-0.9}=\dfrac{1^2-(0.9)^2}{0.1}=\dfrac{1-0.81}{0.1}=1.9$.

On [1, 1.1], the average rate of change is $\dfrac{f(1.1)-f(1)}{1.1-1}=\dfrac{(1.1)^2-1^2}{0.1}=\dfrac{1.21-1}{0.1}=2.1$.

The rates of change appear to approach 2 as the $x$-interval centered about 1 gets smaller.

The average rate of change of $f(x)$ on the interval $[a, b]$ is given by the ratio

$$\frac{\Delta y}{\Delta x}=\frac{f(b)-f(a)}{b-a}.$$

This ratio is called **the difference quotient** of $f(x)$ for $x$ in the interval $[a, b]$. The difference quotient can take several forms, depending on how you name the values $a$ and $b$. For example, the difference quotient of $f(x)$ on the interval $[x, x+h]$ becomes

$$\frac{\Delta y}{\Delta x}=\frac{f(x+h)-f(x)}{(x+h)-x}=\frac{f(x+h)-f(x)}{h}.$$

Or, to find the average rate of change of $f(x)$ on the interval $[x-h, x+h]$, the difference quotient becomes

$$\frac{\Delta y}{\Delta x}=\frac{f(x+h)-f(x-h)}{(x+h)-(x-h)}=\frac{f(x+h)-f(x-h)}{2h}.$$

We will use this last form of the difference quotient.

**Example 5:** For this example, we will use the non-linear function $g(x)=x^3-2x+1$.

a. Use the difference quotient

$$\frac{g(x+h)-g(x-h)}{(x+h)-(x-h)}$$

to symbolically represent the average rate of change of $g$ on the interval $[x-h, x+h]$. Simplify the difference quotient for $g$ as much as possible.

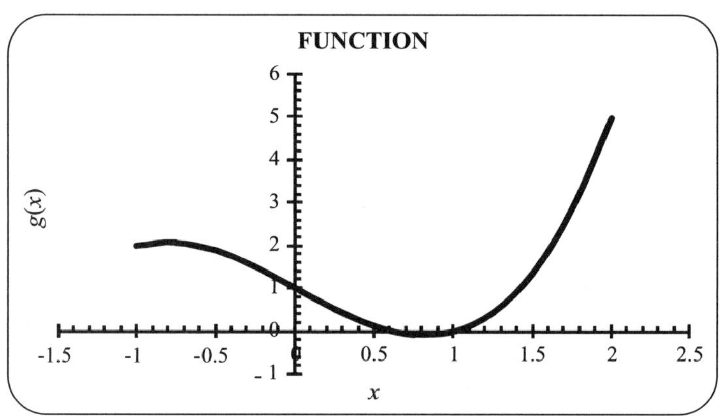


b. Use your answer from Part a to compute the average rate of change of $g$ on the following intervals: [0.9, 1.1], [0.99, 1.01], and [0.999, 1.001]. Identify $x$ and $h$ in each case.

c. What happens to the different quotient as $h$ gets closer and closer to 0?

**SOLUTION**

a. $$\frac{g(x+h)-g(x-h)}{(x+h)-(x-h)}=\frac{\left[(x+h)^3-2(x+h)+1\right]-\left[(x-h)^3-2(x-h)+1\right]}{2h}$$
$$=\frac{(x^3+3x^2h+3xh^2+h^3-2x-2h+1)-(x^3-3x^2h+3xh^2-h^3-2x+2h+1)}{2h}$$
$$=\frac{6x^2h+2h^3-4h}{2h}$$
$$=\frac{2h(3x^2+h^2-2)}{2h}$$
$$=3x^2+h^2-2$$

b. On [0.9, 1.1], $x=1$ and $h=0.1$, so the difference quotient, $3x^2+h^2-2$, is
$$3(1)^2+(0.1)^2-2=1.01.$$
On [0.99, 1.01], $x=1$ and $h=0.01$, so the difference quotient is
$$3(1)^2+(0.01)^2-2=1.001.$$
On [0.999, 1.001], $x=1$ and $h=0.001$, so the difference quotient is
$$3(1)^2+(0.001)^2-2=1.000001.$$

c. As $h$ gets closer to 0, the difference quotient appears to be approaching 1.

The word *derivative* means something that comes from something else, or is *derived* from something else. In mathematics, the ***derivative*** of one function is a second function which is obtained from the first function by finding the difference quotient and seeing what happens as $h$ gets smaller and smaller, that is, closer and closer to 0. This process of taking smaller and smaller values of $h$ is called taking the *limit as h approaches 0.*

We know that a line, $y=mx+b$, has constant slope, $m$. The slope can be thought of as the rate of change of the output with respect to the input. For a linear function, it doesn't matter whether you start at $x=1$ or $x=100$. If you increase $x$ by 1, then the output, often called $y$, will change by that constant rate of change, or slope.

For nonlinear functions, the rate of change depends on where you are; in other words, it is itself a function of $x$. The function that gives the "slope" of a nonlinear function is the derivative; it is a "slope-giving function." Thus the derivative is a generalization of the concept of the slope of a line to nonlinear functions. You can use the derivative to find the "slope of a curve" at a given input value.

The symbol $f'(x)$ means the function which is the derivative of $f(x)$. If we want the derivative of the function $f(x)$ at a specific value $c$, we use the notation $f'(c)$.

**Example 6:** Use ***Differentiating.xls*** to plot the derivative of the function from **Example 5**

$$g(x) = x^3 - 2x + 1 \ .$$

Find the derivative at 1, $g'(1)$.

**SOLUTION**

Enter the function in *Cell C24* and plot on the interval $[-1, \ 2]$. The value of the derivative at 1 is 1, which is the same value we found in **Example 5**.

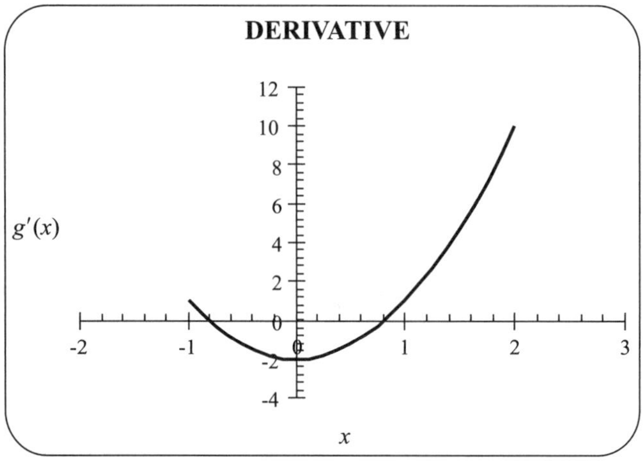


Imagine that you are zooming in on a curve a particular point. As you get closer, the curve begins to look more and more like a line. The line that matches the curve at that point is called the *tangent line*. The tangent line to a curve at a point shares, at that point, the coordinates of the point and the slope.

The equation of any line can be found using the slope-intercept form of the line. If the slope of the line is *m*, and $(x_1, \ y_1)$ is any point on the line, then the equation of the line is

$$y - y_1 = m(x - x_1).$$

In the case of a tangent line to the curve at $x = x_1$, the *y*-value $y_1$ is given by $y_1 = f(x_1)$. The slope is given by the derivative of the function at that point, that is, $f'(x_1)$. Then the equation of the tangent line becomes

$$y - f(x_1) = f'(x_1)(x - x_1) \ .$$

**Example 7:** Find the derivative of $f(x) = \dfrac{1}{x}$ and use it to find the equation of the line tangent to the graph of *f* at $x = 1$. Sketch the graphs of *f* and the tangent line.

**SOLUTION**

The difference quotient is

$$\frac{\Delta y}{\Delta x} = \frac{f(x+h) - f(x-h)}{2h}.$$

Since

$$f(x)=\frac{1}{x},\quad f(x+h)=\frac{1}{x+h}\quad \text{and}\quad f(x-h)=\frac{1}{x-h}.$$

The difference quotient is computed and simplified below.

$$\begin{aligned}\frac{\Delta y}{\Delta x}&=\frac{\frac{1}{x+h}-\frac{1}{x-h}}{2h}\\&=\frac{\frac{1}{x+h}-\frac{1}{x-h}}{2h}\cdot\left[\frac{(x+h)(x-h)}{(x+h)(x-h)}\right]\\&=\frac{(x-h)-(x+h)}{2h(x+h)(x-h)}\\&=\frac{-2h}{2h(x+h)(x-h)}\\&=-\frac{1}{(x+h)(x-h)}\end{aligned}$$

Then we let $x=1$, and the difference quotient becomes

$$-\frac{1}{(1+h)(1-h)}.$$

Next, we substitute smaller and smaller values of $h$, say $h=0.01$, $h=0.001$, $h=0.0001$, and $h=0.00001$. We find that the difference quotient is getting closer and closer to $-1$. That is, $f'(1)=-1$. So the equation of the tangent line is

$$\begin{aligned}y-f(1)&=f'(1)(x-1)\\y-1&=-1(x-1)\\y&=-x+2.\end{aligned}$$

The function $f(x)=\dfrac{1}{x}$ has the following graph:

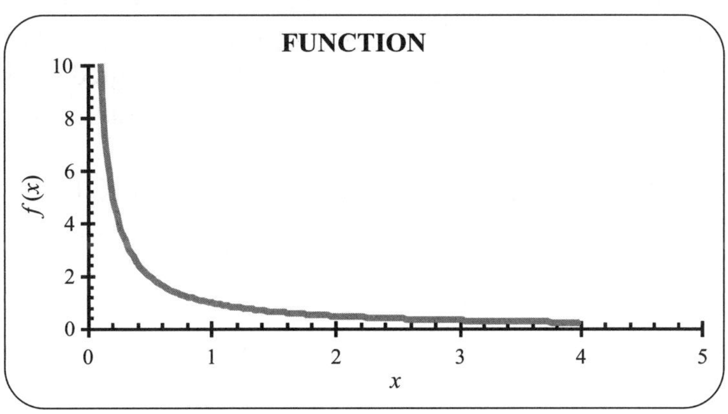

Drawing in the tangent line to the curve at $x = 1$, the function with the tangent line has the following graph:

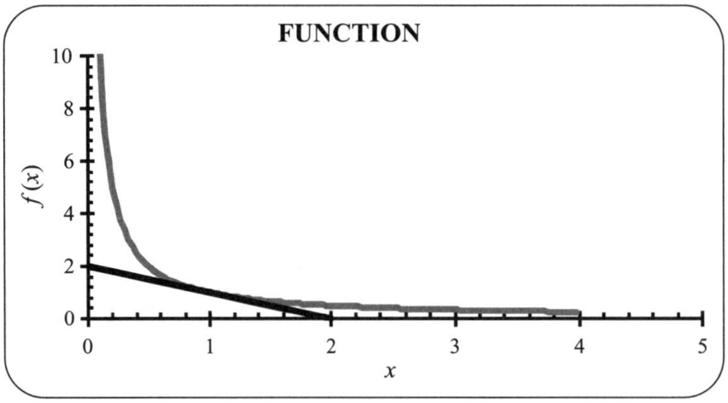


The process of evaluating the difference quotient

$$\frac{\Delta y}{\Delta x} = \frac{f(x+h) - f(x-h)}{2h}$$

at smaller and smaller values of $h$ (that is, values of $h$ very close to 0), is called ***taking the limit***. We write

$$f'(x) = \lim_{h \to 0} \frac{f(x+h) - f(x-h)}{2h}.$$

Before diving into the applications of marginal cost, marginal revenue, and marginal profit functions, we should first note that these functions are exactly the same as the derivatives of the cost, revenue, and profit functions, respectively. In other words, $MC(q) = C'(q)$, $MR(q) = R'(q)$, and $MP(q) = P'(q)$. Also recall that derivatives are functions that give information about the rate of change of the output variable with respect to the input variable; in this case, the input variable is $q$, the production/sales level.

If the rate of change of the output variable (cost, revenue, or profit) with respect to the input variable (production/sales level) is positive, then the output is increasing as the production/sales level increases. If the rate of change is negative, then the output is decreasing as the production/sales level increases.

The main idea to keep in mind is a simple one — you want to maximize the profit. What is profit at a given value of $q$? Profit is the difference between the revenue earned from the sale of $q$ units and the cost of producing those $q$ units. As most cost functions have both fixed and variable costs, the fixed costs must be overcome before any positive profit (i.e., not a loss) is obtained.

Once the revenue catches up with the cost function (this is called the break-even point), then, provided that revenue is increasing more rapidly than production costs as the production/sales level increases, the difference between the revenue and cost (i.e., the profit) will increase. This is expressed mathematically using the derivatives (or marginals) as follows:

$$\begin{aligned} MR(q) &> MC(q) \\ MR(q) - MC(q) &> 0 \\ R'(q) - C'(q) &> 0 \\ P'(q) &> 0 \end{aligned}$$

Since the slope of the profit function is always positive, the profit function is increasing. That is, $P'(q)$ is positive whenever marginal revenue is greater than marginal cost, or when revenue is increasing faster than costs are.

When $MR(q) < MC(q)$, the profit, $P(q) = R(q) - C(q)$, is decreasing.

The $q$-value at which $MR(q) = MC(q)$ is thus the natural place to look when one is looking for the production/sales level that will yield the maximum profit.

**Example 8:** In ***Demand, Revenue, Cost, and Profit*** we discussed the revenue and cost functions for Don's Tofu Tarts. The formulas for these functions are given below.

$$R(q) = -0.0125q^2 + 2.50q \text{ and } C(q) = 15 + 0.50q.$$

a. Find an equation for $MR(q)$. Graph both $R(q)$ and $MR(q)$. At what value of $q$ does the revenue reach its maximum value? What is the maximum revenue?

b. Find an equation for $MC(q)$. Graph both $C(q)$ and $MC(q)$. What does the graph of $MC(q)$ tell us?

c. Compare the graphs of $MR(q)$ and $MC(q)$. Where does $MR(q) = MC(q)$?

d. What is the profit function, $P(q)$? Find $MP(q)$ by two methods. Evaluate $MP(q)$ at the value of $q$ you found in Part c. Explain its significance.

e. What is $MP(q)$ at $q = 100$? Explain the significance of this value.

**SOLUTION**

a.
$$\begin{aligned}
MR(q) &= \lim_{h\to 0} \frac{R(q+h) - R(q-h)}{2h} \\
&= \lim_{h\to 0} \frac{\left[-0.0125(q+h)^2 + 2.50(q+h)\right] - \left[-0.0125(q-h)^2 + 2.50(q-h)\right]}{2h} \\
&= \lim_{h\to 0} \frac{-0.05qh + 5h}{2h} \\
&= \lim_{h\to 0} \frac{2h(-0.025q + 2.5)}{2h} \\
&= \lim_{h\to 0} (-0.025q + 2.5) \\
&= -0.025q + 2.5
\end{aligned}$$

The graphs of the revenue function, $R(q)$, and the marginal revenue function, $MR(q)$, are given below.

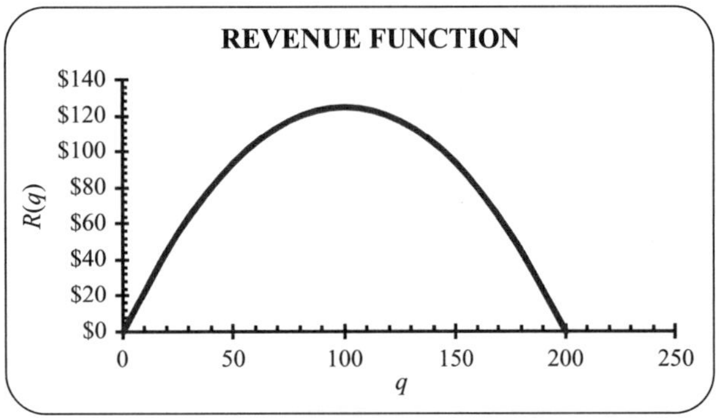


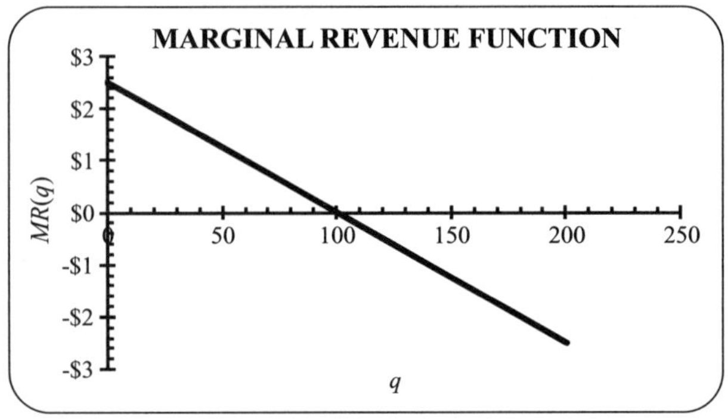


Notice that over the interval $(0,\ 100)$, where $R(q)$ is increasing, $MR(q) > 0$; over the interval $(100,\ 200)$, where $R(q)$ is decreasing, $MR(q) < 0$.

At $q = 100$, $MR(q) = 0$. This is the level of production/sales where the revenue changes from increasing to decreasing; hence, the revenue reaches its maximum value when the daily production /sales level is $q = 100$ tofu tarts. The maximum revenue is

$$R(100) = -0.0125(100)^2 + 2.5(100) = 125.$$

b. The cost function $C(q) = 15 + 0.5q$ is linear, so $MC(q) = C'(q)$ is the slope. That is, $MC(q) = 0.5$. The graphs of the cost function, $C(q)$, and the marginal cost function, $MC(q)$, are given below.

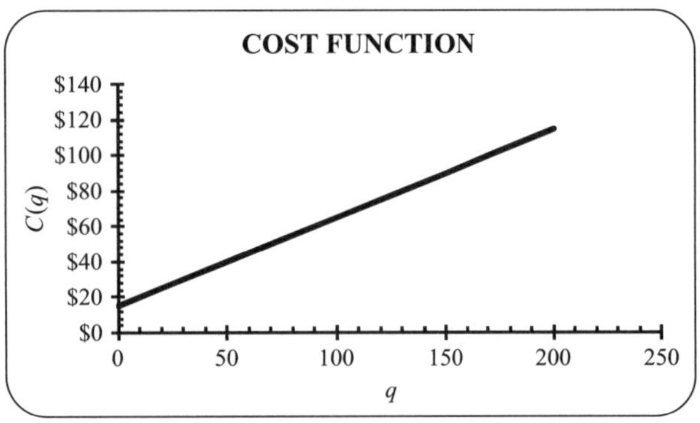


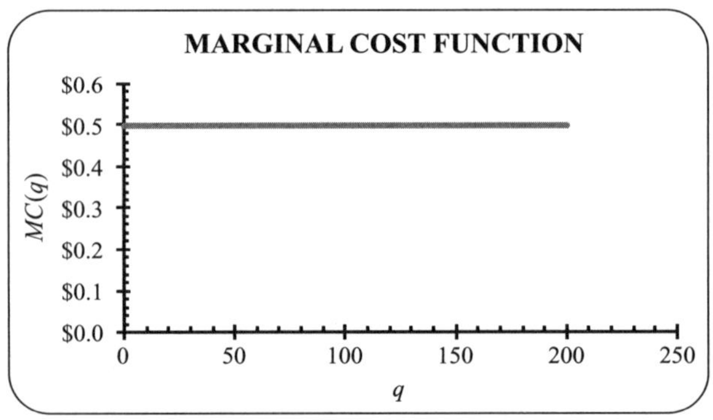


Notice that the cost is increasing at the constant rate of \$0.50 per tart, resulting from the variable cost of 50 cents per tart. The derivative of the cost function is constant.

c. We graph $MR(q)$ and $MC(q)$ on the same set of axes.

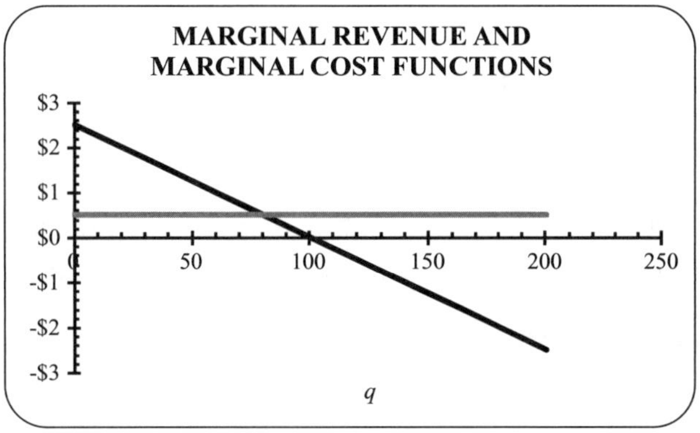


Solve $MR(q) = MC(q)$ for $q$:

$$\begin{aligned} -0.025q + 2.5 &= 0.5 \\ 2 &= 0.025q \\ 80 &= q \end{aligned}$$

d. The profit function is

$$\begin{aligned} P(q) &= R(q) - C(q) \\ &= (-0.0125q^2 + 2.5q) - (15 + 0.5q) \\ &= -0.0125q^2 + 2q - 5. \end{aligned}$$

Therefore

$$\begin{aligned} MP(q) &= \lim_{h\to 0} \frac{P(q+h) - P(q-h)}{2h} \\ &= \lim_{h\to 0} \frac{\left[-0.0125(q+h)^2 + 2(q+h) - 5\right] - \left[-0.0125(q-h)^2 + 2(q-h) - 5\right]}{2h} \\ &= \lim_{h\to 0} \frac{-0.05q + 4h}{2h} \\ &= \lim_{h\to 0} \frac{2h(-0.025q + 2)}{2h} \\ &= \lim_{h\to 0} (-0.25q + 2) \\ &= -0.025q + 2. \end{aligned}$$

Alternately, we could find the marginal profit as the difference between the marginal revenue and the marginal cost.

$$\begin{aligned} MP(q) &= MR(q) - MC(q) \\ &= (-0.025q + 2.5) - (0.5) \\ &= -0.025q + 2 \end{aligned}$$

Both methods of finding $MP(q)$ give the same result.

At $q = 0$, $MP(q) = -0.025(80) + 2 = 0$.

Since $MP(80)=0$, $MR(80)=MC(80)$. The value of $q$ at which $MR(q)=MC(q)$ corresponds to the maximum difference between the revenue and cost functions, that is, the maximum profit. In this case, it is the point at which the profit changes from increasing to decreasing. Prior to this point of intersection, $MR(q)>MC(q)$; hence, revenue is increasing more rapidly than the cost. After this point of intersection, $MR(q)<MC(q)$; that is, the cost is increasing more rapidly and thus closing the profit gap.

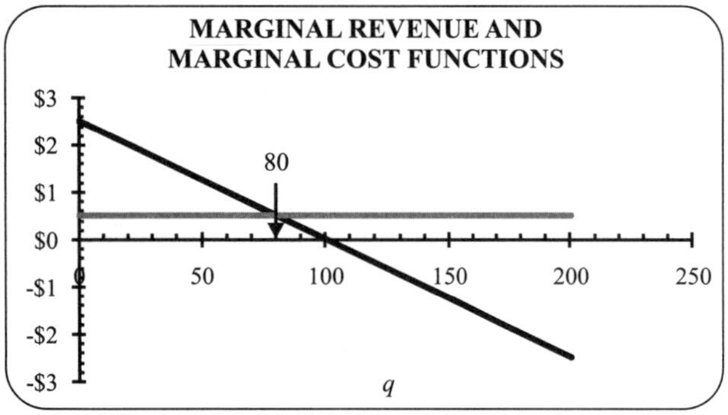


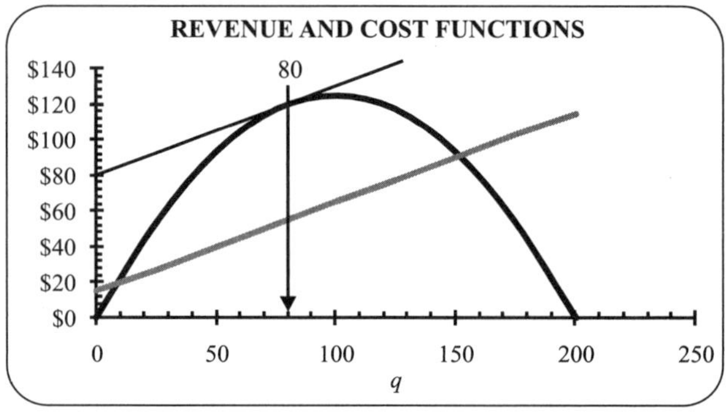


e. $MP(100)=-0.025(100)+2=-0.5$. Since $MR(100)=0$, the maximum **revenue** occurs at $q=100$. But since $MP(100)$ is negative, the **profit** is decreasing at $q=100$.

**Example 9:** Graphs of the marginal revenue and marginal cost functions for a product are given below.

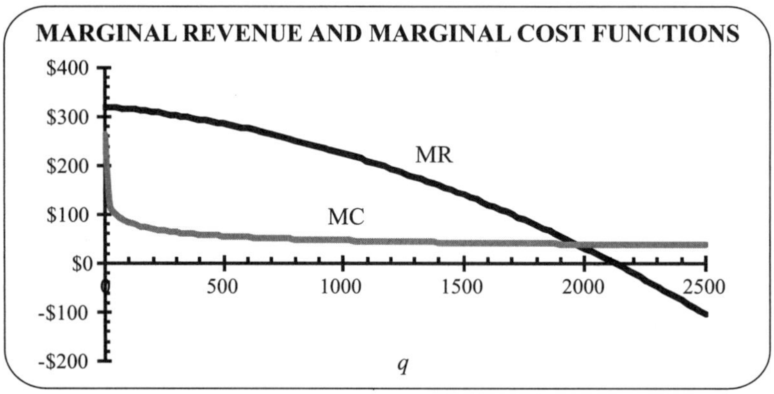


a. Is the revenue function increasing or decreasing at $q=500$? Explain.

b. Is the cost function increasing or decreasing at $q=500$? Explain.

c. Is the revenue function increasing or decreasing at $q=2500$? Explain.

d. Is the cost function increasing or decreasing at $q=2500$? Explain.

e. Is the profit function increasing or decreasing at $q=2500$? Explain.

f. Is the profit function increasing or decreasing at $q=1000$? Explain.

g. For what value of $q$ is profit maximized? Explain.

h. For what value of $q$ is revenue maximized? Explain.

**SOLUTION**

a. $MR(500) > 0$, which means that the slope of the revenue curve is positive at $q = 500$; therefore, the revenue function is increasing at $q = 500$.

b. $MC(500) > 0$, which means that the slope of the cost curve is positive at $q = 500$; therefore, the cost function is increasing at $q = 500$.

c. $MR(2500) < 0$, which means that the slope of the revenue curve is negative at $q = 2500$; therefore, the revenue function is decreasing at $q = 2500$.

d. $MC(2500) > 0$, which means that the slope of the cost curve is positive at $q = 2500$; therefore, the cost function is increasing at $q = 2500$.

e. $MP(2500) = MR(2500) - MC(2500) < 0$, which means that the slope of the profit curve is negative at $q = 2500$; therefore, the profit function is decreasing at $q = 2500$.

f. $MP(1000) = MR(1000) - MC(1000) > 0$, which means that the slope of the profit curve is positive at $q = 1000$; therefore, the profit function is increasing at $q = 1000$.

g. Profit is maximized when $MP(q) = 0$ or $MR(q) = MC(q)$; therefore, profit is maximized at $q \cong 1950$.

h. Revenue is maximized when $MR(q) = 0$; therefore, revenue is maximized at $q \cong 2100$.

## *Properties and Shortcuts*

There are several notations which mean "the derivative of $f(x)$." Two of the most useful ones are $f'(x)$ and $\frac{d}{dx}f(x)$ or $\frac{dy}{dx}$, where $y = f(x)$. These mean exactly the same thing.

The properties of derivatives can be used to simplify differentiation. Three basic properties are given here.

1. $\frac{d}{dx}[k \cdot f(x)] = k \cdot f'(x)$

2. $\frac{d}{dx}[f(x) \pm g(x)] = f'(x) \pm g'(x)$

3. $\frac{d}{dx}[k] = 0$

**Example 10:** Given that $\frac{d}{dx}(x^2) = 2x$ and that $\frac{d}{dx}(x) = 1$, use these and the properties above to find the derivative of $f(x) = 3x^2 + 5x + 7$.

**SOLUTION**

$$\begin{aligned} f'(x) &= \frac{d}{dx}(3x^2+5x+7) \\ &= \frac{d}{dx}(3x^2)+\frac{d}{dx}(5x)+\frac{d}{dx}(7) \quad \text{by Property 2} \\ &= 3\frac{d}{dx}(x^2)+5\frac{d}{dx}(x)+0 \qquad \text{by Properties 1 and 3} \\ &= 3(2x)+5(1) \\ &= 6x+5 \end{aligned}$$

**Example 11:** Suppose that $g(x)=4\cdot f(x)$ and that $f'(x)=\dfrac{x^3}{x+6}$. What is $g'(x)$?

**SOLUTION**

$$\begin{aligned} g'(x) &= 4\cdot f'(x) \\ &= 4\left(\frac{x^3}{x+6}\right) \\ &= \frac{4x^3}{x+6} \end{aligned}$$

**Example 12**: Suppose that $h(x)=f(x)-g(x)$, $f'(x)=-3x^2+2$, and $g'(x)=\frac{1}{6}x^3+1$. Find $h'(x)$.

**SOLUTION**

$$\begin{aligned} h'(x) &= f'(x)-g'(x) \\ &= (-3x^2+2)-\left(\frac{1}{6}x^3+1\right) \\ &= -\frac{1}{6}x^3-3x^2+1 \end{aligned}$$

Rather than having to use the limit of the difference quotient

$$f'(x)=\lim_{h\to 0}\frac{f(x+h)-f(x-h)}{2h}$$

every time to calculate derivatives, there are several formulas, or shortcuts, which can be used. All of these shortcuts come from the limit of the difference quotient, but using the shortcuts will speed things up.

1. $\dfrac{d}{dx}(k)=0$ where $k$ is any constant.

2. $\frac{d}{dx}x^n = nx^{n-1}$. This is known as the Power Rule.

3. $\frac{d}{dx}e^x = e^x$. Other than the zero function, the exponential function is the only function whose derivative is equal to itself.

4. $\frac{d}{dx}b^x = b^x \ln b$. The base $b$ is any positive number that is different from 1.

5. $\frac{d}{dx}\ln x = \frac{1}{x}$.

**Example 13:** Use the properties and shortcuts for derivatives to find the equation of the tangent line to $f(x) = (2x-1)(3x+4)$ at $x = -1$.

**SOLUTION**

Using algebra, we multiply to get

$$f(x) = (2x-1)(3x+4) = 6x^2 + 5x - 4$$

so that

$$f'(x) = 6(2x) + 5(1) + 0 = 12x + 5.$$

Then $f'(-1) = 12(-1) + 5 = -7$. Since $f(-1) = -3$, the equation of the tangent line is

$$\begin{aligned} y - f(-1) &= f'(-1)\left(x - (-1)\right) \\ y - (-3) &= (-7)\left(x - (-1)\right) \\ y + 3 &= -7x - 7 \\ y &= -7x - 10. \end{aligned}$$

**Example 14:** The cost to manufacture $q$ notebook computers is given by

$$C(q) = 0.0002q^3 - 0.06q^2 + 120q + 5000.$$

What is the actual cost incurred in manufacturing the 401st computer? What is the marginal cost when $q = 400$?

**SOLUTION**

The actual cost to produce the 401st computer is the difference between the costs to manufacture 401 computers and 400 computers.

$$C(401) - C(400) = 56368.18 - 56200 = 168.18$$

The marginal cost at $q = 400$ is the value of the derivative at $q = 400$.

$$\begin{aligned} MC(q) &= C'(q) \\ &= \frac{d}{dq}(0.0002q^3 - 0.06q^2 + 120q + 5000) \\ &= (0.0002)(3q^2) - (0.06)(2q) + 120 \\ &= 0.0006q^2 - 0.12q + 120 \\ MC(400) &= 0.0006(400)^2 - 0.12(400) + 120 = 168 \end{aligned}$$

Notice that the marginal cost at $q = 400$ and the actual cost at $q = 400$ to produce the next computer are essentially the same.

In addition to the shortcuts, there are three rules which give us ways to find derivatives of functions which are products, or quotients, or compositions, of simpler functions.

1. The Product Rule: $\frac{d}{dx}\left(f(x) \cdot g(x)\right) = f'(x) \cdot g(x) + f(x) \cdot g'(x)$

2. The Quotient Rule: $\frac{d}{dx}\left(\frac{f(x)}{g(x)}\right) = \frac{f'(x) \cdot g(x) - f(x) \cdot g'(x)}{\left(g(x)\right)^2}$

3. The Chain Rule: $\frac{d}{dx}\left(f(g(x))\right) = f'(g(x)) \cdot g'(x)$

The Chain Rule is the one that we use when the function we are differentiating is a composition of simpler functions. The function $f(x)$ is the "outside" function and the function $g(x)$ is the "inside" function. Sometimes you have a try a few combinations to figure out which choices for $f(x)$ and $g(x)$ give you the correct composition $f(g(x))$.

**Example 15:** Use the shortcuts and rules for differentiation to find derivatives of each of the following functions.

a. $F(x) = 2x \cdot e^x$

b. $G(x) = \sqrt{x} \cdot (x^2 - 3x)$

c. $h(x) = \dfrac{6 - 7x}{6 + 7x}$

d. $k(x) = \dfrac{x^5 + 5x^4 - x}{x^2}$

**SOLUTION**

a. Use the Product Rule:

$$\begin{aligned}F'(x) &= \frac{d}{dx}(2x\cdot e^x)\\ &= \left(\frac{d}{dx}(2x)\right)\cdot e^x + (2x)\cdot\left(\frac{d}{dx}(e^x)\right)\\ &= 2e^x + 2xe^x\end{aligned}$$

b. Although you could use the Product Rule, a little algebra will make its use unnecessary.

$$\begin{aligned}G(x) &= \sqrt{x}\cdot(x^2-3x)\\ &= (x^{1/2})(x^2-3x)\\ &= x^{1/2}x^2 - 3x^{1/2}x\\ &= x^{5/2} - 3x^{3/2}\end{aligned}$$

Then using the Power Rule,

$$\begin{aligned}G'(x) &= \frac{5}{2}x^{5/2-1} - 3\cdot\left(\frac{3}{2}x^{3/2-1}\right)\\ &= \frac{5}{2}x^{3/2} - \frac{9}{2}x^{1/2}.\end{aligned}$$

c. Use the Quotient Rule.

$$\begin{aligned}h'(x) &= \frac{\left(\frac{d}{dx}(6-7x)\right)\cdot(6+7x) - (6-7x)\cdot\left(\frac{d}{dx}(6+7x)\right)}{(6+7x)^2}\\ &= \frac{(-7)(6+7x)-(6-7x)(7)}{(6+7x)^2}\\ &= \frac{-42-49x-42+49x}{(6+7x)^2}\\ &= \frac{-84}{(6+7x)^2}\end{aligned}$$

d. We could use the Quotient Rule, but a little algebra will simplify the function.

$$\begin{aligned}k(x) &= \frac{x^5+5x^4-x}{x^2}\\ &= \frac{x^5}{x^2} + \frac{5x^4}{x^2} - \frac{x}{x^2}\\ &= x^3 + 5x^2 - x^{-1}\end{aligned}$$

Using the Power Rule, the derivative is

$$\begin{aligned} k'(x) &= 3x^2 + 5(2x) - (-1)x^{-2} \\ &= 3x^2 + 10x + x^{-2}. \end{aligned}$$

**Example 16:** Use the Chain Rule to find the derivative of the each function.

a. $s(x) = (x^3 + 3)^{50}$

b. $t(x) = e^{x^2}$

c. $H(x) = \ln(e^x + 6)$

**SOLUTION**

a. $s(x) = (x^3 + 3)^{50} = (f(g(x))$ where $f(x) = x^{50}$ and $g(x) = x^3 + 3$. Since $f'(x) = 50x^{49}$ and $g'(x) = 3x^2$, we get

$$\begin{aligned} s'(x) &= f'(g(x)) \cdot g(x) \\ &= 50(x^3 + 3)^{49} \cdot 3x^2 \\ &= 150x^2 (x^3 + 3)^{49}. \end{aligned}$$

b. $t(x) = e^{x^2} = f(g(x))$ where $f(x) = e^x$ and $g(x) = x^2$. Since $f'(x) = e^x$ and $g'(x) = 2x$, we get

$$\begin{aligned} t'(x) &= f'(g(x)) \cdot g'(x) \\ &= e^{x^2} \cdot 2x. \end{aligned}$$

c. $H(x) = \ln(e^x + 6) = f(g(x))$ where $f(x) = \ln(x)$ and $g(x) = e^x + 6$. Since $f'(x) = \dfrac{1}{x}$ and $g'(x) = e^x$, we get

$$\begin{aligned} H'(x) &= f'(g(x)) \cdot g'(x) \\ &= \frac{1}{e^x + 6} \cdot e^x \\ &= \frac{e^x}{e^x + 6}. \end{aligned}$$

**Example 17:** The revenue from selling $x$ items is given by $R(x) = 100x \cdot e^{-0.0001x}$. What is the marginal revenue when $x = 5{,}000$? Give an explanation of $MR(5000)$.

**SOLUTION**

$$\begin{aligned} MR(x) &= R'(x) \\ &= \left(\frac{d}{dx}(100x)\right)\cdot(e^{-0.0001x}) + (100x)\cdot\left(\frac{d}{dx}(e^{-0.0001x})\right) \\ &= (100)e^{-0.0001x} + (100x)(e^{-0.0001x})(-0.0001) \\ &= e^{-0.0001x}(100 - 0.01x) \end{aligned}$$

Note that we used the Product Rule and then needed to use the Chain Rule to find the derivative of the factor $e^{-0.0001x}$. Evaluating the marginal revenue at $x = 5000$, we get

$$MR(5000) = e^{-0.0001(5000)}\left(1000 - 0.01(5000)\right) = 30.33.$$

At $x = 5000$ items, the revenue is increasing at a rate of \$30.33 per item. We can also say that the approximate increase in revenue from the 5000th item to the 5001th item is \$30.33.

If a function has a maximum or a minimum value at $x = x_0$, and if $f'(x_0)$ exists, then we must have $f'(x_0) = 0$. That is, the function must have a horizontal tangent at $x = x_0$. Consider the function $f(x) = x^3 - 9x^2 + 15x + 30$. Its derivative is

$$\begin{aligned} f'(x) &= 3x^2 - 18x + 15 \\ &= 3(x^2 - 6x + 5) \\ &= 3(x-5)(x-1). \end{aligned}$$

Since $f'(5) = 0$ and $f'(1) = 0$, $f(x)$ has horizontal tangents at both $x = 5$ and $x = 1$. Which one gives a maximum and which one gives a minimum?

The Second Derivative Test gives us a method to distinguish between $x$ values which give a maximum and those which give a minimum. The Second Derivative Test says that if $f'(x_0) = 0$ and if $f''(x_0) > 0$, then the function has a minimum at $x = x_0$; the minimum value of the function is $f(x_0)$. Further, if $f'(x_0) = 0$ and if $f''(x_0) < 0$, then the function has a maximum at $x = x_0$; the maximum value of the function is $f(x_0)$. Note that the second derivative, $f''(x)$, is the derivative of the first derivative, $f'(x)$.

The second derivative of $f(x) = x^3 - 9x^2 + 15x + 30$ is the derivative of $f'(x) = 3x^2 - 18x + 15$, or $f''(x) = 6x - 18$. Since $f''(5) = 30 - 18 = 12 > 0$, there is a minimum at $x = 5$; this minimum value is $f(5) = 5$. And, since $f''(1) = 16 - 28 = -12 < 0$, there is a maximum at $x = 1$; this maximum value is $f(1) = 37$.

**Example 18:** The monthly cost to manufacture $x$ camcorders is

$$C(x) = 0.0025x^2 + 80x + 10{,}000.$$

a. What is the average cost $\overline{C}(x)$?
b. At what level is the average cost minimized? What is the minimum cost?

**SOLUTION**

a. The average cost is the cost per item, that is,

$$\begin{aligned}\overline{C}(x) &= \frac{C(x)}{x} \\ &= \frac{0.0025x^2 + 80x + 10{,}000}{x} \\ &= 0.0025x + 80 + 10{,}000x^{-1}.\end{aligned}$$

b. The minimum average cost occurs where the derivative of the average cost is 0.

$$\begin{aligned}\overline{C}'(x) &= 0.0025 - 10{,}000x^{-2} \\ &= 0.0025 - \frac{10{,}000}{x^2} \\ &= \frac{0.0025x^2 - 10{,}000}{x^2}\end{aligned}$$

$\overline{C}'(x) = 0$ where $0.0025x^2 - 10{,}000 = 0$, that is, where the numerator is 0. This happens at $x = 2000$. Since

$$\overline{C}''(x) = 20{,}000x^{-3},$$

$\overline{C}''(2000) > 0$, so, by the Second Derivative Test, $\overline{C}(x)$ has a minimum at $x = 2000$. This minimum average cost is $\overline{C}(2000) = 90$.

**Example 19:** A hotel has 10 luxury units that it can rent at \$300 per day. For each \$50 increase in cost, one unit will become vacant.

a. How much should the hotel charge to maximize revenue?

b. If it costs \$100 per day to clean and maintain each room, how much should they charge to maximize profit?

**SOLUTION**

a. First, we have to figure out the formula for the revenue function in terms of $x$, the number of \$50 increases.

| Rental amount per day | Number of units rented | Revenue |
|---|---|---|
| 300 | 10 | 300(10) |
| $300 + 50 = 350$ | $10 - 1 = 9$ | 350(9) |
| $300 + 50(2) = 400$ | $10 - 2 = 8$ | 400(8) |
| $300 + 50x$ | $10 - x$ | $(300 + 50x)(10 - x)$ |

The formula for the revenue function is $R(x) = (300+50x)(10-x) = 3,000+200x-50x^2$. The derivative, $MR(x) = R'(x) = 200-100x$, is zero at $x = 2$. The second derivative, $R''(x) = -100$, is negative at $x = 2$, which means by the Second Derivative Test that the revenue function has a maximum at $x = 2$. The maximum revenue occurs if they charge $\$300+\$50(2) = \$400$ per day. At that rate, they will rent $10-2 = 8$ units, for a maximum revenue of $R(2) = \$3,200$.

b. Since the daily cost to maintain each room is \$100, the formula for the profit function is

$$\begin{aligned} P(x) &= R(x) - C(x) \\ &= (3000+200x-50x^2) - 100(10-x) \\ &= 2000+300x-50x^2. \end{aligned}$$

The derivative, $MP(x) = P'(x) = 300-100x$, is zero at $x = 3$. The second derivative, $P''(x) = -100$, is negative at $x = 3$, so the profit function has a maximum at $x = 3$. The maximum profit is realized when they charge \$450 per room and rent $10-3 = 7$ rooms, for a maximum profit of \$2,450.

## Exercises

1. *Basic linear equations.*
   a. Given the slope $m = 2$ and the $y$-intercept $(0, \ -3)$, find the slope-intercept form of the line.
   b. Using this equation, find three other points on the line.
   c. Using the point $(0, \ -3)$ and the slope, find, without using the equation, three other points on this line.
   d. Again, just using the point $(0, \ -3)$ and the slope, determine the $y$-coordinate of the point on this line that has an $x$-coordinate of 100.

**SOLUTION**

a. $y = 2x-3$
b. Possible solutions (answers may vary): $(1, \ -1)$, $(2, \ 1)$, $(-1, \ -5)$
c. $x_0 = 0$; $y_0 = -3$; $x_1 = 1$; $y_1 = -1$; $x_2 = 2$, $y_2 = 1$; $x_3 = 3$; $y_3 = 3$
d. $y = 200$

2. *Linear and nonlinear functions.*
   a. Sketch the graphs of $f(x) = -0.4x^2+4x$ and $g(x) = \frac{9}{5}x$ over the interval $[0, \ 10]$.
   b. Use the graph to estimate where the rates of change are the same and estimate the vertical distance between the two graphs at that point.

c. Find the quadratic function representing the distance between the two curves and its vertex to check the accuracy of your visual approximation.

**SOLUTION**

a.

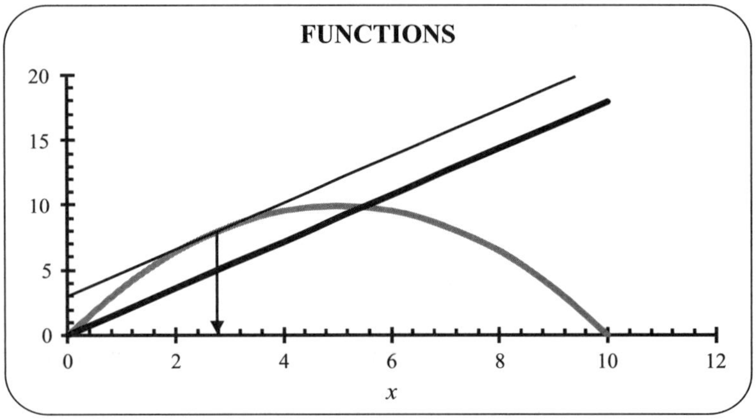


b. Approximately at $x = 3$; approximately 3 units apart.

c. $d(x) = -0.4x^2 + 2.2x$; vertex = (2.75, 3.025)

3. Given the function $f(x) = -x^2 + 16x - 44$, do the computations and answer the questions below.

a. Use the definition of the derivative, $f'(x) = \lim_{h \to 0} \dfrac{f(x+h) - f(x-h)}{2h}$, to find $f'(x)$.

b. Find the value of $x$ where $f(x)$ has a maximum.

c. What is the maximum value of $f(x)$?

d. Find the equation of the tangent line at $x = 10$.

e. Graph the function; on the same axes, graph the tangent line to the curve at $x = 10$.

**SOLUTION**

a. $f'(x) = -2x + 16$

b. $x = 8$

c. $y = 20$

d. $y = -4x + 56$

e. 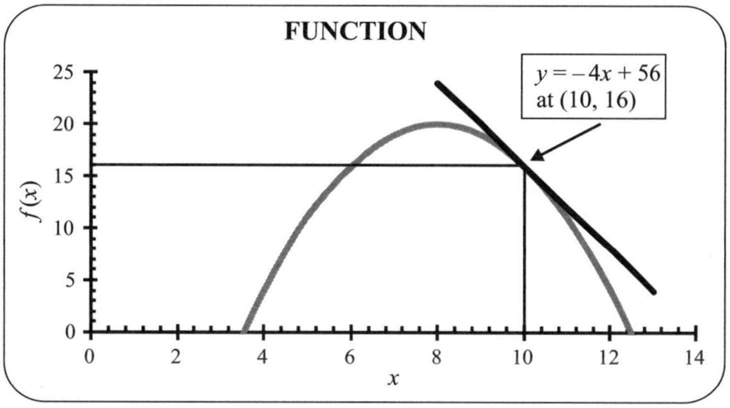


4. Given $f(x) = -\frac{2}{3}x^2 + 4x + 4$, answer the following questions.

   a. What is $f'(x)$?

   b. What is the slope of the graph at $x = 2$?

   c. What is the equation of the tangent line at $x = 2$?

   d. Draw the tangent line you found in Part c on the graph given below.

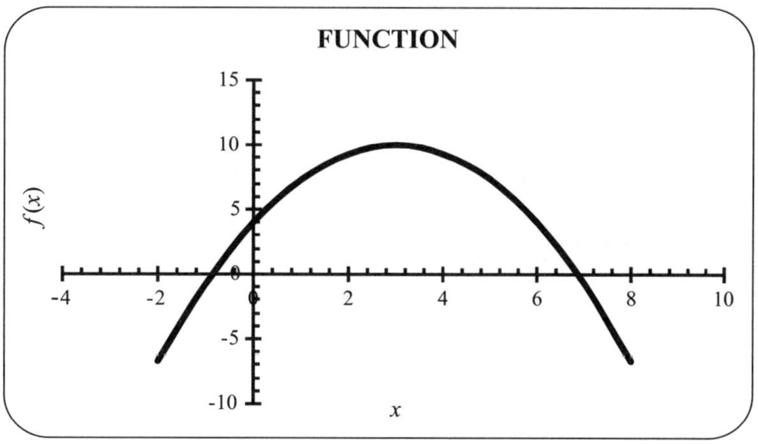


   e. Use the graph in Part d to estimate the value of $x$ at which the tangent line is horizontal.

   f. Use algebra to find the value of $x$ on the interval [0, 10] at which the tangent line is horizontal.

**SOLUTION**

a. $f'(x) = -\frac{4}{3}x + 4$

b. $\frac{4}{3}$

c. $y=\dfrac{4}{3}x+\dfrac{20}{3}$

d.

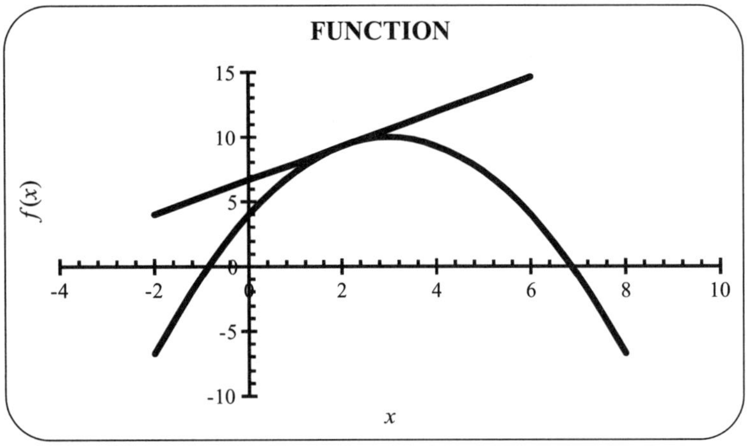


e. $x=3$

f. $x=3$

5. Graphs of the demand and marginal profit functions for a good are shown below.

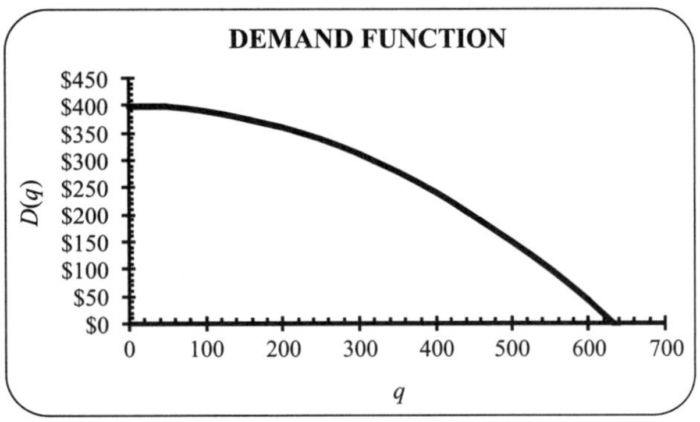


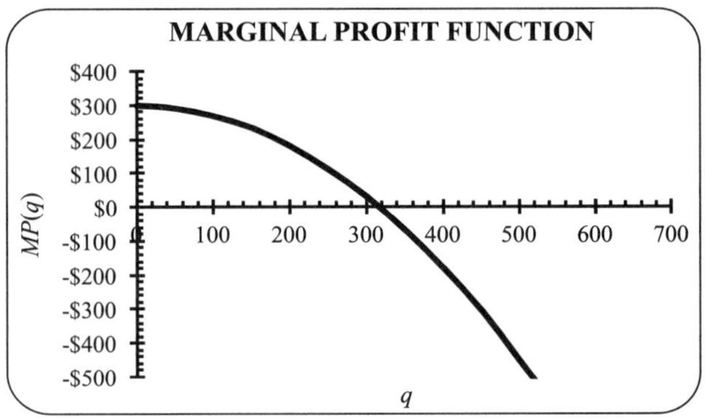


Answer the questions based on these graphs.

a. Estimate the number of units that should be produced and sold in order to maximize profit.

b. Estimate the price per unit for the number of units that you found in Part a.

c. What is the revenue at this sales level? Is this the profit as well? Why or why not?

d. Describe the graph of the profit function specifying the intervals over which it is increasing and those over which it is decreasing.

## SOLUTION

The following answers are approximate. Yours may differ somewhat.

a. 320 units

b. $300/unit

c. $96,000. No, the profit must also take into account the cost.

d. The profit function will be increasing steeply at first. It will continually slow down until it reaches its maximum at approximately $q = 320$ units. Without information about the cost function, we cannot tell what that maximum is. The interval over which the profit is increasing is (0, 320). After reaching its maximum at $q=320$, the profit begins to decrease, slowly at first and then progressively more rapidly.

6. Suppose a demand function for a given product is $D(q) = -0.4q + 800$. Also, suppose that the fixed cost of production is \$40,000 and the variable cost is \$40 per unit.

   a. Graph the demand function.

   b. Find the formulas for the revenue and the cost functions.

   c. Find the formulas for the profit and marginal profit functions.

   d. Find the number of units that yields a maximum profit.

   e. What is the maximum possible profit?

**SOLUTION**

a.

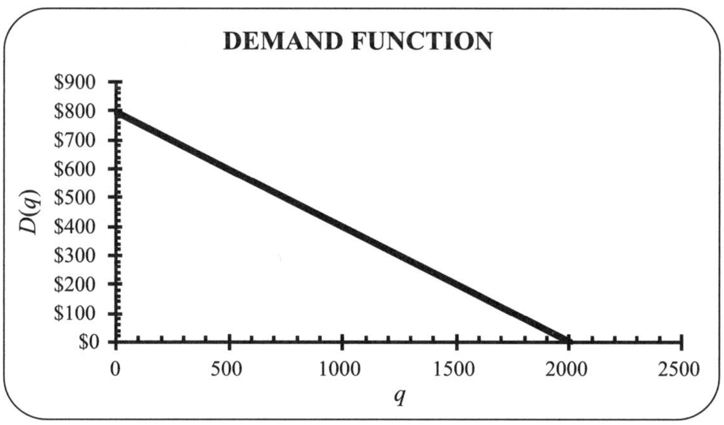


b. $R(q) = -0.4q^2 + 800q$

$C(q) = 40q + 40,000$

c. $P(q) = -0.4q^2 + 760q - 40,000$

$MP(q) = -0.8q + 760$

d. 950 units

e. \$321,000

7. Suppose the marginal profit is given by

$$MP(q) = -0.0005q^2 + 0.25q + 152.5.$$

   a. For what value of $q$ is $MR = MC$?

   b. For what values of $q$ is $MR > MC$?

   c. For what values of $q$ is $MR < MC$?

   d. For what value of $q$ is profit maximized?

**SOLUTION**

a. At $q \cong 856$.

b. For $0 < q < 856$.

c. For $q > 856$.

d. At $q \cong 856$.

8. Use the properties, shortcuts and rules for differentiation to find the derivative of each function.

a. $y = x^{3/2}(2+\sqrt{x})$

b. $y = x^2 \cdot 3^x$

c. $y = \sqrt{x^7 - 6}$

d. $y = e^{3x}(5^x - x^4)$

e. $y = e - e^{-x/2}$

f. $y = \dfrac{3x^2 - 1}{4 - x}$

**SOLUTION**

a. $y' = 3x^{1/2} + 2x$

b. $y' = x3^x(2 + x \ln 3)$

c. $y' = \dfrac{7x^6}{2\sqrt{x^7 - 6}}$

d. $y' = e^{3x}(5^x - 4x^3 - 3x^4 + 5^x \ln 5)$

e. $y' = \frac{1}{2}e^{-x/2}$

f. $y' = \dfrac{-3x^2 + 24x - 1}{(4-x)^2}$

9. Suppose the cost function is given by $C(q) = 0.000001q^3 - 0.01q + 50$ and the revenue function is given by $R(q) = -0.02q^2 + 150q$, for $0 \le q \le 7500$.

a. Where does $MR(q) = MC(q)$? What is this common value?

b. What is the formula for the profit function, $P(q)$?

c. At what value of $q$ is the profit maximized?

**SOLUTION**

a. At $q = 3333$. $MR(3333) = MC(3333) = 16.67$.

b. $P(q) = 0.000001q^3 - 0.01q^2 + 100q$

c. At $q = 3333$. $MP(3333) = 185{,}185.18$.

10. The weekly demand for big-screen televisions is $D(q) = 600 - 0.05q$ and the weekly cost to produce those televisions is $C(q) = 0.000002q^3 - 0.03q^2 + 400q + 80000$.

a. What is the formula for the revenue function, $R(q)$?

b. What is $MR(2000)$? Explain what this means.

c. What is the formula for the profit function, $P(q)$?

d. What is $MP(2000)$? Explain what this means.

**SOLUTION**

a. $R(q) = 600q - 0.05q^2$

b. $MR(2000) = 400$. At 2,000 TVs, the weekly revenue is increasing at a rate of \$400 for each additional TV.

c. $P(q) = -0.000002q^3 - 0.02q^2 + 200q - 80{,}000$

d. $MP(2000) = 136$. At 2,000 TVs, the weekly profit is increasing at a rate of \$136 for each additional TV.

11. *Inventory Control*: The demand for motorcycle tires is 40,000 per year, with a uniform distribution. The cost to order each shipment is \$400. It costs \$2 per tire per year to store each tire. What is the cost per day? How many times per year should orders be placed in order to minimize both ordering and storing costs?

**SOLUTION**

If $t$ is the number of days between orders, and $T$ is the cost of each tire ($T$ is a constant), then the total cost per day is

$$C(t) = \left(400 \cdot \frac{365}{t} + 40{,}000T\right) + \left(2 \cdot \frac{20{,}000t}{365}\right).$$

The minimum cost occurs at $t = 36.5$ days; place the order 10 times per year.

# Using *Solver*

This section will help familiarize you with the ***Solver*** tool in Microsoft *Excel* and how to use it to solve business applications. You will begin with a simple application.

**Example 1:** Daniel's Super Gamer Store has a revenue function given by $R(q) = -(q-7.071)^2 + 50$. The quantity $q$ represents the number of video games sold in thousands of units and $R(q)$ represents the revenue in thousands of dollars.

a. Use ***Graphing.xls*** to plot the revenue function over the interval $[0, 15]$.

b. Use ***Solver*** to find the values of $q$ where $R(q) = 0$. Round your answers to the nearest unit.

c. Use ***Solver*** to find the number of video games that should be sold in order to maximize revenue. Round your answer to the nearest video game.

d. What is the maximum revenue? Round your answer to the nearest dollar.

## SOLUTION

Remember that, if you are using Microsoft *Office* 2003, the macro security level must be set to medium or low in order to use ***Graphing.xls***. If you are using Microsoft *Office* 2007, the macro must be enabled. Remember also that *Excel* must be set for automatic computation and that the variable must be called $x$.

a. Open ***Graphing.xls*** and enter either of the formulas =-((x-7.071)^2)+50 or =-1*(x-7.071)^2+50 in *Cell C18* and the values 0 and 15 in *Cells H18* and *I18*, respectively. Then click the **Graph** button to run the macro.

| | A | B | C | D | E | F | G | H | I | J | K | L |
|---|---|---|---|---|---|---|---|---|---|---|---|---|
| 16 | | **Definition** | | | **Computation** | | | **Plot Interval** | | | **Constants** | |
| 17 | | **Formula for *f*(*x*)** | | | *x* | *f*(*x*) | | *a* | *b* | | *s* | |
| 18 | | | 0.000959 | | | 0.000959 | | 0 | 15 | | *t* | |
| 19 | | | | | | | | | | | *u* | |
| 20 | | | | | | | | | | | *v* | |
| 21 | | | | | | | | | | | *w* | |

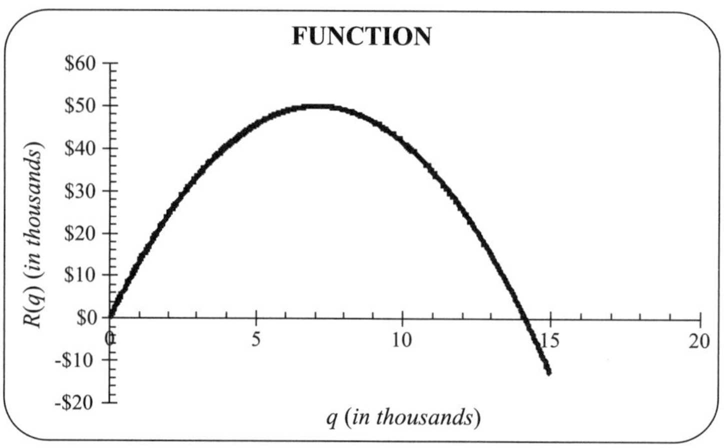

b. To open ***Solver*** in Microsoft *Excel* 2003, pull down the **Tools** menu and select ***Solver***. Click the menu expansion arrows if you do not see ***Solver***. If ***Solver*** does not appear on the menu, you will need to load it. Pull down the **Tools** menu and select **Add-Ins**. Check the **Solver Add-In** check box and click **OK**. (You may be prompted to insert the *Excel* installation disk.)

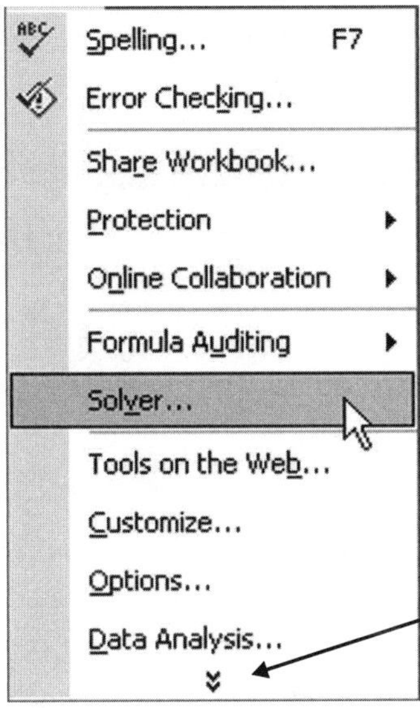


Click the menu expansion arrows if you do not see Solver. If it does not appear on the menu, follow the instructions in the paragraph above.

To open ***Solver*** in Microsoft *Excel* 2007, click the **Data** tab and then click **Solver**.

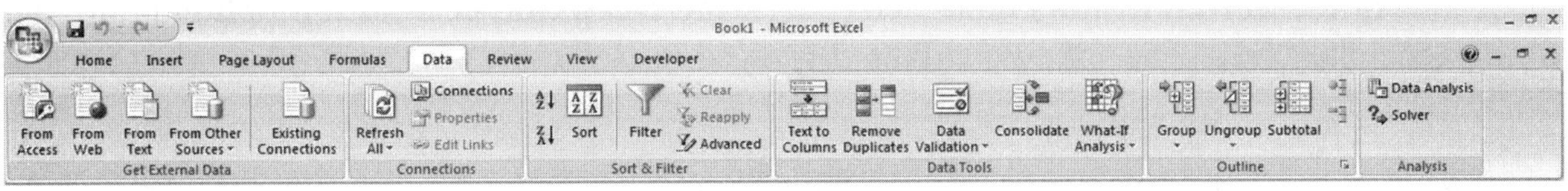


If ***Solver*** does not appear on the **Data** tab, you will need to load it. Click the **Microsoft Office Button** and then click **Excel Options**. Click **Add-Ins**, select **Excel Add-Ins** (in the Manage box), and click **Go**.

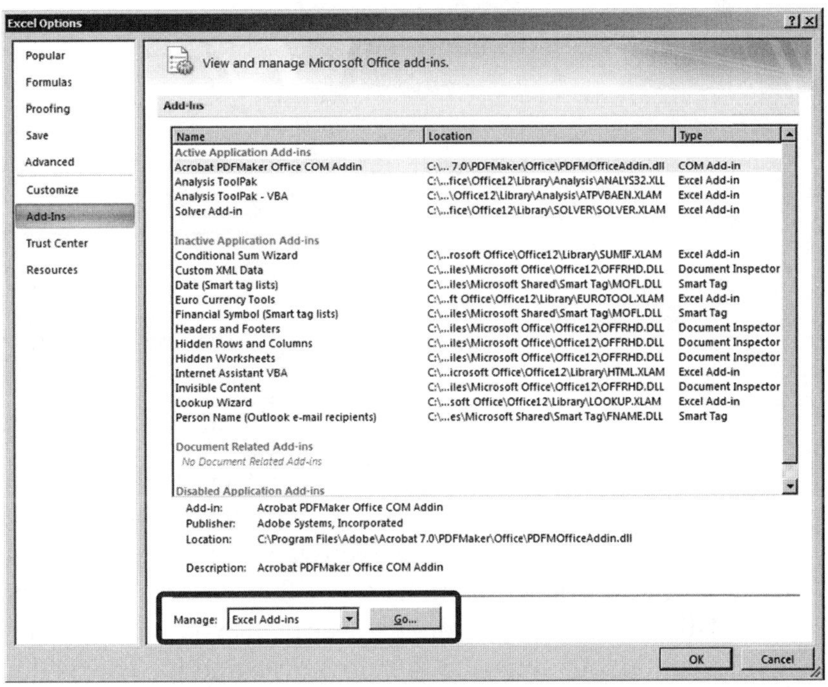

Check the **Solver Add-in** check box and click **OK**. (You may be prompted to insert the *Excel* installation disk.)

The target cell is the cell that contains the formula that you want to maximize, minimize, or set to a predetermined value. In this example, the target cell is *Cell F18*, the cell that contains the formula for the revenue function. Click in the **Set Target Cell** box and select *Cell F18*. Then click the **Value of** option button and enter the value 0 in the box. The value of the revenue function will change depending on the quantity $q$ in *Cell E18*. Click in the **By Changing Cells** box and select *Cell E18*. Then click **Solve**.

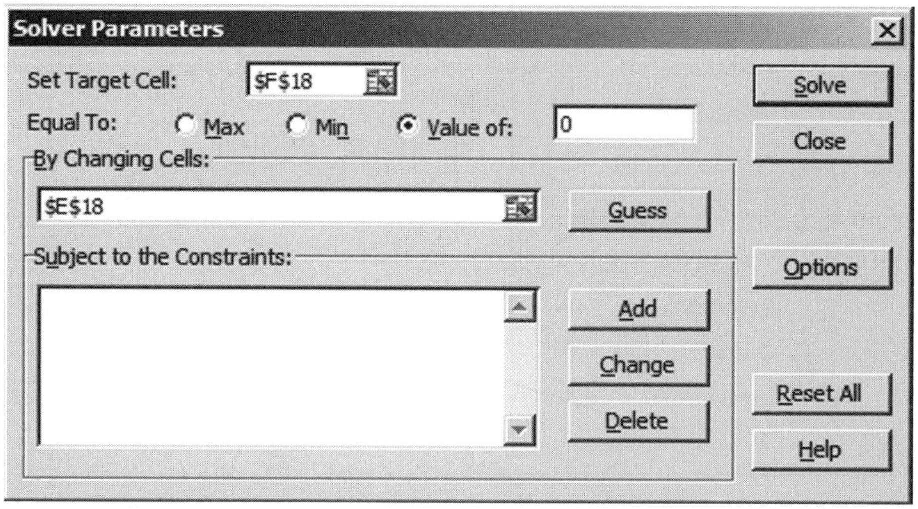


In the **Solver Results** dialog box, click the **Keep Solver Solution** option button and then click **OK**. The solution will be displayed in *Cell E18*.

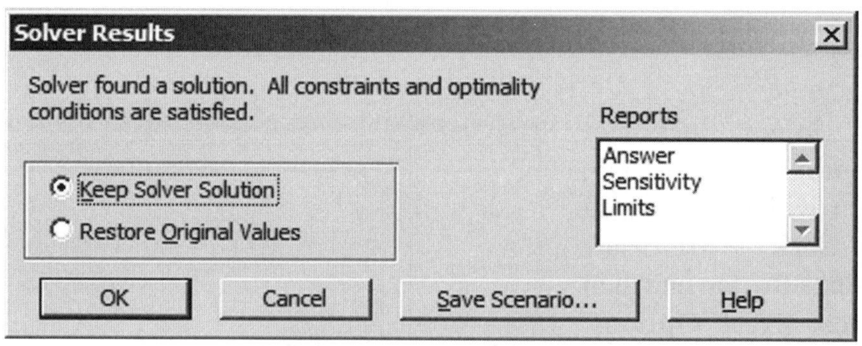


| | E | F |
|---|---|---|
| 16 | **Computation** | |
| 17 | *x* | *f*(*x*) |
| 18 | -6.8E-05 | -2.3E-11 |

The values in *Cells E18* and *F18* are displayed in scientific notation: -6.8E-05 for $x$ (quantity) and -2.3E-11 for $f(x)$ (revenue). In decimal form, these values are –0.000068 and –0.000000000023, respectively, which are approximately equal to zero.

The computation cells show that no revenue is generated when zero units are sold. This is a trivial solution, which can also be seen from the graph. There is also a nontrivial solution, but you need to help ***Solver*** find it. This is done by starting with an initial value of $q$ that is close to the other solution. From the graph of the revenue function, it appears that $R(q) = 0$ when $q \cong 0$ and $q \cong 14$. Note that ***Solver*** found the first solution when the initial value of $q$ was 0. ***Solver*** would find the same solution, $q = 0$, with any initial value for $q$ between 0 and 7. However, if an initial value for $q$ of 8 or more is used, ***Solver*** will find the other solution. *The bottom line is that* ***Solver*** *will find the solution that is closest to the initial value.*

Now find the other solution. Enter the value 10 in *Cell E18* and run ***Solver*** as previously described. The computation cells below show that no revenue is generated when 14,142 units are sold. Note that -6.5E-07 is approximately equal to zero.

| | E | F |
|---|---|---|
| 16 | **Computation** | |
| 17 | $x$ | $f(x)$ |
| 18 | 14.14207 | -6.5E-07 |

c. Open ***Solver*** and use the following screen captures as a reference.

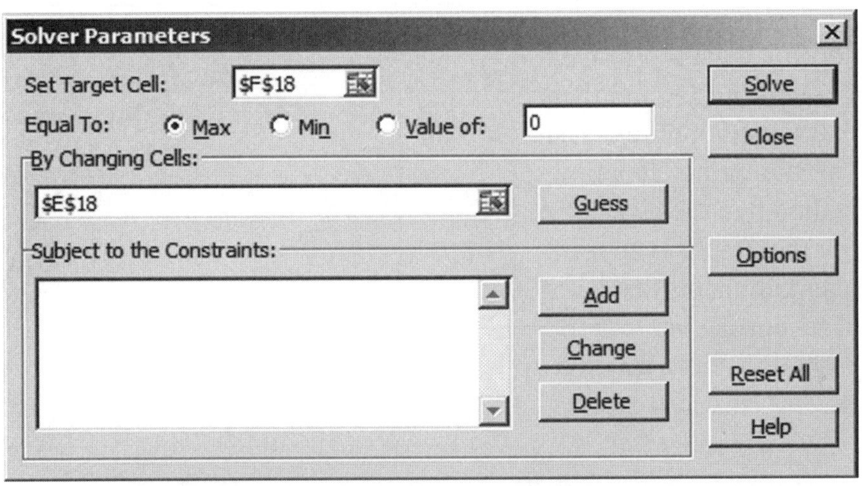


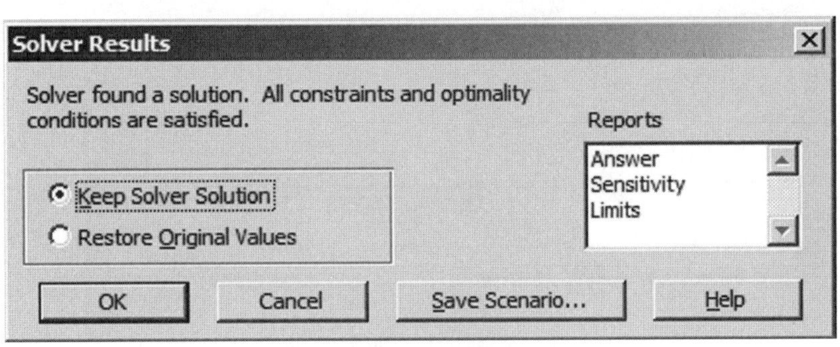


**Computation**

| $x$ | $f(x)$ |
|---|---|
| 7.071 | 50 |

The computation cells show that 7,071 video games should be sold in order to maximize revenue.

d. The computation cells from Part c show that the maximum revenue is $50,000.

**Example 2:** Daniel's Super Gamer Store has a cost function given by $C(q) = 12 + 2q$. The quantity $q$ represents the number of video games sold in thousands of units and $C(q)$ represents the cost in thousands of dollars.

a. Use ***Solver*** to find the number of video games that should be produced and sold in order to maximize profit. Round your answer to the nearest video game.

b. What profit would be generated from the sale of the quantity in Part a. Round your answer to the nearest dollar.

**SOLUTION**

a. The profit function is given by

$$\begin{aligned} P(q) &= R(q) - C(q) \\ &= -(q-7.071)^2 + 50 - (12+2q) \\ &= -(q-7.071)^2 - 2q + 38. \end{aligned}$$

Enter this equation in *Cell C18* of ***Graphing.xls*** and then click the **Graph** button. The graph of the profit function, $P(q)$, is shown to the right.

From the graph, it appears that the maximum profit occurs at $q \cong 6$, but ***Solver*** can be used to obtain a more accurate solution.

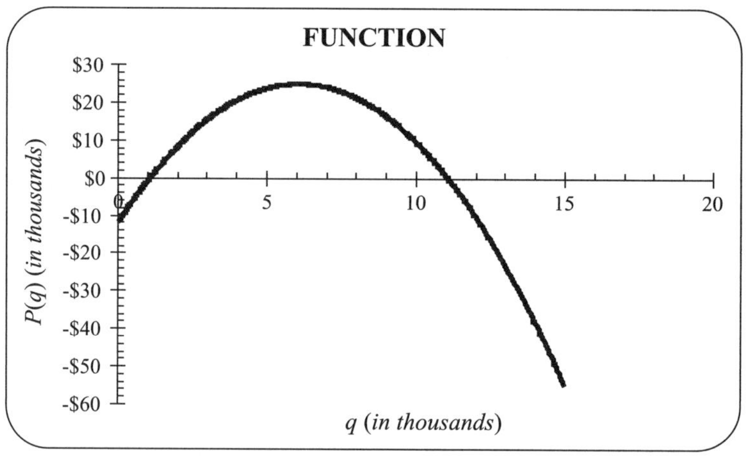


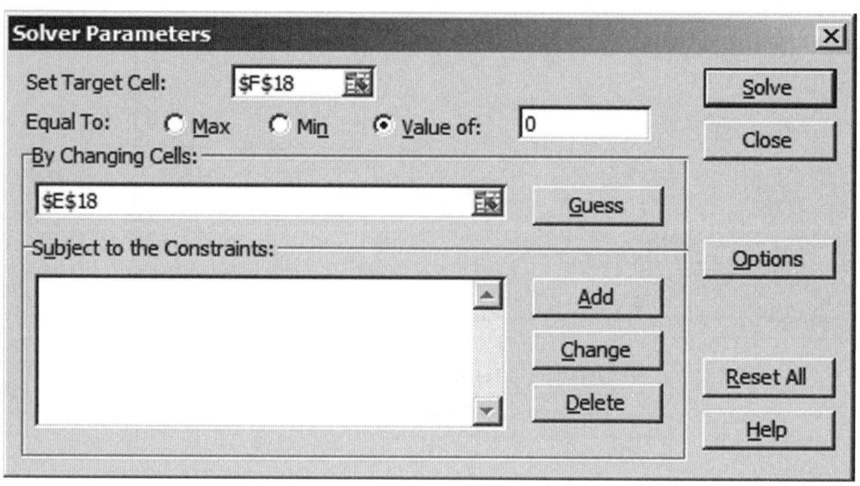


**Computation**

| $x$ | $f(x)$ |
|---|---|
| 6.071 | 24.858 |

Approximately 6,071 video games should be sold in order to maximize profit.

b. The computation cells from Part a show that a profit of \$24,858 would be generated from the sale of 6,071 video games.

**Example 3:** For Daniel's Super Gamer Store answer the following questions.

a. Use *Excel* to plot $R(q)$ and $C(q)$ on one graph. Plot the functions over the interval [0, 15] using 76 equally spaced points.

b. Use ***Solver*** to find the values of $q$ where the company breaks even. Round your answers to the nearest unit.

c. On what interval(s) does loss occur?

d. On what interval(s) does profit occur?

**SOLUTION**

a.

| $q$ | $R(q)$ | $C(q)$ |
|---|---|---|
| 0 | 0.00 | 12.00 |
| 0.05 | 0.71 | 12.10 |
| 0.1 | 1.41 | 12.20 |
| 0.15 | 2.10 | 12.30 |
| 0.2 | 2.79 | 12.40 |
| ⋮ | ⋮ | ⋮ |
| 14.8 | –9.74 | 41.60 |
| 14.85 | –10.51 | 41.70 |
| 14.9 | –11.29 | 41.80 |
| 14.95 | –12.08 | 41.90 |
| 15 | –12.87 | 42.00 |

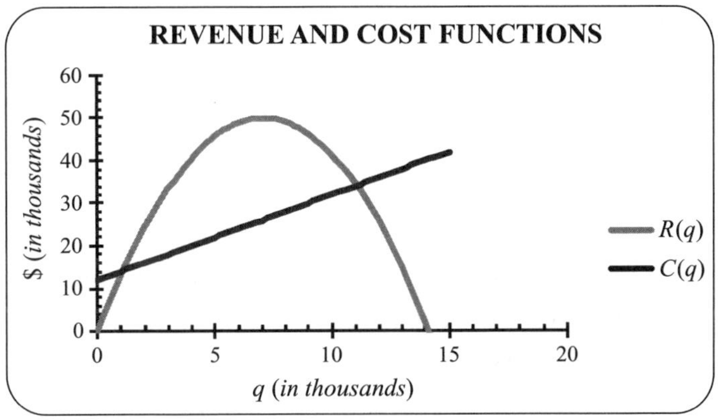


b. The company breaks even when $R(q) = C(q)$. Using ***Solver*** to find the corresponding values of $q$ will require the use of a constraint. Copy the labels and first row of data to *Cells A80: C81* of your *Excel* spreadsheet. Open ***Solver***, and enter the information shown below.

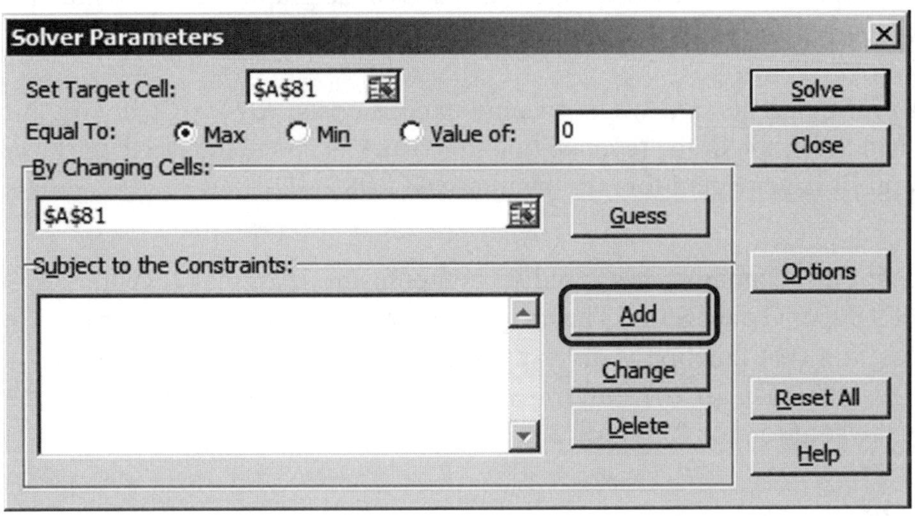


For this example it does not matter whether you select *Cell A81*, *Cell B81*, or *Cell C81* as the target cell. It also does not matter which option you select in the **Equal To** section of the **Solver Parameters** dialog box; the constraint will override all of the options.

To add the constraint, click **Add** in the **Solver Parameters** dialog box. Click in the **Cell Reference** box and select *Cell B81*. Then click the down arrow at the right of the middle box and select =. Click in the **Constraint** box and select *Cell C81*. Then click **OK**.

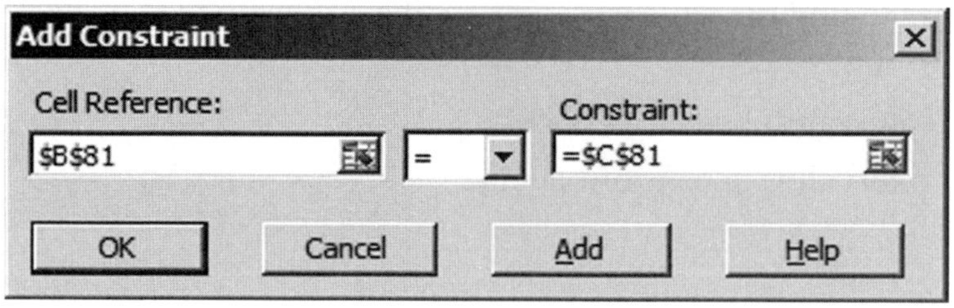


In the **Solver Parameters** dialog box, click **Solve**. The solution will be displayed in *Cell A81*.

| | A | B | C |
|---|---|---|---|
| 80 | $q$ | $R(q)$ | $C(q)$ |
| 81 | 1.08522 | 14.17044 | 14.17044 |

Since the initial value of $q$ was 0, the first solution that is displayed is the break-even point that is closest to zero. The first break-even point occurs at $q = 1.085$.

From the graph, it appears that the second break-even point occurs at $q \cong 10$. Enter the value 10 in *Cell A81* and repeat the previous steps to find the second break-even point.

| | A | B | C |
|---|---|---|---|
| 80 | $q$ | $R(q)$ | $C(q)$ |
| 81 | 11.05678 | 34.11356 | 34.11356 |

The second break-even point occurs at $q = 11.057$.

c. A loss is incurred when costs exceed revenue, $C(q) > R(q)$. Graphically this happens when the graph of the cost function, $C(q)$, is above the graph of the revenue function, $R(q)$. A loss is incurred for $q$ in the interval (0, 1.085) or (11.057, $\infty$).

d. A profit is generated when revenue exceed costs, $R(q) > C(q)$. Graphically this happens when the graph of the revenue function, $R(q)$, is above the graph of the cost function, $C(q)$. A profit is generated for $q$ in the interval (1.085, 11.057).

**Example 4:** An economist was hired to compute the marginal revenue and marginal cost for Daniel's Super Gamer Store. The marginal revenue function is given by $MR(q) = -2(q - 7.071)$, the marginal cost function is given by $MC(q) = 2$, and the marginal profit function is given by $MP(q) = MR(q) - MC(q) = -2(q - 7.071) - 2$. Use ***Solver*** along with marginal analysis to verify the solution for **Example 2**.

**SOLUTION**

Maximum profit occurs when $MR(q) = MC(q)$ or when $MP(q) = 0$. The following screen captures illustrate how to use ***Solver*** to find where $MR(q) = MC(q)$.

| | A | B | C | D | E | F | G |
|---|---|---|---|---|---|---|---|
| 80 | $q$ | $R(q)$ | $C(q)$ | $P(q)$ | $MR(q)$ | $MC(q)$ | $MP(q)$ |
| 81 | 0 | 0.000959 | 12 | -11.999 | 14.142 | 2 | 12.142 |

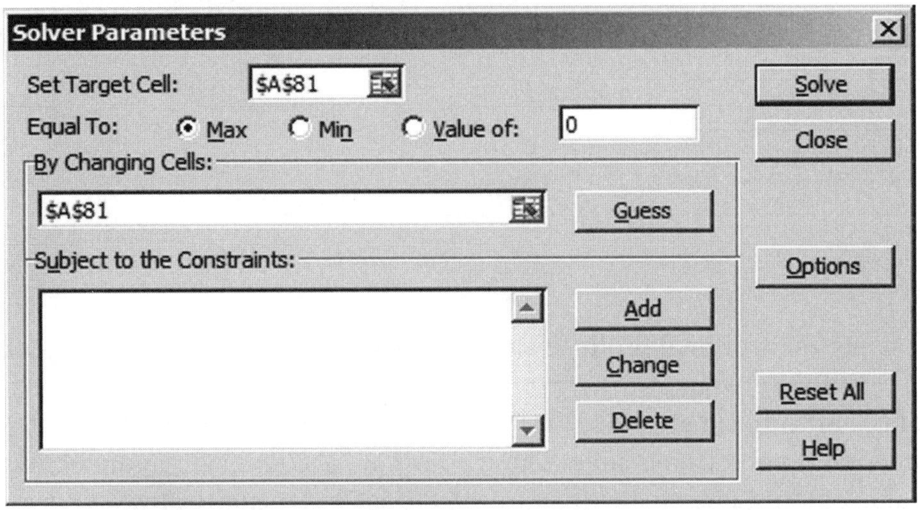


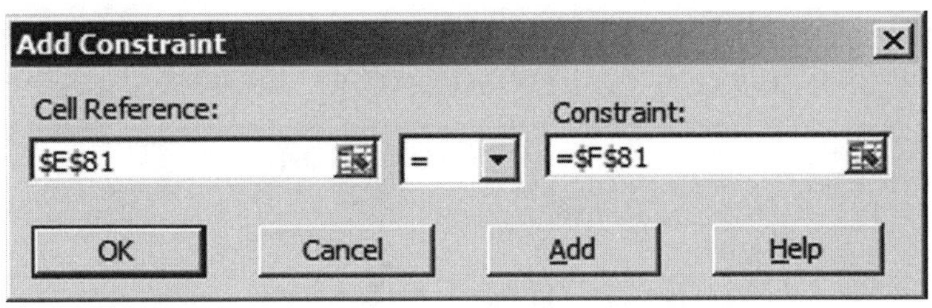


| | A | B | C | D | E | F | G |
|---|---|---|---|---|---|---|---|
| 80 | $q$ | $R(q)$ | $C(q)$ | $P(q)$ | $MR(q)$ | $MC(q)$ | $MP(q)$ |
| 81 | 6.071 | 49 | 24.142 | 24.858 | 2 | 2 | 0 |

As shown in **Example 2**, 6,071 video games should be produced and sold in order to maximize profit, and the maximum profit is $24,858. Note that the solution for **Example 2** could also have been verified by using ***Solver*** to find the value of $q$ for which $MP(q) = 0$.

**Example 5:** The graph of the profit function for Pocket PC Computers is given below.

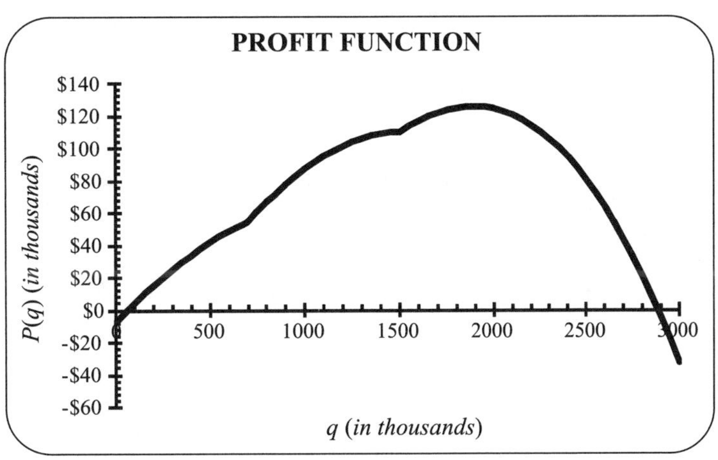

You want to use ***Solver*** to find the maximum profit. Which initial value of $q$ would give the correct solution after running ***Solver***?

A.

| $q$ | $MP(q)$ |
|---|---|
| 0 | \$119.21 |

B.

| $q$ | $MP(q)$ |
|---|---|
| 1000 | \$87.38 |

C.

| $q$ | $MP(q)$ |
|---|---|
| 2000 | –\$26.40 |

D.

| $q$ | $MP(q)$ |
|---|---|
| 2600 | –\$182.02 |

E. C or D

F. All

G. None

## SOLUTION

You are using the marginal profit function, $MP(q)$, to find the quantity that maximizes profit. Profit is maximized when $MP(q) = 0$. From the graph of the profit function, $P(q)$, it appears that profit is maximized at $q \cong 1{,}900$. When using ***Solver***, *Excel* will stop at points where the profit function is not differentiable. These points are shown in the figure below. Therefore, the initial value of $q$ must be greater than 1,500, the second quantity at which the profit function is not differentiable. Also, it does not matter whether the initial value of $q$ is greater than or less than the quantity that gives the maximum. Therefore, the answer is choice E.

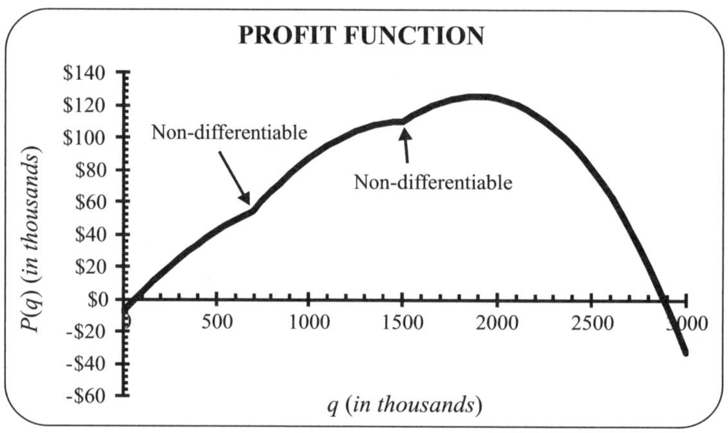


**Example 6:** The graph of the marginal profit function for Pocket PC Computers is given below.

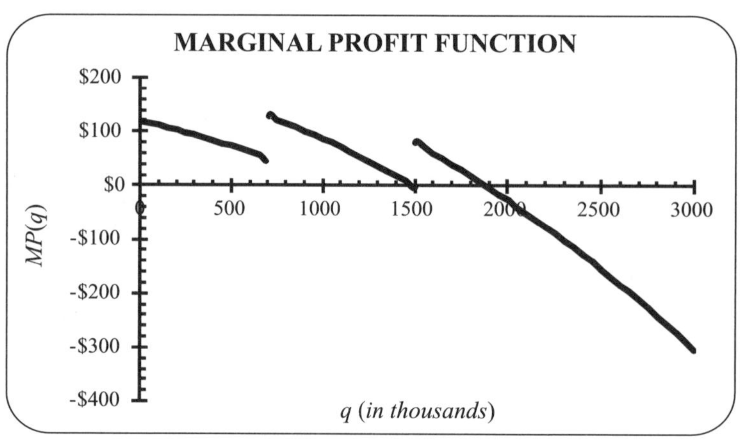

You want to use ***Solver*** to find the maximum profit. Which initial value of $q$ would give the correct solution after running ***Solver***?

A.

| $q$ | $MP(q)$ |
|---|---|
| 5000 | $73.54 |

B.

| $q$ | $MP(q)$ |
|---|---|
| 1650 | $50.75 |

C.

| $q$ | $MP(q)$ |
|---|---|
| 1000 | $87.38 |

D. All

E. None

**SOLUTION**

You are using the marginal profit function, $MP(q)$, to find the maximum profit. $MP(q)$is not a continuous function which causes a problem for ***Solver***. The initial value of $q$ needs to be close to where the graph of the marginal profit function crosses the horizontal axis. This is where $MP(q) = 0$ and thus is where maximum profit occurs. For this particular example it *appears* that $MP(q) = 0$ at two different values of $q$ ($q \cong 1{,}500$ and $q \cong 1{,}900$). From the graph of the profit function in **Example 5**, it appears that profit is maximized at $q \cong 1{,}900$. Therefore, the initial value of $q$ must be greater than 1,500. The answer is choice B.

It is worth noting that profit is maximized where $MR(q) = MC(q)$. The marginal revenue function is continuous, but the marginal cost function is not continuous. This is shown in the graph below. It looks as if $MR(q) = MC(q)$ at two different values of $q$. Therefore, you need to use the graph of the profit function to estimate the quantity that maximizes profit and then use that estimate as the initial value of $q$ when using ***Solver***.

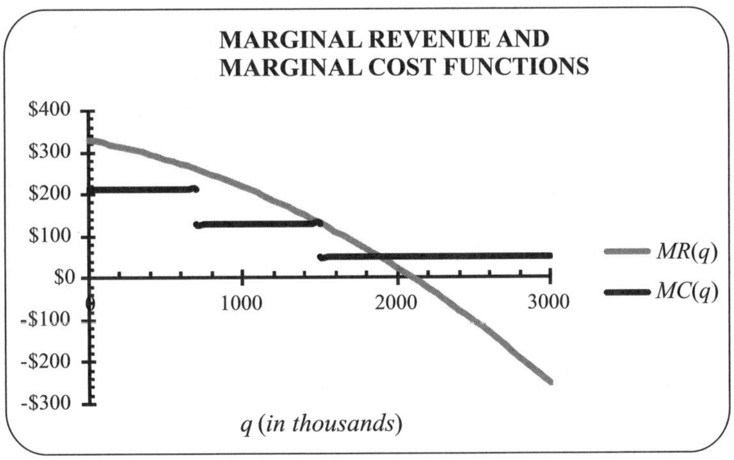


## *Exercises*

1. Gustavo's Gas Grills has a revenue function given by $R(q) = -0.3646q^2 + 3.208q$. The quantity $q$ represents the number of gas grills sold in thousands of units and $R(q)$ represents the revenue in millions of dollars.
   a. Use ***Graphing.xls*** to plot the revenue function over the interval [0, 10].
   b. Use ***Solver*** to find the values of $q$ where $R(q) = 0$. Round your answers to the nearest unit.
   c. Use ***Solver*** to find the number of grills that should be sold in order to maximize revenue. Round your answer to the nearest grill.
   d. What is the maximum revenue? Round your answer to the nearest dollar.

## SOLUTION

a.

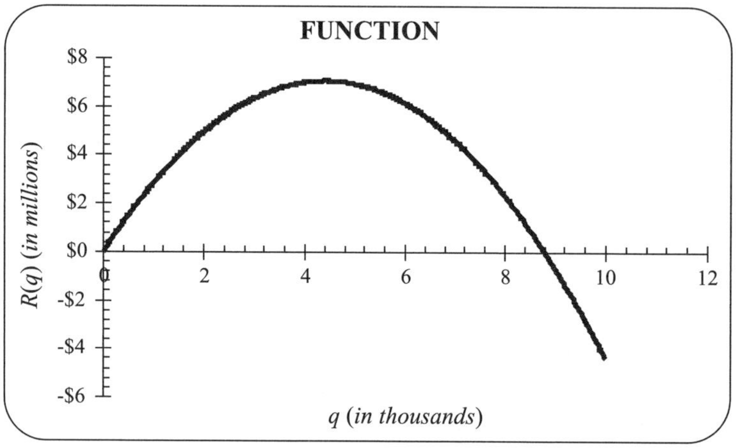


b. $R(q) = 0$ when $q = 0$ and $q = 8.799$.

c. 4,399 grills should be sold in order to maximize revenue.

d. Maximum revenue is $7,056,544.

2. Gustavo's Gas Grills has a cost function given by $C(q) = 2 + 0.2q$. The quantity $q$ represents the number of gas grills sold in thousands of units and $C(q)$ represents the cost in millions of dollars.

   a. Use *Excel* to graph $R(q)$ and $C(q)$ on one graph. Plot the functions over the interval [0, 10] using 41 equally spaced points.

   b. Use ***Solver*** to find the values of $q$ where the company breaks even. Round your answers to the nearest unit.

   c. On what interval(s) does loss occur?

   d. On what interval(s) does profit occur?

## SOLUTION

a.

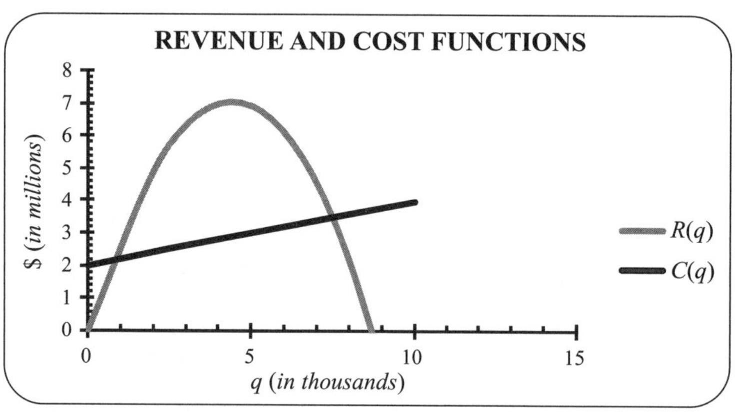


b. The break-even points occur at $q = 0.729$ and $q = 7.521$.

c. Loss occurs on the intervals $(0, 0.729)$ and $(7.521, \infty)$.

d. Profit occurs on the interval $(0.729, 7.521)$.

3. Gustavo, the owner of the company, wants to know how many grills the company needs to sell in order to maximize profit.
   a. Find the formula for the profit function, $P(q)$.
   b. Use ***Solver*** to find the number of grills that should be produced and sold in order to maximize profit.
   c. What is the maximum profit?

**SOLUTION**

a. $P(q) = -0.3646q^2 + 3..008q - 2$
b. 4,125 grills should be produced and sold in order to maximize profit.
c. Maximum profit is \$4,204,103.

4. Gustavo has asked that marginal analysis be used to confirm the solution for **Exercise 3**. The marginal revenue function is given by $MR(q) = -0.7292q + 3.208$, and the marginal cost function is given by $MC(q) = 0.2$. Use ***Solver*** along with marginal analysis to verify the answers for Parts b and c of **Exercise 3**.

**SOLUTION**

4,125 grills should be produced and sold in order to maximize profit, and the maximum profit is \$4,204,103.

5. The graphs of the profit function and the marginal revenue and marginal cost functions for a good are given below.

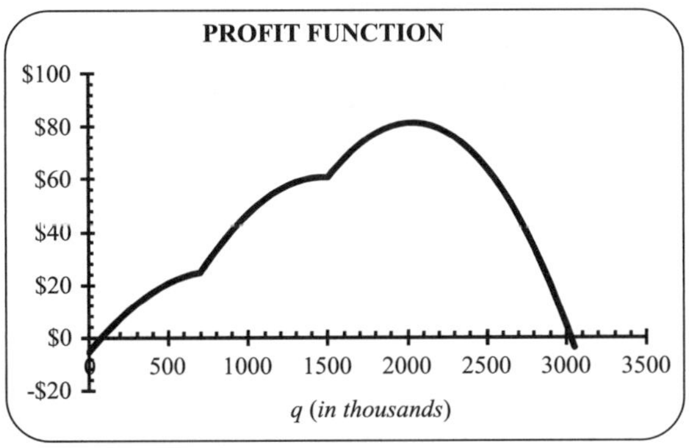


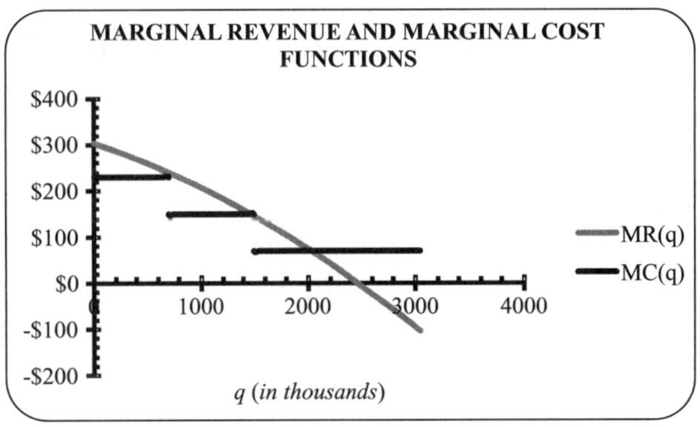


You want to use ***Solver*** to find the maximum profit. Which initial value of $q$ would give the correct solution after running ***Solver***?

a. Using the profit function.
b. Using the marginal revenue and marginal cost functions.

A. $q = 300$ B. $q = 1{,}100$ C. $q = 3{,}000$ D. All E. None

**SOLUTION**

a. The answer is choice D.
b. The answer is choice C.

# Integration

## *Area Under the Curve*

Basic calculus is comprised of two major subfields — differential calculus and integral calculus. Both subfields use the mathematical construct of the limit to make their studies rigorous, but both certainly can be understood, at least in part, from intuitive, graphical, and numerical points of view. We have already seen that differential calculus is based on the idea of the rate of change, or slope, of a function. As nonlinear functions have nonconstant slopes, a function, the derivative of the original function, is needed to express the slope of a nonlinear function. We have seen applications of the derivative primarily through marginal cost, marginal revenue, and marginal profit functions used in market analysis.

We now turn to the other side of calculus — integral calculus. Integral calculus is probably why calculus took on the name "calculus." A calculus is something that is built up in layers, such as the tartar on one's teeth. Integral calculus is essentially an accumulation process, albeit a mathematical one. Initially, it will seem that the two divisions of calculus have little in common, except that they both use the construct of a limit. Later, we will see that they are quite strongly related through the remarkable Fundamental Theorem of Calculus.

First, however, we will start with an accumulation process with which we are already familiar — the accumulation of distance traveled over time. For this, let's retrieve our walker, hiker Jeff.

**Example 1:** Compute the total distance traveled for each of the following scenarios:

a. Jeff walks for two hours at a constant rate of 2.5 mph. Draw a sketch over time, using rectangles to represent distance traveled.

b. Jeff walks for 0.5 hour at 1.5 mph, one hour at 3 mph, and 0.5 hour at 2 mph. Draw a sketch over time, using rectangles to represent distance traveled.

c. Jeff walks for 0.5 hour at 1.5 mph, 0.5 hour at 2 mph, 0.5 hour at 2.5 mph, and 0.5 hour at 3 mph.

**SOLUTION**

Let $D$ represent the total distance traveled.

a. $D = Rate \cdot Time = 2.5\dfrac{\text{miles}}{\text{hour}} \cdot 2 \text{ hours} = 5 \text{ miles}$

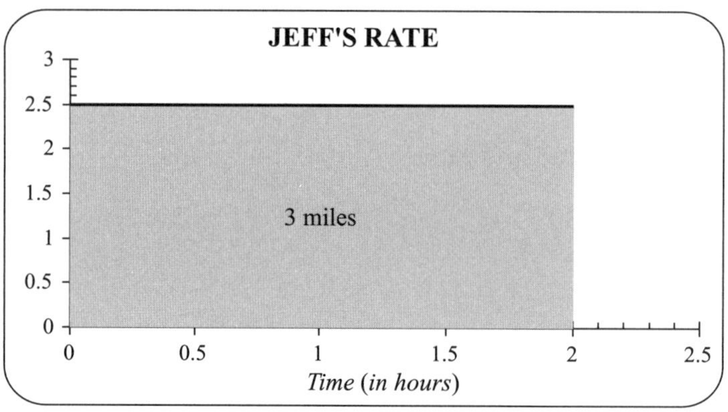

b. $D = 1.5(0.5) + 3(1) + 2(0.5) = 4.75$ miles

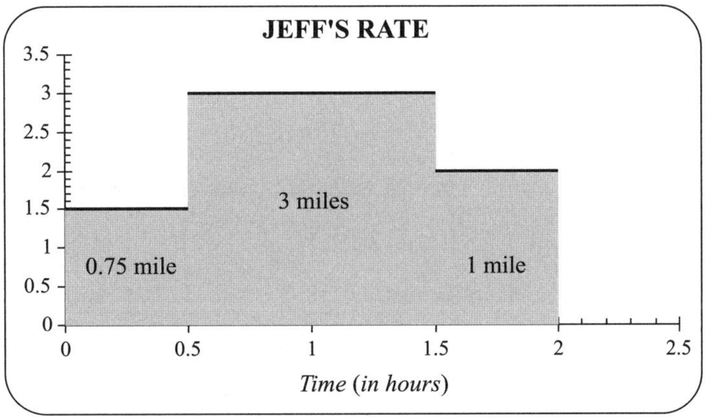


c. $D = 1.5(0.5) + 2(0.5) + 2.5(0.5) + 3(0.5) = 5$ miles (no graph provided)

From this example, it is easy to see how area can be used to represent something of interest, such as distance traveled or profit earned. Rates, such as speed of travel or income streams, are seldom constant; thus, it is often necessary to divide (mathematically, we say partition) our time interval of interest into small subintervals of time over which the rates do not vary greatly. We then sum up, or accumulate, the total distance traveled (or profit earned) over all of the smaller subintervals of time. This is the basic idea behind integral calculus.

**Example 2:** Suppose that Jeff starts out from rest (0 mph) and increases his walking speed steadily over the course of two hours to a brisk 5 mph. Find the total distance that he has traveled. Note that "steadily" need not necessarily imply constant acceleration, but we will assume that it does here. Constant acceleration implies that his walking rate is a linear function of time.

**SOLUTION**

Jeff's walking rate is given by a linear function $r = mt + b$. Since he starts from rest, $b = 0$. Since he accelerates from 0 mph to 5 mph in two hours, $m = (5 - 0)/2 = 2.5$ miles per hour. Thus $r(t) = 2.5t$.

The graph illustrates Jeff's rate, $r$, over time.

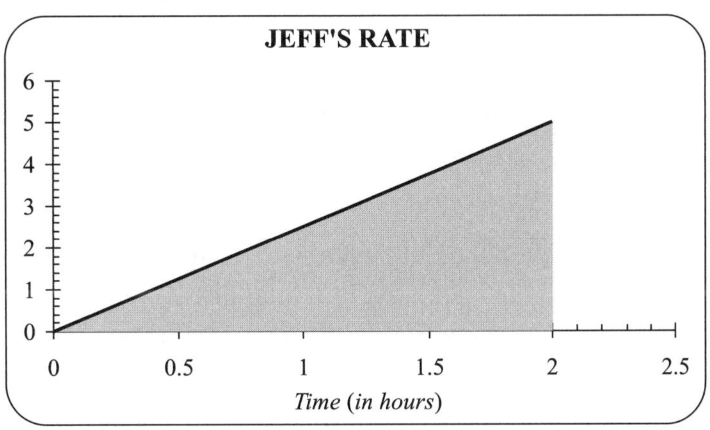

We still need to use the formula *Distance* = *Rate* · *Time*, but as Jeff's rate is never constant, we will break up his two hours of walking time into four 0.5-hour intervals and treat his rate as constant (using the midpoint rates) over each subinterval. This process, generally speaking, will give us an approximation of the total distance traveled. First, we find the rates.

Jeff's rate at 0.25 hours is 2.5(0.25) = 0.625 mph.

Jeff's rate at 0.75 hours is 2.5(0.75) = 1.875 mph.

Jeff's rate at 1.25 hours is 2.5(1.25) = 3.125 mph.

Jeff's rate at 1.75 hours is 2.5(1.75) = 4.375 mph.

Now, we multiply the rate times the time interval to obtain distance. Recall from the first example, we used the area of rectangles to represent distance.

$$\text{Distance \#1} = \text{Area of rectangle \#1} = \text{height}\cdot\text{width} = \text{rate}\cdot\text{time} = 0.625(0.5) = 0.3125 \text{ miles}$$

$$\text{Distance \#2} = \text{Area of rectangle \#2} = 1.875(0.5) = 0.9375 \text{ miles}$$

$$\text{Distance \#3} = \text{Area of rectangle \#3} = 3.125(0.5) = 1.5625 \text{ miles}$$

$$\text{Distance \#4} = \text{Area of rectangle \#4} = 4.375(0.5) = 2.1875 \text{ miles}$$

$$\begin{aligned}\text{Total Distance} &= \text{Sum of the areas of all four rectangles}\\ &= 0.3125 + 0.4375 + 1.5625 + 2.1875 = 5 \text{ miles}\end{aligned}$$

Notice that Jeff's rates, which are given by the heights of the rectangles, are in units of miles per hour and the time intervals, the widths of the rectangles, are in units of hours; therefore, the units represented by the areas of the rectangles are in units given by

$$\text{height}\cdot\text{time} = \frac{\text{miles}}{\text{hour}}\cdot\text{hours} = \text{miles}.$$

You may also have noticed that Jeff's total distance traveled, 5 miles, could have been easily found in this case by taking his average rate, 2.5 mph, multiplied by 2 hours. This is only because the acceleration (the rate of change of Jeff's speed) was itself constant.

This is also why the rectangles built over the midpoints of the time subintervals gave the exact answer. In general, we cannot expect the numbers to work out so nicely, but if we build enough rectangles over small enough subintervals, we can obtain an approximation that is as close to the area as needed.

One final note about this example: The area under the straight line and above the $t$-axis, which represented the total distance that Jeff traveled, is a simple geometric shape — a triangle. Hopefully, you remember from your high school days that the area of a triangle is $\frac{1}{2}\cdot\text{base}\cdot\text{height}$. In this case, $\frac{1}{2}(2)(5) = 5$ miles.

Depending on the function graphed, the area under the graph can represent a variety of interesting quantities, such as total distance traveled, total profit made, consumer surplus, and even the probability that a continuous random variable takes on a value in a given interval. The moral of the story is that we want to be able to find that area. If the area is something for which we already have a formula, then we do not need to use calculus. Many of the curves that are used in business applications, however, do *not* describe simple geometric shapes.

**Example 3:** Use midpoint sums to compute the area of the region between graph of $f(x)=x^2-4x+7$ and the $x$-axis over the interval [0, 4]. Divide the interval into $n=8$ subintervals.

**SOLUTION**

To obtain the areas of our approximating rectangles, we need the height and width of each rectangle. If we divide the interval [0, 4] into 8 subintervals of equal width, then we will obtain the width of each rectangle as shown below. Note that we use the notation $\Delta x$ to represent the width of the rectangles. We will also need $x$-values at which we will evaluate our function to obtain the heights of the rectangles. First, we will partition the interval, using 9 partition points equally spaced (we always need exactly 1 more partition point than the desired number of subintervals). Second, we will average adjacent partition points to obtain midpoint values at which we will evaluate our function. All of this is carefully shown below.

$a=0;\ b=4$

$$\Delta x=\frac{b-a}{n}=\frac{4-0}{8}=0.5$$

$x_i=a+i\cdot\Delta x,\ i=0,1,\ldots,8$ $\qquad$ $m_i=\dfrac{x_{i-1}+x_i}{2},\ i=1,2,\ldots,8$

$x_0=0+0(0.5)=0.0$

$x_1=0+1(0.5)=0.5$ $\qquad$ $m_1=\dfrac{x_{1-1}+x_1}{2}=\dfrac{x_0+x_1}{2}=\dfrac{0.0+0.5}{2}=0.25$

$x_2=0+2(0.5)=1.0$ $\qquad$ $m_2=\dfrac{x_{2-1}+x_2}{2}=\dfrac{x_1+x_2}{2}=\dfrac{0.5+1.0}{2}=0.75$

$x_3=0+3(0.5)=1.5$ $\qquad$ $m_3=\dfrac{x_{3-1}+x_3}{2}=\dfrac{x_2+x_3}{2}=\dfrac{1.0+1.5}{2}=1.25$

$x_4=0+4(0.5)=2.0$ $\qquad$ $m_4=\dfrac{x_{4-1}+x_4}{2}=\dfrac{x_3+x_4}{2}=\dfrac{1.5+2.0}{2}=1.75$

$x_5=0+5(0.5)=2.5$ $\qquad$ $m_5=\dfrac{x_{5-1}+x_5}{2}=\dfrac{x_4+x_5}{2}=\dfrac{2.0+2.5}{2}=2.25$

$x_6=0+6(0.5)=3.0$ $\qquad$ $m_6=\dfrac{x_{6-1}+x_6}{2}=\dfrac{x_5+x_6}{2}=\dfrac{2.5+3.0}{2}=2.75$

$x_7=0+7(0.5)=3.5$ $\qquad$ $m_7=\dfrac{x_{7-1}+x_7}{2}=\dfrac{x_6+x_7}{2}=\dfrac{3.0+3.5}{2}=3.25$

$x_8=0+8(0.5)=4.0$ $\qquad$ $m_8=\dfrac{x_{8-1}+x_8}{2}=\dfrac{x_7+x_8}{2}=\dfrac{3.5+4.0}{2}=3.75$

$$f(m_1) = f(0.25) = (0.25)^2 - 4(0.25) + 7 = 6.0625$$

$$f(m_2) = f(0.75) = (0.75)^2 - 4(0.75) + 7 = 4.5625$$

$$f(m_3) = f(1.25) = (1.25)^2 - 4(1.25) + 7 = 3.5625$$

$$f(m_4) = f(1.75) = (1.75)^2 - 4(1.75) + 7 = 3.0625$$

$$f(m_5) = f(2.25) = (2.25)^2 - 4(2.25) + 7 = 3.0625$$

$$f(m_6) = f(2.75) = (2.75)^2 - 4(2.75) + 7 = 3.5625$$

$$f(m_7) = f(3.25) = (3.25)^2 - 4(3.25) + 7 = 4.5625$$

$$f(m_8) = f(3.75) = (3.75)^2 - 4(3.75) + 7 = 6.0625$$

Finally, we have all of our required building blocks; that is, all of our heights and all of our widths. We multiply our heights times our widths and add them all up to obtain our total area approximation. The notation used below is the same as that which is used in the text.

$$S_n(f,[a,b]) = \sum_{i=1}^{n} f(m_i) \cdot \Delta x$$

$$\begin{aligned} S_8(f,[0,\ 4]) &= \sum_{i=1}^{8} f(m_i) \cdot 0.5 \\ &= 0.5 \sum_{i=1}^{8} f(m_i) \\ &= 0.5(6.0625 + 4.5625 + 3.5625 + 3.0625 + 3.0625 + 3.5625 + 4.5625 + 6.0625) \end{aligned}$$

$$S_8(f,[0,\ 4]) = 0.5(34.50) = 17.25$$

The area of the region between graph of $f(x) = x^2 - 4x + 7$ and the $x$-axis over the interval [0, 4] is approximately 17.25 square units.

**Example 4:** For $f(x) = 2x^5 - 3x^2 + 15$, use *Excel* to calculate $S_{30}(f,[-2,\ 1])$. Your worksheet should contain $i$, $x_i$, $m_i$, and $f(m_i)$.

**SOLUTION**

| $n$ | $a$ | $b$ | $\Delta x$ |
|---|---|---|---|
| 4 | 1 | 9 | 2 |
| | | | |
| $i$ | $x_i$ | $m_i$ | $f(m_i)$ |
| 0 | -2.0 | | |
| 1 | -1.9 | -1.95 | -52.7976 |
| 2 | -1.8 | -1.85 | -38.6075 |
| 3 | -1.7 | -1.75 | -27.0137 |
| 4 | -1.6 | -1.65 | -17.6217 |
| ⋮ | ⋮ | ⋮ | ⋮ |
| 26 | 0.6 | 0.55 | 14.1932 |
| 27 | 0.7 | 0.65 | 13.9646 |
| 28 | 0.8 | 0.75 | 13.7871 |
| 29 | 0.9 | 0.85 | 13.7199 |
| 30 | 1 | 0.95 | 13.8401 |
| | | Sum | 150.6996 |
| | | | |
| | | Area | 15.06996 |

**ANSWER**

**Example 5:** Use ***Midpoint Sums.xls*** to approximate the area of the region between graph of $f(x)=e^{-x}$ and the $x$-axis over the interval $[0,\ 5]$. Divide the interval into $n=4$ subintervals. Do the same for $n=10$ and $n=100$. What pattern do you notice?

**SOLUTION**

| **Intervals** | | **Midpoint Sum** |
|---|---|---|
| $n$ | $\Delta x$ | $\sum_{i=1}^{n} f(m_i)\cdot\Delta x$ |
| 4 | 1.25000 | 0.93143 |

| **Intervals** | | **Midpoint Sum** |
|---|---|---|
| $n$ | $\Delta x$ | $\sum_{i=1}^{n} f(m_i)\cdot\Delta x$ |
| 10 | 0.50000 | 0.98299 |

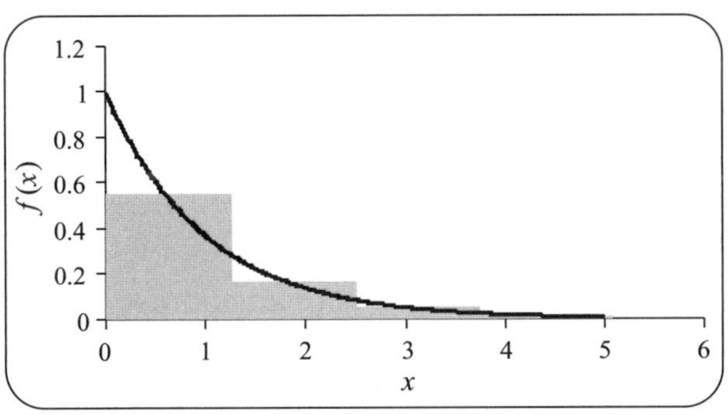


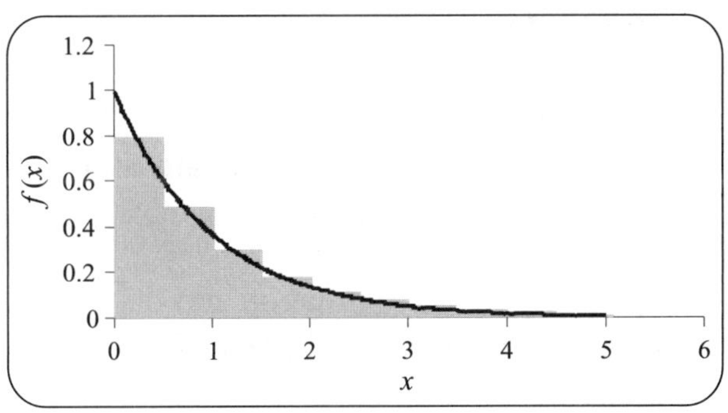

For $n = 4$, the result is 0.93143 square units, pictured above. For $n = 10$, the result is 0.98299 square units, also pictured above. For $n = 100$, the result is 0.99316; the picture for this is not included. It appears that the area of the region between graph of $f(x) = e^{-x}$ and the $x$-axis over the interval [0, 5] is close to 0.99. Note that the function, $f(x) = e^{-x}$, is the 1st-quadrant portion of the probability density function for an exponential random variable. You will learn more about these in the next section. If we were to integrate from 0 to infinity, the area under the curve would be exactly 1.

The definite integral is used to express the area of the region between the graph of a nonnegative function and the $x$-axis over an interval $[a, b]$. Thus, if $f$ is nonnegative, $\int_a^b f(x)\,dx$ is the area of the region between graph of $f$ and the $x$-axis over the interval $[a, b]$.

**Example 6:** Use ***Integrating.xls*** to approximate the area of the region between graph of $f(x) = e^x$ and the $x$-axis over the interval [0, 5]. Do you think that $f(x) = e^x$ over the interval [0, 5] could be a probability density function for a continuous random variable? Explain your answer.

**SOLUTION**

Enter the function in *Cell C20*, plot on the interval [0, 5], and integrate over the same interval.

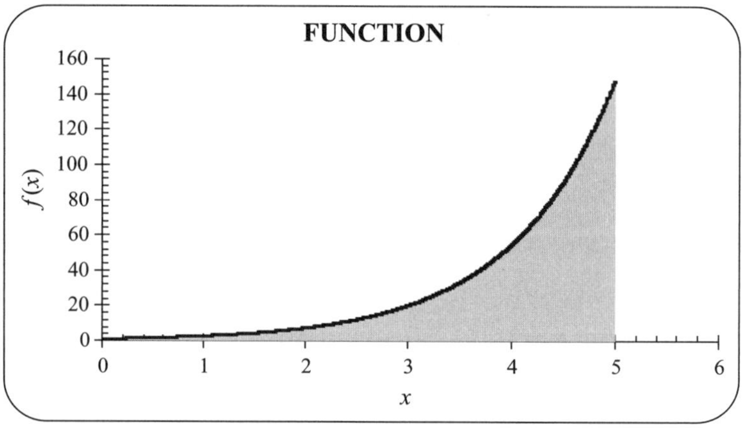


The approximate area of the region between graph of $f(x) = e^x$ and the $x$-axis over the interval [0, 5] is 147.4132 square units. There is no way that $f(x) = e^x$ over the interval [0, 5] could be a *p.d.f.* for a continuous random variable. The reason for this is simple. The area between the graph of a *p.d.f* and the $x$-axis over an interval represents the probability that the random variable takes on a value in that interval. The maximum possible value of any probability is 1.0, so 147.4132 is way out of range.

You may have noticed that in ***Midpoint Sums.xls***, summation notation, $\sum_{i=1}^{n} f(m_i)\cdot\Delta x$, is used to represent the approximate area under a curve; whereas in ***Integrating.xls***, the definite integral notation, $\int_a^b f(x)\,dx$, is used. Technically, the difference is that the summation notation represents an approximation and the definite integral represents an exact value; however, both ***Midpoint Sums.xls*** and ***Integrating.xls*** yield approximate values of the exact area. We saw in **Examples 1** and **2** that we could find the area exactly with the help of geometry. We will now look at a couple more examples of definite integrals that we can compute exactly using geometry. Later we will see in *Closed Form Integration* a technique to find the exact area of many definite integrals that do not have such nicely formed geometric boundaries.

**Example 7:** The graph of $f(x)=3x+4$ is given below. Write the shaded area as an integral and find the area using geometry. Hint: The area formula for a trapezoid is $A=\frac{1}{2}h(B+b)$.

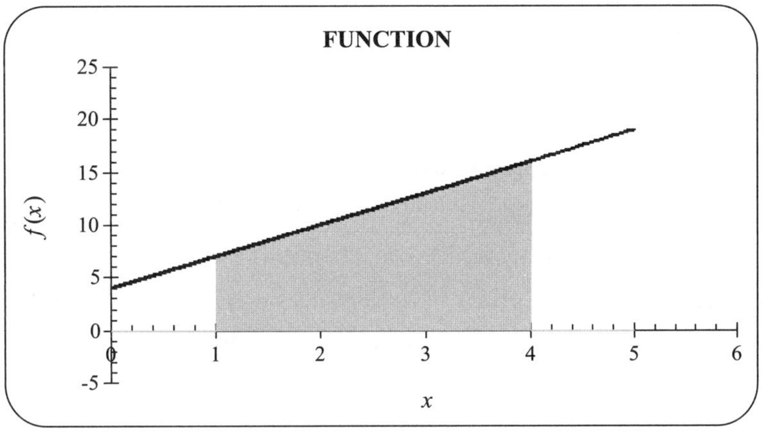


**SOLUTION**

$$\text{Area} = \int_{1}^{4} (3x+4)\, dx .$$

To use the area formula for a trapezoid, we will assume that the trapezoid is on its side as the big base and little base must be parallel. So, $h=4-1=3$, $b=3(1)+4=7$, and $B=3(4)+4=16$. Notice that we are using the function, $f(x)=3x+4$, to find the two bases of the trapezoid. Putting all of these numbers into the formula, we obtain $A=\frac{1}{2}(3)(7+16)=34.5$ square units.

**Example 8:** The graph of $f(x)=-|x|+4$ is given to the right. Write the shaded area as an integral and find the area using geometry.

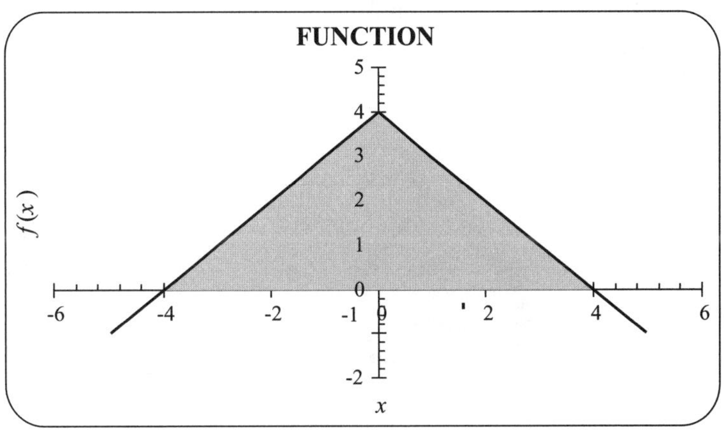


**SOLUTION**

$$\text{Area} = \int_{-4}^{4} (-|x|+4)\, dx .$$

If we split the shaded region into two congruent triangles, we need only find the area of one triangle and double it. The triangle formed in the 1st-quadrant by the $x$-axis, $y$-axis, and downward-sloping side of the graph of the function is a right triangle with *height* = 4 and *base* = 4. Thus the area of the right-hand triangle is $A = \frac{1}{2}(4)(4) = 8$. Remember that the area shaded is twice that, so the final answer is $\int_{-4}^{4} \left(-|x| + 4\right)\, dx = 16$ square units.

## *Applications*

In the market analysis that we have been doing, we have been trying to find a single optimal price, and the corresponding maximum number of units that can be sold at that price. This optimal price is determined by a demand function, which is estimated via market sampling and test marketing techniques; the goal is to maximize profit. Not all sales environments establish a single fixed price for their products. Some sales environments will establish a minimum price and will not sell for under that price.

In sales environments such as car sales and time-shares, the sales personnel may try to get each buyer to pay the maximum price that he or she is willing to pay. If every possible buyer paid the maximum price that they were willing to pay and no minimum (other than \$0) was set, then the total amount of revenue that could be earned in this manner is called, not surprisingly, the total possible revenue. Recall that the revenue earned from the sale of $q$ units at \$$p$/unit, $p \cdot q$, can be visualized as the area of the rectangle inscribed under the demand curve with diagonal endpoints at (0, 0) and ($q$, $p$). The total possible revenue can likewise be visualized as an area under the curve — namely, the total area under the curve.

As described in your text, as soon as the selling price is determined, the buyers who would have been willing to pay more can be thought of as saving money. The total amount of money saved by buyers in this way is called the **consumer surplus**. Of course, fixing the selling price also cuts some buyers out of the market. Revenue not realized from items not sold to these buyers is simply referred to as "not sold."

Your book gives a nice graphical illustration of the breakdown of the total possible revenue into the three non-overlapping regions, revenue, consumer surplus, and not sold. In this section, we give you a bit more practice finding these values for given demand curves.

**Example 9:** Recall that the demand function for Don's tofu tarts is $D(q) = -0.0125q + 2.50$. Using $q = 80$ find the values below.

a. Total possible revenue.

b. Revenue.

c. Consumer surplus.

d. Revenue not realized from items not sold.

e. Show that the sum of the consumer surplus, the revenue, and the revenue not realized from items not sold is equal to the total possible revenue.

**SOLUTION**

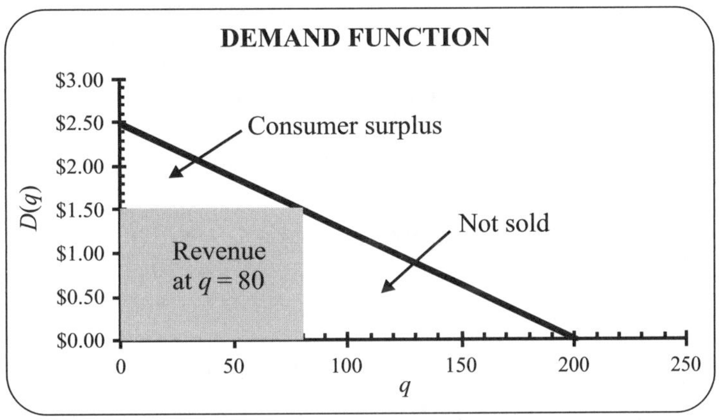


a. The total possible revenue is the area under the entire curve.

$$\text{Total possible revenue } = \frac{1}{2}(200)(2.5) = \$250$$

b. The revenue is the area of the revenue at $q = 80$ region.

$$R(80) = 80(1.5) = \$120$$

c. The consumer surplus is the area of the consumer surplus region.

$$\text{Consumer surplus } = \frac{1}{2}(80)(1) = \$40$$

d. The revenue not realized from items not sold is the area of the not sold region.

$$\text{Revenue not realized from items not sold } = \frac{1}{2}(200 - 80)(1.5) = \$90$$

e. The toal possible revenue is the sum of all the areas.

$$\begin{aligned}\text{Total possible revenue} &= \text{consumer surplus} + \text{revenue} + \text{not sold} \\ &= \$40 + \$120 + \$90 \\ &= \$250\end{aligned}$$

Refer to your text for a more advanced presentation on integration and on finding the area under nonlinear curves.

**Example 10:** The demand function for Pocket PC Computers is given by

$$D(q) = -0.00001366q^2 - 0.03542396q + 329.21286693.$$

The graph of $D(q)$ is given below.

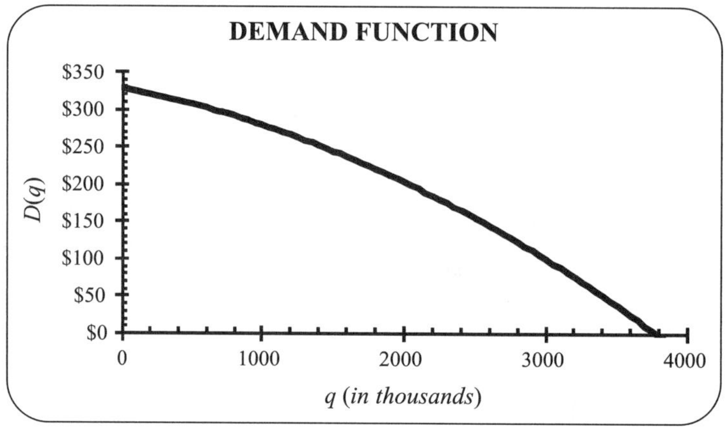


a. Use ***Solver*** in ***Graphing.xls*** to find the value of $q$ for which $D(q) = 0$.
b. Find a formula for the revenue function, $R(q)$.
c. Find the revenue at $q = 2000$.
d. Use ***Integrating.xls*** to find the consumer surplus at $q = 2000$.
e. Use ***Integrating.xls*** to find the total possible revenue.
f. Use ***Integrating.xls*** to find the area of the "not sold" region for $q = 2000$.
g. Find the total amount of lost revenue at $q = 2000$.

Round all your answers to the nearest whole number.

**SOLUTION**

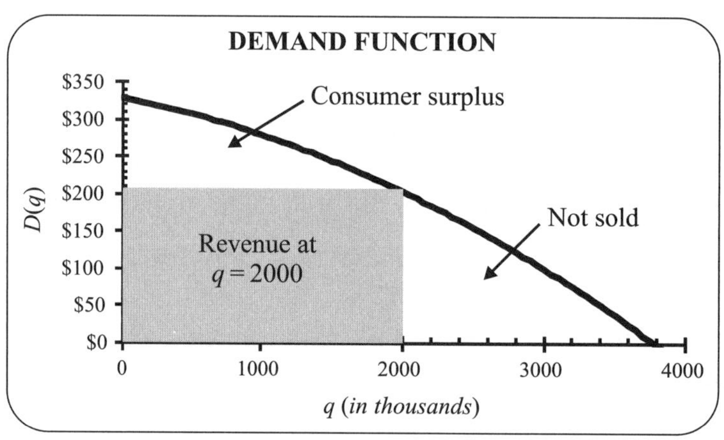


a.

**Definition**

| Formula for $f(x)$ |
|---|
| 329.2129 |

**Computation**

| $x$ | $f(x)$ |
|---|---|
| 3780.943 | 6.54E-07 |

**Plot Interval**

| $a$ | $b$ |
|---|---|
| 0 | 4000 |

$q \cong 3781$ ⟶

b. $R(q) = q(-0.00001366q^2 - 0.03542396q + 329.21286693)$
$= -0.00001366q^3 - 0.03542396q^2 + 329.21286693q$

c. $R(2000) = \$407,450$

d. $$\text{Consumer surplus} = \int_0^{q_0} D(q)\,dq - q_0 \cdot D(q_0)$$
$$= \int_0^{q_0} D(q)\,dq - R(q_0)$$
$$= \int_0^{2000} (-0.00001366q^2 - 0.03542396q + 329.21286693)\,dq - 407,450$$
$$= 551,151 - 407,450$$
$$= \$143,701$$

e. $$\int_0^{3781} (-0.00001366q^2 - 0.03542396q + 329.21286693)\,dq = \$745,423$$

f. $$\int_{2000}^{3781} (-0.00001366q^2 - 0.03542396q + 329.21286693)\,dq = \$194,271$$

g. The total amount of lost revenue is the sum of the consumer surplus and the revenue not realized from items not sold.
$$\text{Total lost revenue} = \$143,701 + \$194,271 = \$337,972$$

## *Closed Form Integration*

If an antiderivative can be found quickly for a given integrand, then direct use of the Fundamental Theorem of Calculus (FTC) is often the quickest way to find a desired area. The FTC tells us that for a continuous function, or a piecewise continuous function, *f*, over a closed interval [*a*, *b*],

$$\int_a^b f(x)\,dx = F(b) - F(a),$$

where $F(x)$ is an antiderivative of *f*. An antiderivative of a function, *f*, is a function, *F*, such that $F'(x) = f(x)$.

*If* we can find such an *F*, then all we have to do is evaluate the function for two input values, *a* and *b*, and perform a subtraction. No ***Midpoint Sums.xls***, no ***Intgrating.xls***, no building of lengthy spreadsheets, no fuss, no muss, no bother! So, you are most certainly wondering, *what is this FTC magic*? The formulas and properties given below are also in your text, but for your convenience, we include them here as well.

*Formulas for Antiderivatives*

1. $\int k \; dx = kx + C$

2. $\int x^n \; dx = \dfrac{x^{n+1}}{n+1} + C, \; n \neq -1$

3. $\int e^{kx} dx = \dfrac{1}{k} e^{x} + C$

4. $\int b^x \; dx = \dfrac{b^x}{\ln b} + C$

5. $\int \dfrac{1}{x} \; dx = \ln|x| + C$

In each case, $C$ is an arbitrary constant, called the constant of integration.

*Properties of Integrals*

1. $\int_a^b k \cdot f(x) \; dx = k \cdot \int_a^b f(x) \; dx$

2. $\int_a^b [f(x) \pm g(x)] \; dx = \int_a^b f(x) \; dx \pm \int_a^b g(x) \; dx$

3. $\int_a^b f(x) \; dx = \int_a^c f(x) \; dx + \int_c^b f(x) \; dx$, for any $c$ with $a < c < b$

When we use closed-form integration to evaluate definite integrals, we will first find the indefinite integral, or antiderivative, and then apply the FTC, that is, evaluate $F(x)$ at the limits of integration, $a$ and $b$, and take the difference, $F(b) - F(a)$. A couple quick notes before we embark on our collection of examples, however, are in order. First, the provided list of formulas is miniscule. There are books published with literally thousands of such formulas! Second, usually when we find antiderivatives, or indefinite integrals, we include a "+$C$". We will continue to do so, because it is good mathematical form, but we will not bother with it when we invoke the FTC as it will always subtract off. We will include it in the first example only.

**Example 11:** Evaluate $\int_1^4 (3x+4) \; dx$ using closed-form integration.

**SOLUTION**

First we find the antiderivative or indefinite integral, and next we evaluate it at the limits of integration according to the FTC. Thus we have

$$\int (3x+4) \; dx = 3\int x \; dx + \int 4 \; dx = 3 \cdot \frac{x^2}{2} + 4x + C.$$

Thus, by the FTC, we have

$$\int_1^4 (3x+4)\,dx = \left(3\cdot\frac{(4)^2}{2}+4(4)+C\right)-\left(3\cdot\frac{(1)^2}{2}+4(1)+C\right)=(40+C)-(5.5+C)=34.5.$$

Notice how the $C$'s subtracted off. We will do a few more exercises leaving off the $C$'s when we use the FTC.

**Example 12:** Evaluate $\int_0^2 (x^2+2e^{3x})\,dx$ using closed-form integration.

**SOLUTION**

$$F(x)=\int (x^2+2e^{3x})\,dx=\int x^2\,dx+2\int e^{3x}\,dx=\frac{x^3}{3}+\frac{2}{3}e^{3x}+C.$$

Thus, by the FTC,

$$\int_0^2 (x^2+2e^{3x})\,dx=\left(\frac{2^3}{3}+\frac{2}{3}e^{3(2)}\right)-\left(\frac{0^3}{3}+\frac{2}{3}e^{3(0)}\right)=2+\frac{2}{3}e^6.$$

**Example 13:** Evaluate $\int_1^e \left(\frac{1}{x}-1\right)dx$ using closed-form integration.

**SOLUTION**

$$F(x)=\int\left(\frac{1}{x}-1\right)dx=\int\frac{1}{x}\,dx-\int 1\,dx=\ln|x|-x+C.$$

Thus, by the FTC,

$$\int_1^e\left(\frac{1}{x}-1\right)dx=\left(\ln|e|-e\right)-\left(\ln|1|-1\right)=(1-e)-(0-1)=2-e.$$

From *Properties of Integrals*, we know that the integral of the sum is the sum of the integrals and that the integral of the difference is the difference between the integrals. These results follow from the analogous property of derivatives. You may recall that the shortcuts for derivatives of products and quotients were not so easy. For instance, the derivative of the product is *not* the product of the derivatives. The same applies here. You *cannot* find the integral of a product or quotient simply by taking the product or quotient of the integrals. Many mathematics instructors or professors find this particular common mistake so mathematically offensive that they will not award any partial credit points at all for a problem in which a student tries to take an integral of a product by erroneously taking the product of the integral!

**Example 14:** Evaluate $\int_1^2 (2x+1)(2x-1)\,dx$ using closed-form integration.

**SOLUTION**

$$F(x)=\int (2x+1)(2x-1)\,dx=\int (4x^2-1)\,dx=4\int x^2\,dx-\int 1\,dx=4\cdot\frac{x^3}{3}-x+C.$$

Thus, by the FTC,

$$\int_1^2 (2x+1)(2x-1)\,dx=\left(4\cdot\frac{2^3}{3}-2\right)-\left(4\cdot\frac{1^3}{3}-1\right)=\frac{25}{3}.$$

Notice that we *first* multiplied out the integrand and *then* integrated, *not* the other way around.

**Example 15:** Find $\int_1^2 \frac{(2x+1)}{x}\,dx$ using closed-form integration.

**SOLUTION**

$$F(x)=\int \frac{(2x+1)}{x}\,dx=\int\left(2+\frac{1}{x}\right)dx=\int 2\,dx+\int\frac{1}{x}\,dx=2x+\ln|x|+C.$$

Thus, by the FTC,

$$\int_1^2 \frac{(2x+1)}{x}\,dx=\left(2(2)+\ln|2|\right)-\left(2(1)-\ln|1|\right)=2+\ln 2.$$

Notice that we *first* divided out the integrand and *then* integrated, *not* the other way around.

## *Exercises*

For exercises 1 through 5, sketch the graph of the given function, shade the region between the graph of the function and the $x$-axis over the indicated interval, and compute the area of the region.

1. $f(x)=3x-3$; $[1, 5]$.

**SOLUTION**

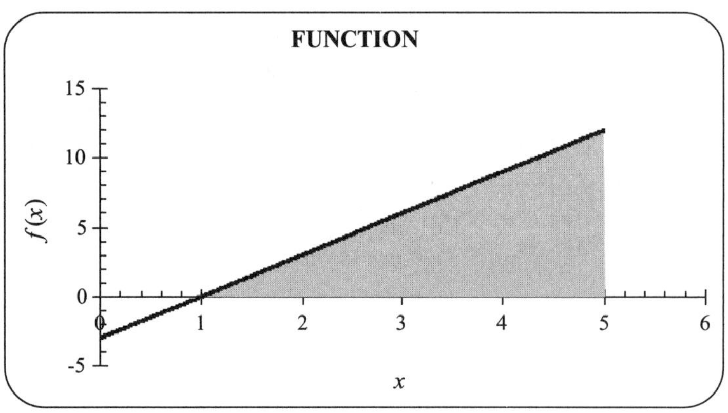


Area = 24 square units.

2. $g(x) = 2$; $[2, 5]$.

**SOLUTION**

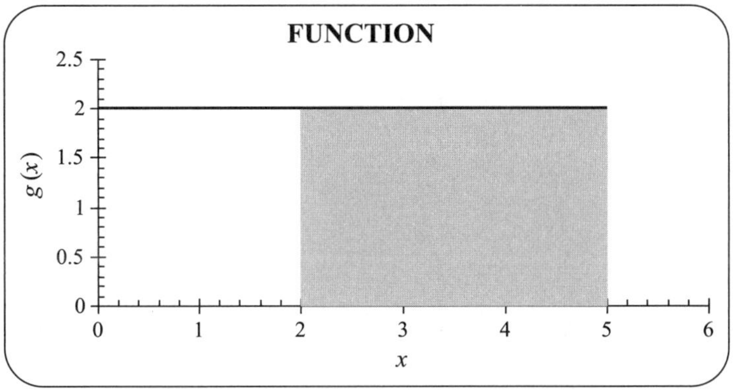


Area = 6 square units.

3. $h(x) = 3x$; $[1, 5]$.

**SOLUTION**

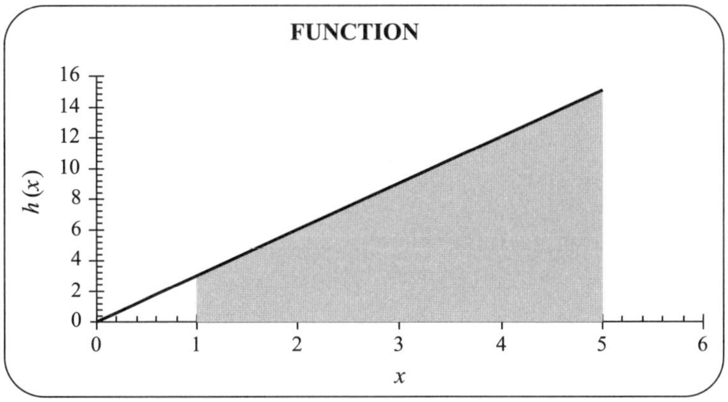


Area = 36 square units.

4. $y = \begin{cases} \frac{1}{2} & \text{if } -2 \le x \le 0 \\ x & \text{if } 0 < x \le 3 \end{cases}$; $[-2, 3]$.

**SOLUTION**

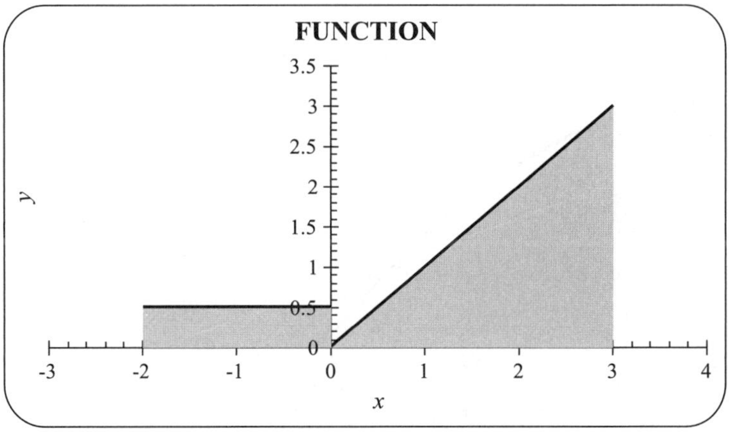


Area = 5.5 square units.

5. $y = |x| + 1$; [–1, 1].

**SOLUTION**

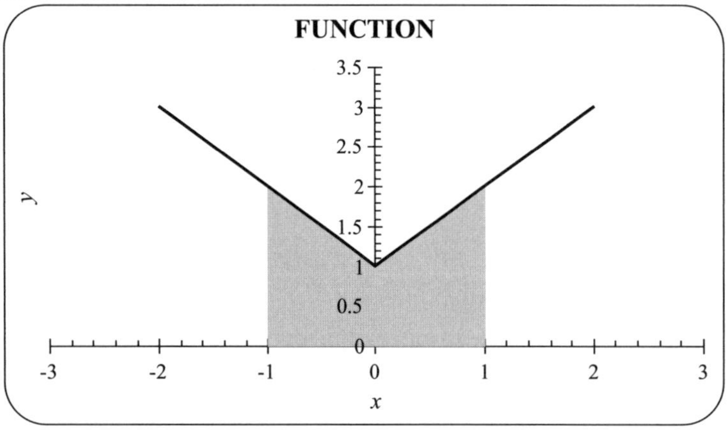


Area = 3 square units

6. Let $f(x) = -(x+2)^2 + 36$. You are to approximate the area of the region between the graph of *f* and the *x*-axis over the interval [–2, 4]. Include a graph of *f*.

   a. Find points $x_0$, $x_1$, $x_2$, and $x_3$ that divide the interval [–2, 4] into three subintervals of equal width.

   $x_0 =$ ____, $x_1 =$ ____, $x_2 =$ ____, $x_3 =$ ____.

   b. Find the midpoints $m_1$, $m_2$, and $m_3$ of the subintervals.

   $m_1 =$ ____, $m_2 =$ ____, $m_3 =$ ____.

   c. Compute the midpoint sum $S_3(f,[-2,\ 4])$.

**SOLUTION**

   a. $x_0 = -2$, $x_1 = 0$, $x_2 = 2$, $x_3 = 4$.

   b. $m_1 = -1$, $m_2 = 1$, $m_3 = 3$.

   c. $S_3(f,[-2,\ 4]) = 146$ square units.

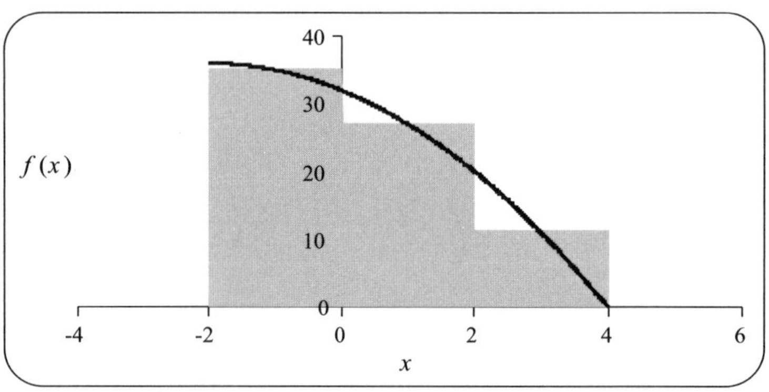

7. Let $f(x) = 2x^2$. Approximate the area of the region between the graph of $f$ and the $x$-axis over the interval [1, 9]. Divide the interval into $n = 4$ subintervals.

   a. Use midpoint sums to compute the answer by hand.
   b. Use ***Midpoint Sums.xls*** to compute the answer.

**SOLUTION**

a. 480
b. 480

8. Use ***Midpoint Sums.xls*** to approximate the area under the graph of $f(x) = \frac{1}{4}e^{-x/4}$ over the interval [0, 20] for $n = 4$, $n = 10$, and $n = 100$. What pattern do you notice?

**SOLUTION**

For $n = 4$, $\sum_{i=1}^{n} f(m_i) \cdot \Delta x = 0.93143$.

For $n = 10$, $\sum_{i=1}^{n} f(m_i) \cdot \Delta x = 0.98299$.

For $n = 100$, $\sum_{i=1}^{n} f(m_i) \cdot \Delta x = 0.99316$.

The results appear to be close to 1.0.

9. Use ***Intgrating.xls*** to approximate the area under the graph of $f(x) = \frac{1}{4}e^{-x/4}$ over the interval [0, 100]. Could the function

$$f(x) = \begin{cases} 0 & \text{if } x < 0 \\ \frac{1}{4}e^{-x/4} & \text{if } x \geq 0 \end{cases}$$

be a probability density function for a continuous random variable?

**SOLUTION**

Approximately 1.0. Yes, because the function values are always non-negative, the function is piece-wise continuous, and the total area between the curve and the $x$-axis is 1.0.

10. Use ***Intgrating.xls*** to approximate the area under the graph of $f(x) = -x^2 + 4$ over the interval [0, 2]. Could the function

$$f(x) = \begin{cases} 0 & \text{if } x < 0 \\ -x^2 + 4 & \text{if } 0 \leq x \leq 2 \end{cases}$$

be a probability density function for a continuous random variable?

**SOLUTION**

Approximately 5.33. No, because the total area between the curve and the $x$-axis exceeds 1.

11. Repeat **Example 9** for $q = 100$.

**SOLUTION**

a. \$250
b. \$62.50
c. \$125
d. \$62.50
e. $\$62.50 + \$125 + \$62.50 = \$250$

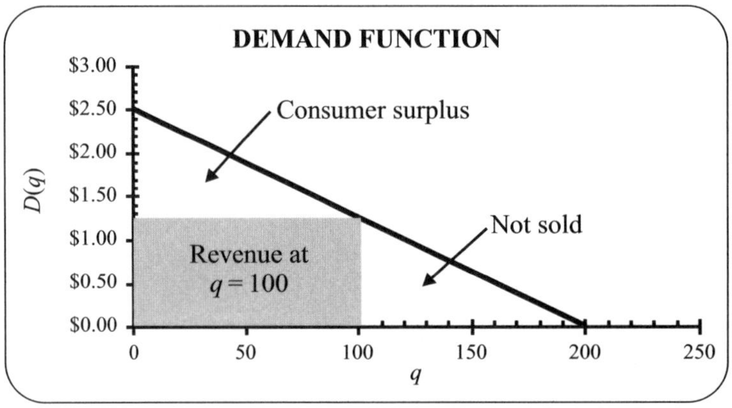


12. The graph of $f(x) = 2x + 5$ is given below. Write the shaded area as an integral and find the area using geometry. Hint: The area formula for a trapezoid is $A = \frac{1}{2}h(B + b)$.

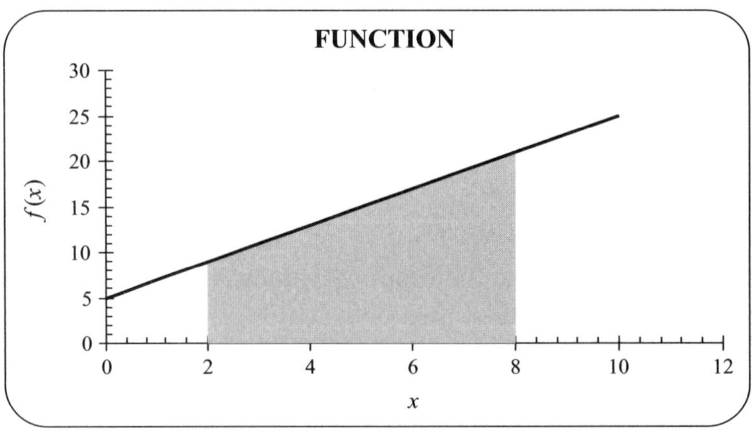


**SOLUTION**

$$\int_2^8 (2x + 5)\, dx = 90$$

13. The graph of $f(x) = 3 - |x|$ is given below. Write the shaded area as an integral and find the area using geometry.

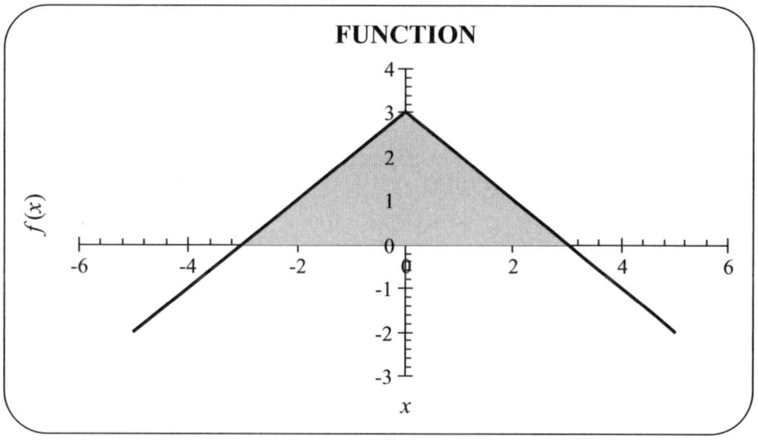


**SOLUTION**

$$\int_{-3}^{3} (3-|x|)\, dx = 9$$

14. Let $X$ be a continuous random variable with probability density function

$$f_X(x) = \begin{cases} 0.2 & \text{if } 0 \le x \le 5 \\ 0 & \text{elsewhere} \end{cases}.$$

Use basic geometry to find the shaded region illustrated below. Interpret this region as a definite integral and as a probability.

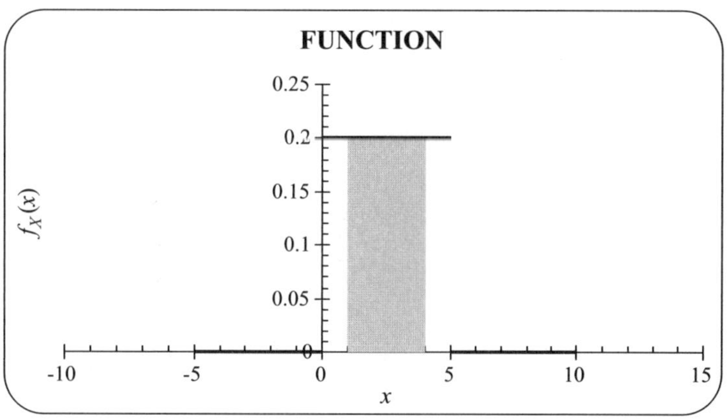


**SOLUTION**

$$\int_{1}^{4} 0.2\, dx = 0.6 = P(1 \le X \le 4).$$

15. Use closed form integration to evaluate $\int_{-1}^{4} (3-5x)\, dx$.

**SOLUTION**

$$\int_{-1}^{4} (3-5x)\,dx = -22.5$$

16. Use closed form integration to evaluate $\int_{0}^{1} (e^{2x}-1)\,dx$. Give an exact answer.

**SOLUTION**

$$\int_{0}^{1} (e^{2x}-1)\,dx = \frac{1}{2}(e^{2}-3)$$

17. Use closed form integration to evaluate $\int_{-1}^{1} (x-3)^{2}\,dx$. Give an exact answer.

**SOLUTION**

$$\int_{-1}^{1} (x-3)^{2}\,dx = \frac{56}{3}$$

18. Use closed form integration to evaluate $\int_{1}^{e} \left(\frac{3-2x}{x}\right) dx.$ Give an exact answer.

**SOLUTION**

$$\int_{1}^{e} \left(\frac{3-2x}{x}\right) dx = 5-2e$$

# Project 2: Bidding on an Oil Lease

## Business Background

An auction is a mechanism that facilitates the transfer of goods and services between buyers and sellers. The objective of the seller is to obtain the highest possible price for the item being auctioned; the objective of the bidder is to obtain the item at the lowest price. Generally the bidder that submits the highest bid wins the auction. However, the seller often specifies a reserve price for the item being auctioned. If the highest bid is not higher than the reserve price, then the item is not sold.

Auctions may be categorized based upon how much information the bidders have about the amounts that the other bidders are willing to pay. In a sealed-bid auction the bids are submitted privately, so bidders have no information about the amounts that other bidders are willing to pay. In an open auction the bids are submitted publicly, so bidders have much more information.

Sealed-bid auctions may be further categorized based upon the way in which the price paid is determined. In a first-price auction the price paid is equal to the amount of the highest bid submitted. This type of auction is used in the United States for offshore oil and gas leases and in Switzerland for construction contracts. In addition, this type of auction has been used in the United States Treasury for U.S. Treasury securities. In a second-price (or Vickrey) auction the price paid is equal to the amount of the second highest bid submitted. This type of auction is used in Guinea, Nigeria, and Uganda for foreign exchange. In addition, this type of auction has been used in Czechoslovakia to refinance credit.

Open auctions may be further categorized based upon how the price changes during the bidding. In an English (or ascending-price) auction the bidding starts at the reserve price or some other low opening price. The bidding increases until there are no more bids. As is the case with the first-price, sealed-bid auction, the winning bidder pays the amount of the highest bid. This type of auction is used in auction houses such as Sothebys and Christie's as well as in online auctions such as E-bay. In a Dutch (or descending-price) auction the bidding starts at an extremely high price. The bid is progressively lowered until some bidder claims the item by indicating a willingness to pay the amount of the current bid. This type of auction is used in the Netherlands for produce and flowers, in England and Israel for fish, in Bolivia, Jamaica, and Zambia for foreign exchange, and in Rumania to refinance credit.

As was mentioned previously, the United States government sells offshore oil leases in first-price, sealed-bid auctions. The bidders are major oil companies such as Chevron, ConocoPhillips, ExxonMobile, Marathon Oil, and Shell. Participating companies may submit bids individually or with a bidding ring (a consortium of companies). Companies participating in a bidding ring submit one bid on behalf of the ring and must agree on the allocation of cost and profits among the companies in the ring before the bid is submitted. The lease gives the owner the right to drill for oil in tracts of the floor of the ocean off the Alaska, California, Texas, or Louisiana coast. The bidding companies hire geologists to conduct assays to estimate the amount of oil in and the difficulty of extracting oil from each tract. This information is reflected in the geologist's signal or estimate of the amount that should be paid in order to receive a fair and reasonable profit.

For example, suppose that ExxonMobile is planning to participate in the upcoming auction for an offshore oil lease in the Gulf of Mexico. The geologist has estimated that the revenue from the sale of the oil will be \$150,000,000. Furthermore, the company has estimated that the cost of operating the wells will be \$125,000,000. If the desired profit margin is 10%, then the company should be willing to pay \$10,000,000 for the lease. It is this amount that when subtracted from Net Operating Income will produce a 10% profit. The abbreviated income statement for this example is given below.

| | |
|---|---|
| Revenue | \$150,000,000 |
| Operating Costs | - 125,000,000 |
| Net Operating Income | \$ 25,000,000 |
| Lease Payment | - 10,000,000 |
| Net Income (Profit) | \$ 15,000,000 |

Additional information about offshore oil and gas leases in the Gulf of Mexico may be found at `http://www.gomr.mms.gov`.

# Distributions

## *Finite Random Variables*

Random variables were first introduced in *Mathematics for Business Decisions Part 1: Probability and Simulations*. Further properties will be discussed here.

A **finite random variable** is a random variable that can assume only a finite number of distinct values. For example, finite random variables can represent the number of heads obtained in 20 tosses of a fair coin, the number of cars in an automobile dealership's inventory at the end of the month, the number of students that are logged on to a university's computer network during one hour, etc.

The **probability mass function** (*p.m.f.*) of a finite random variable $X$ is given by

$$f_X(x) = P(X = x).$$

A probability mass function must satisfy the two properties given below.

1. $0 \le f_X(x) \le 1$
2. $\sum_{\text{all } x} f_X(x) = 1$

The **cumulative distribution function** (*c.d.f.*) of any random variable $X$ is given by

$$F_X(x) = P(X \le x).$$

The domain of any cumulative distribution function is the set of real numbers, and a *c.d.f.* must satisfy the two properties given below.

1. $0 \le F_X(x) \le 1$
2. $F_X(x) \to 0$ as $x \to -\infty$, and $F_X(x) \to 1$ as $x \to \infty$

The expected value or mean of a finite random variable $X$ is given by

$$E(X) = \mu_X = \sum_{\text{all } x} x \cdot f_X(x).$$

**Example 1**: Jack has three pennies, four nickels, two dimes, and one quarter in his pocket. Let $X$ be the denomination (in cents) of a coin selected at random from his pocket.

a. Find the values of the *p.m.f.* of $X$.

b. Plot the *p.m.f.* of $X$.

c. Find a formula for the *c.d.f.* of $X$.

d. Plot the *c.d.f.* of $X$.

e. Use the *p.m.f.* of $X$ to compute the probability that Jack selects a quarter from his pocket.

f. Use the *p.m.f.* of $X$ to compute the probability that Jack selects at most five cents from his pocket.

g. Use the *c.d.f.* of $X$ to compute the probability that Jack selects at most five cents from his pocket.
h. Compute the expected value of $X$.

**SOLUTION**

a. The *p.m.f.* of $X$ is given below.

| $x$ | 1 | 5 | 10 | 25 |
|---|---|---|---|---|
| $f_X(x)$ | $3/10=0.3$ | $4/10=0.4$ | $2/10=0.2$ | $1/10=0.1$ |

b.

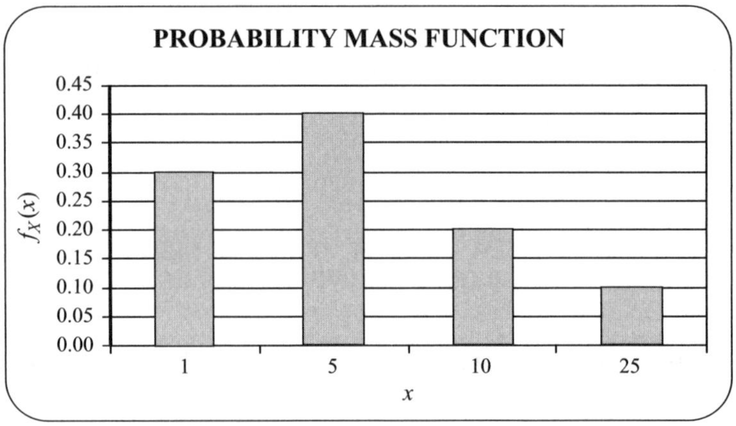


c. The *c.d.f.* of $X$ is given by

$$F_X(x)=\begin{cases}0 & \text{if } x<1\\ 0.3 & \text{if } 1\le x<5\\ 0.7 & \text{if } 5\le x<10\\ 0.9 & \text{if } 10\le x<25\\ 1.0 & \text{if } x\ge 25.\end{cases}$$

d.

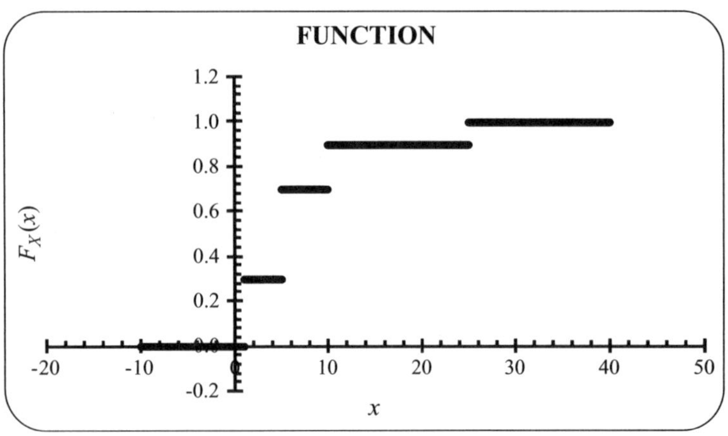


e. The probability that Jack selects a quarter is given by $P(X=25)=f_X(25)=0.1$.

f. The probability that Jack selects at most five cents is given by

$$P(X \le 5) = P(X = 1) + P(X = 5) = f_X(1) + f_X(5) = 0.3 + 0.4 = 0.7.$$

g. The probability that Jack selects at most five cents is given by

$$P(X \le 5) = F_X(5) = 0.7.$$

Notice that it is easier to use the *c.d.f.* rather than the *p.m.f.* to compute this probability.

h. $E(X) = \mu_X = \sum_{\text{all } x} x \cdot f_X(x) = 1(0.3) + 5(0.4) + 10(0.2) + 25(0.1) = 6.8$ cents.

**Example 2:** The probability mass function of a finite random variable *Y* is given below.

| $y$ | 0 | 1 | 2 | 3 |
|---|---|---|---|---|
| $f_Y(y)$ | 0.6141 | 0.3251 | 0.0574 | 0.0034 |

a. Find $P(Y = 1)$.

b. Find $P(Y \le 1)$.

c. Find $F_Y(2)$.

d. Find $P(Y \ge 2)$.

e. Find $f_Y(2)$.

f. Find $E(Y)$.

**SOLUTION**

a. $P(Y = 1) = f_Y(1) = 0.3251$

b. 
$$\begin{aligned} P(Y \le 1) &= P(Y = 0) + P(Y = 1) \\ &= f_Y(0) + f_Y(1) \\ &= 0.6141 + 0.3251 \\ &= 0.9392 \end{aligned}$$

c. 
$$\begin{aligned} F_Y(2) &= P(Y \le 2) \\ &= f_Y(0) + f_Y(1) + f_Y(2) \\ &= P(Y = 0) + P(Y = 1) + P(Y = 2) \\ &= 0.6141 + 0.3251 + 0.0574 \\ &= 0.9966 \end{aligned}$$

d. 
$$\begin{aligned} P(Y \ge 2) &= P(Y = 2) + P(Y = 3) \\ &= f_Y(2) + f_Y(3) \\ &= 0.0574 + 0.0034 \\ &= 0.0608 \end{aligned}$$

or

$$P(Y \geq 2) = 1 - P(Y \leq 1)$$
$$= 1 - 0.9392$$
$$= 0.0608$$

e. $f_Y(2) = 0.0574$

f. $E(Y) = \sum_{\text{all } y} y \cdot f_Y(y) = 0(0.6141) + 1(0.3251) + 2(0.0574) + 3(0.0034) = 0.4501$

**Example 3:** The *c.d.f.* of *Y*, the number of years among the next five in which a stock index falls, is given below.

$$F_Y(y) = \begin{cases} 0 & \text{if } y < 0 \\ 0.12 & \text{if } 0 \leq y < 1 \\ 0.40 & \text{if } 1 \leq y < 2 \\ 0.75 & \text{if } 2 \leq y < 3 \\ 0.95 & \text{if } 3 \leq y < 4 \\ 0.99 & \text{if } 4 \leq y < 5 \\ 1 & \text{if } y \geq 5 \end{cases}$$

a. Find the values of the *p.m.f.* of *Y*.
b. Plot the *p.m.f.* of *Y*.
c. What value of *Y* is most likely?
d. What value of *Y* is least likely?
e. Find $P(Y = 4)$.
f. Find $P(Y \leq 3)$.

**SOLUTION**

a.

| $y$ | $f_Y(y)$ |
|---|---|
| 0 | $0.12 - 0.00 = 0.12$ |
| 1 | $0.40 - 0.12 = 0.28$ |
| 2 | $0.75 - 0.40 = 0.35$ |
| 3 | $0.95 - 0.75 = 0.20$ |
| 4 | $0.99 - 0.95 = 0.04$ |
| 5 | $1.00 - 0.99 = 0.01$ |

b.

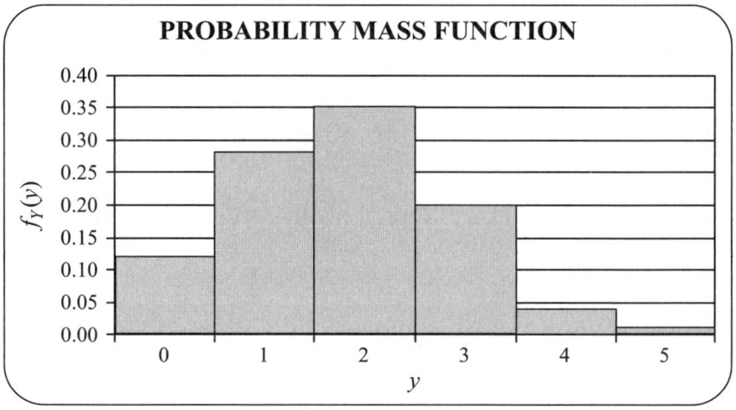


c. The maximum probability is 0.35 for $Y = 2$.

d. The minimum probability is 0.01 for $Y = 5$.

e. $P(Y = 4) = f_Y(4) = 0.04$

f. $P(Y \leq 3) = F_Y(3) = 0.95$

## *Binomial Random Variables*

A Bernoulli trial is an experiment that has exactly two outcomes: success, with probability $p$, and failure, with probability $1 - p$. A **Bernoulli random variable** assumes the value 1 for a success and the value 0 for a failure. The *p.m.f.* of a Bernoulli random variable $X$ is given in the table to the right.

| $x$ | $f_X(x)$ |
|---|---|
| 1 | $p$ |
| 0 | $1-p$ |

The expected value of a Bernoulli random variable $X$ with parameter $p$ is given by

$$E(X) = \sum_{\text{all } x} x \cdot f_X(x) = 1 \cdot p + 0 \cdot (1 - p) = p.$$

**Example 4:** A person goes into an old-fashioned ice cream parlor to get a single scoop of ice cream. There are only three choices: vanilla, chocolate, and strawberry. Let $X = 1$ if the person orders a scoop of vanilla ice cream and let $X = 0$ if the person does not order a scoop of vanilla ice cream. Based on prior sale receipts, the probability of ordering a scoop of vanilla ice cream is 40%.

a. Find the values of the *p.m.f.* of $X$.

b. Use the special formula for the expected value of a Bernoulli random variable to find $E(X)$, and interpret the result.

**SOLUTION**

a. The *p.m.f.* of $X$ is given below.

| $x$ | $f_X(x)$ |
|---|---|
| 1 | 0.40 |
| 0 | 0.60 |

b. $E(X) = p = 0.40$
We can verify the result using the formula for the expected value of any finite random variable.

$$E(X) = \sum_{\text{all } x} x \cdot f_X(x) = 1(0.40) + 0(0.60) = 0.40$$

On average, for a large number of customers, the probability that a customer will order a scoop of vanilla ice cream is 40%.

A **binomial random variable** gives the number of successes in $n$ independent Bernoulli trials with fixed probability of success, $p$.

The expected value of a binomial random variable $X$ with parameters $n$ and $p$ is given by

$$E(X) = \mu_X = np.$$

It can also be computed using the formula for the mean of any finite random variable.

$$E(X) = \mu_X = \sum_{\text{all } x} x \cdot f_X(x)$$

**Example 5:** A recent Gallup Poll of investors' confidence showed that 7 out of 10 American investors believe that now is a good time to invest. Three American investors are selected at random. Let $X$ be the number of investors in the sample who believe that now is a good time to invest. $X$ is a binomial random variable with $n = 3$ and $p = 0.70$.

a. Find the values of the *p.m.f.* of $X$.
b. Find the values of the *c.d.f.* of $X$.
c. Use the special formula for the expected value of a binomial random variable to find $E(X)$.

**SOLUTION**

a. Let $G$ be the event that an investor believes that now is a good time to invest and let $B$ be the event that an investor does not believe that now is a good time to invest. Since 7 out of 10 American investors believe that now is a good time to invest, $P(G) = 0.70$ and $P(B) = 1 - 0.70 = 0.30$. Tree diagrams are used to find the outcomes in the sample space.

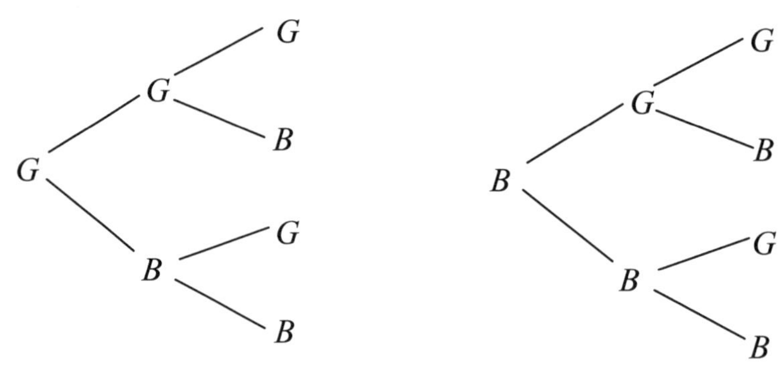


The sample space is given by

$$S = \{GGG,\ GGB,\ GBG,\ GBB,\ BBB,\ BBG,\ BGB,\ BGG\}.$$

The probability that all three investors believe that now is a good time to invest is given by $P(GGG) = P(G \cap G \cap G)$. Since the opinions of the investors are independent,

$$P(G \cap G \cap G) = P(G) \cdot P(G) \cdot P(G) = \left(P(G)\right)^3 = (0.70)^3 = 0.343.$$

$$P(X = 3) = P(GGG) = 0.343.$$

The probability that the first two investors believe that now is a good time to invest and the third investor does not believe that now is a good time to invest is given by $P(GGB) = P(G \cap G \cap B)$. Since the opinions of the investors are independent,

$$P(G \cap G \cap B) = P(G) \cdot P(G) \cdot P(B) = \left(P(G)\right)^2 \cdot P(B) = (0.70)^2 (0.30) = 0.147.$$

The order of the responses does not matter; therefore,

$$P(GGB) = P(GBG) = P(BGG) = 0.147.$$

The events *GGB*, *GBG*, and *BGG* are mutually exclusive, so

$$\begin{aligned} P(X = 2) &= P(GGB \cup GBG \cup BGG) \\ &= P(GGB) + P(GBG) + P(BGG) \\ &= 0.147 + 0.147 + 0.147 \\ &= 0.441. \end{aligned}$$

The probability that the first investor believes that now is a good time to invest and the second and third investors do not believe that now is a good time to invest is given by

$$P(GBB) = P(G \cap B \cap B).$$

Since the opinions of the investors are independent,

$$P(G \cap B \cap B) = P(G) \cdot P(B) \cdot P(B) = P(G) \cdot \left(P(B)\right)^2 = (0.70)(0.30)^2 = 0.063.$$

The order of the responses does not matter; therefore,

$$P(GBB) = P(BBG) = P(BGB) = 0.063.$$

The events *GBB*, *BBG*, and *BGB* are mutually exclusive, so

$$\begin{aligned} P(X = 1) &= P(GBB \cup BGB \cup BBG) \\ &= P(GBB) + P(BGB) + P(BBG) \\ &= 0.063 + 0.063 + 0.063 \\ &= 0.189. \end{aligned}$$

The probability that none of the investors believe that now is a good time to invest is given by $P(BBB) = P(B \cap B \cap B)$. Since the opinions of the investors are independent,

$$P(B \cap B \cap B) = P(B) \cdot P(B) \cdot P(B) = \left(P(B)\right)^3 = (0.30)^3 = 0.027.$$

$$P(X = 0) = P(BBB) = 0.027.$$

Recall that the probability mass function (*p.m.f.*) of a finite random variable $X$ is given by

$$f_X(x) = P(X = x).$$

The computations are summarized in the table below.

| $x$ | 0 | 1 | 2 | 3 |
|---|---|---|---|---|
| $f_X(x)$ | 0.027 | 0.189 | 0.441 | 0.343 |

We can use the fact that $\sum_{\text{all } x} f_X(x)$ must be equal to 1 to verify the results.

$$\begin{aligned}\sum_{\text{all } x} f_X(x) &= f_X(0)+f_X(1)+f_X(2)+f_X(3)\\ &= 0.027+0.189+0.441+0.343\\ &= 1.\end{aligned}$$

b. Recall that $F_X(x) = P(X \le x)$. Therefore,

$$\begin{aligned}F_X(0) &= P(X \le 0) = P(X=0) = 0.027\\ F_X(1) &= P(X \le 1) = P(X=0)+P(X=1) = 0.027+0.189 = 0.216\\ F_X(2) &= P(X \le 2) = P(X=0)+P(X=1)+P(X=2)\\ &= 0.027+0.189+0.441 = 0.657\\ F_X(3) &= P(X \le 2) = P(X=0)+P(X=1)+P(X=2)+P(X=3)\\ &= 0.027+0.189+0.441+0.343 = 1\end{aligned}$$

The computations are summarized in the table below.

| $x$ | 0 | 1 | 2 | 3 |
|---|---|---|---|---|
| $F_X(x)$ | 0.027 | 0.261 | 0.657 | 1.000 |

c. $E(X) = np = 3(0.70) = 2.10.$
We can verify the result using the formula for the expected value of any finite random variable.

$$\begin{aligned}E(X) &= \sum_{\text{all } x} x \cdot f_X(x)\\ &= 0(0.027)+1(0.189)+2(0.441)+3(0.343)\\ &= 2.10\end{aligned}$$

**Example 6:** Use the ***BINOMDIST*** function in *Excel* to verify the results found in Parts a and b of **Example 5**.

**SOLUTION**

Create a table in *Excel* with the same rows and columns as shown to the right.

|  | A | B | C |
|---|---|---|---|
| 1 | $x$ | $f_X(x)$ | $F_X(x)$ |
| 2 | 0 |  |  |
| 3 | 1 |  |  |
| 4 | 2 |  |  |
| 5 | 3 |  |  |

Insert the ***BINOMDIST*** function in *Cell B2*, enter the information shown below, and then click **OK**.

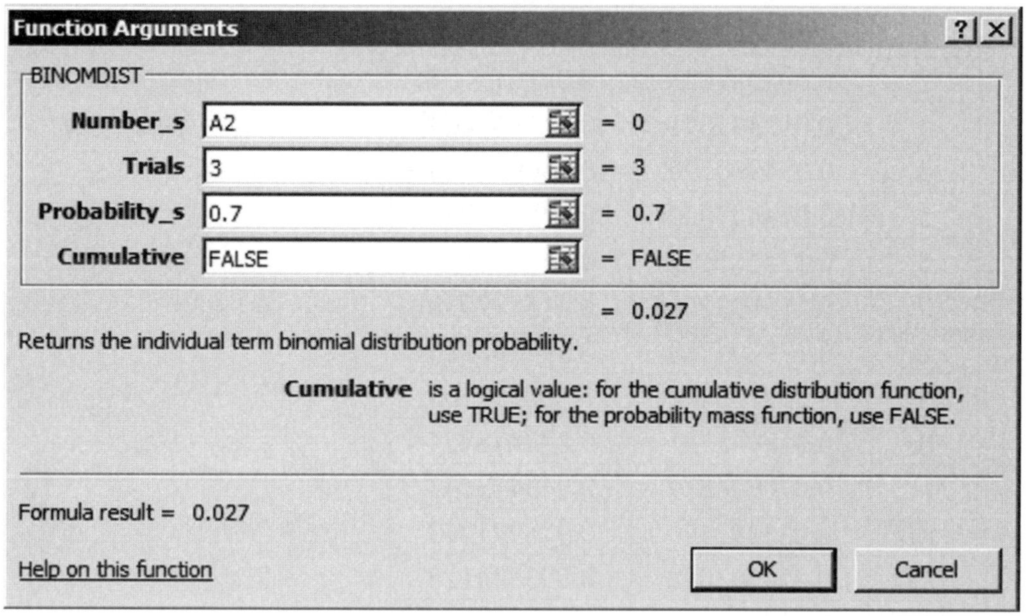


The value of $f_X(0)$ will appear in *Cell B2*. To compute the other values of the *p.m.f.*, copy the formula from *Cell B2* into *Cells B3:B5*. To compute the values of the *c.d.f.*, repeat the above, except in the **Cumulative** box, type "TRUE."

The table shown below gives the values of the *p.m.f.* and *c.d.f.* for the random variable in **Example 5**.

| $x$ | $f_X(x)$ | $F_X(x)$ |
|---|---|---|
| 0 | 0.027 | 0.027 |
| 1 | 0.189 | 0.216 |
| 2 | 0.441 | 0.657 |
| 3 | 0.343 | 1 |

**Example 7:** A coin that is biased so that the probability of tails is 0.40 is tossed 20 times. Let $X$ be the random variable that represents the number of tails observed. $X$ is a binomial random variable with parameters $n = 20$ and $p = 0.40$. Compute $P(X = 15)$ and $P(X \le 10)$.

**SOLUTION**

The results in the table given below were obtained by using ***BINOMDIST*** with $n = 20$ and $p = 0.40$.

| $x$ | $P(X = x)$ | $P(X \leq x)$ |
|---|---|---|
| 0 | 0.000036562 | 0.000036562 |
| 1 | 0.000487488 | 0.000524049 |
| 2 | 0.003087423 | 0.003611472 |
| 3 | 0.012349691 | 0.015961163 |
| 4 | 0.034990790 | 0.050951953 |
| 5 | 0.074647020 | 0.125598973 |
| 6 | 0.124411699 | 0.250010672 |
| 7 | 0.165882266 | 0.415892938 |
| 8 | 0.179705788 | 0.595598725 |
| 9 | 0.159738478 | 0.755337203 |
| 10 | 0.117141551 | 0.872478754 |
| 11 | 0.070994879 | 0.943473633 |
| 12 | 0.035497440 | 0.978971073 |
| 13 | 0.014563052 | 0.993534125 |
| 14 | 0.004854351 | 0.998388475 |
| 15 | 0.001294494 | 0.999682969 |
| 16 | 0.000269686 | 0.999952655 |
| 17 | 0.000042304 | 0.999994959 |
| 18 | 0.000004700 | 0.999999659 |
| 19 | 0.000000330 | 0.999999989 |
| 20 | 0.000000011 | 1.000000000 |

$P(X = 15) \cong 0.001294494$

$P(X \leq 10) \cong 0.872478754$

## *Continuous Random Variables*

A **continuous random variable** is a random variable that can assume any value in some interval of numbers. For example, continuous random variables can represent the interval of time it takes for a balloon to lose all of its air, the depth of the water in a bucket until it completely evaporates, or the length of time until a candle completely melts, etc.

The **probability density function** (*p.d.f.*) of a continuous random variable $X$ is given by $f_X(x)$. The probability that $X$ is between $a$ and $b$ is equal to the area between the graph of $f_X(x)$ and the $x$-axis over the interval $[a, b]$. That is,

$$P(a \leq X \leq b) = \text{area under the graph of } f_X(x) \text{ over the interval } [a, b],$$

which is given by the integral

$$\int_a^b f_X(x)\, dx.$$

A *p.d.f.* must satisfy the two properties given below.

1. $f_X(x) \geq 0$ for all $x$.
2. The total area under the curve must be equal to one.

If $X$ is any continuous random variable, then $P(X = x) = 0$. Why?

Consider $T$, the time it takes for a balloon to lose all of its air. Let's say that it takes between 0 and 10 seconds for the balloon to completely lose all of its air. The probability that the balloon loses all of its air in 6.53 seconds is given by the area between the graph of $f_T(t)$ and the $t$-axis over the point 6.53. Since the point 6.53 is equivalent to the interval [6.53, 6.53] and since the width of this interval is zero, the area of interest is zero as well.

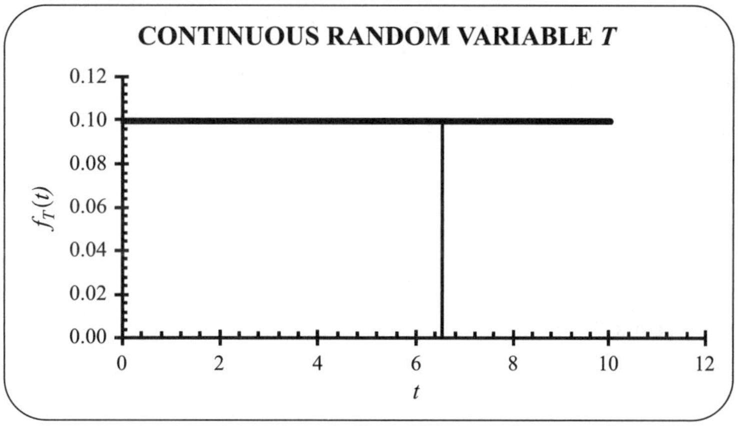


Because the probability of any single value is zero, events are given by intervals of numbers.

Note that the definition of the *c.d.f.* of a continuous random variable $X$ is the same as for a finite random variable — $F_X(x) = P(X \leq x)$. The probability that $X$ is between $a$ and $b$ is given by the difference between $F_X(b)$ and $F_X(a)$. That is, $P(a < X \leq b) = F_X(b) - F_X(a)$.

The expected value or mean of a continuous random variable $X$ is given by

$$E(X) = \mu_X = \int_{-\infty}^{\infty} x \cdot f_X(x)\, dx.$$

**Example 8:** The formulas for and graphs of the *p.d.f.* and *c.d.f.* of a continuous random variable $X$ are given below.

$$f_X(x) = \begin{cases} 0 & \text{if } x < 0 \\ 0.005x & \text{if } 0 \leq x \leq 20 \\ 0 & \text{if } x > 20 \end{cases} \qquad F_X(x) = \begin{cases} 0 & \text{if } x < 0 \\ 0.0025x^2 & \text{if } 0 \leq x < 20 \\ 1 & \text{if } x \geq 20 \end{cases}$$

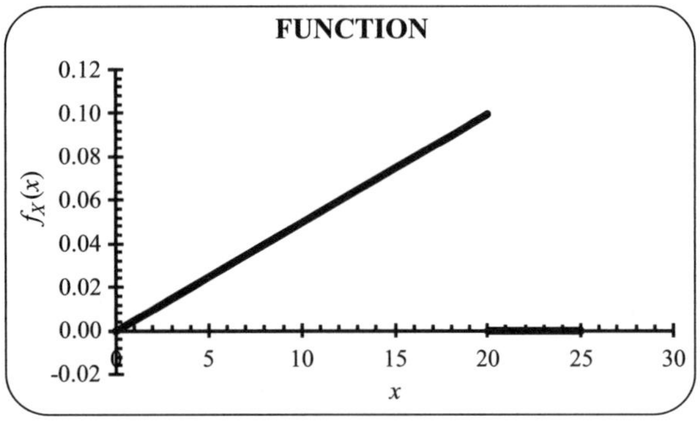


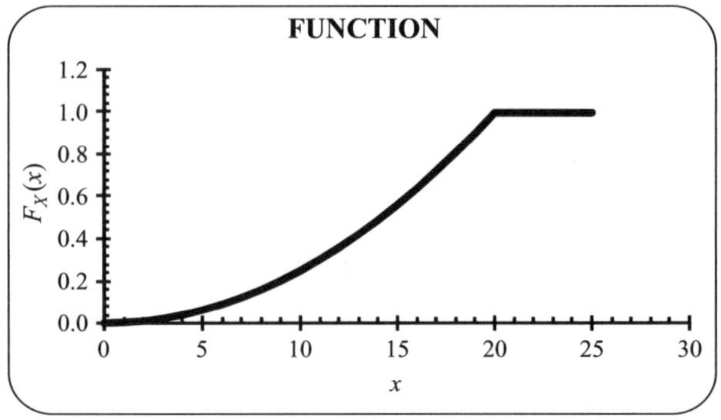


a. Use the graph of the *p.d.f.* to compute $P(X \le 15)$.

b. Use ***Integrating.xls*** to compute $P(X \le 15)$.

c. Evaluate the closed-form integral that gives $P(X \le 15)$.

d. Use the formula for the *c.d.f.* to compute $P(X \le 15)$.

e. Use ***Integrating.xls*** to compute $E(X)$.

f. Evaluate the closed-form integral that gives $E(X)$.

## SOLUTION

a. $P(X \le 15)$ is given by the area between the graph of the *p.d.f.* and the *x*-axis over the interval [0, 15].

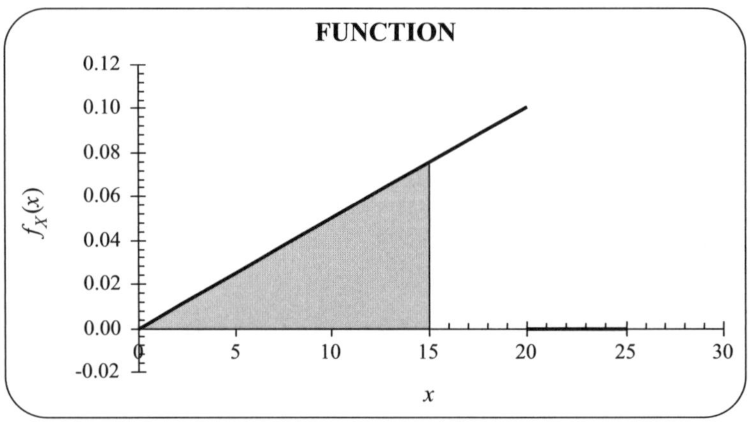


This is the area of a triangle with base $15 - 0 = 15$ and height $f_X(15) = 0.005(15) = 0.075$. Therefore, $P(X \le 15) = \frac{1}{2}(15)(0.075) = 0.5625$.

b. $P(X \le 15) = \int_0^{15} 0.005x \, dx = 0.5625$

| Definition | Computation | | Plot Interval | | Integration Interval | | $\int_a^b f(x)\,dx$ |
|---|---|---|---|---|---|---|---|
| Formula for $f(x)$ | $x$ | $f(x)$ | $A$ | $B$ | $a$ | $b$ | |
| 0 | | 0 | 0 | 20 | 0 | 15 | 0.5625 |

c.
$$
\begin{aligned}
P(X \le 15) &= \int_0^{15} 0.005x \, dx \\
&= 0.005 \int_0^{15} x \, dx \\
&= 0.005 \left( \frac{x^2}{2} \right) \Bigg|_0^{15} \\
&= 0.005 \left( \frac{(15)^2}{2} - \frac{(0)^2}{2} \right) \\
&= 0.005(112.5) \\
&= 0.5625
\end{aligned}
$$

d. $P(X \le 15) = F_X(15) = 0.0025(15)^2 = 0.5625$

e. $E(X) = \int_0^{20} x \cdot 0.005x \, dx = \int_0^{20} 0.005x^2 \, dx = 13.3333$

| Definition | Computation | | Plot Interval | | Integration Interval | | $\int_a^b f(x)\,dx$ |
|---|---|---|---|---|---|---|---|
| Formula for $f(x)$ | $x$ | $f(x)$ | $A$ | $B$ | $a$ | $b$ | |
| 0 | | 0 | 0 | 20 | 0 | 20 | 13.3333 |

f.
$$
\begin{aligned}
E(X) &= \int_0^{20} 0.005x^2 \, dx \\
&= 0.005 \int_0^{20} x^2 \, dx \\
&= 0.005 \left( \frac{x^3}{3} \right) \Bigg|_0^{20} \\
&= 0.005 \left( \frac{(20)^3}{3} - \frac{(0)^3}{3} \right) \\
&= 0.005(2666.667) \\
&= 13.3333
\end{aligned}
$$

**Example 9:** The formulas for and graphs of the *p.d.f.* and *c.d.f.* of a continuous random variable *X* are given below.

$$f_X(x) = \begin{cases} 0 & \text{if } x < 0 \\ 0.25xe^{-x/2} & \text{if } x \geq 0 \end{cases} \qquad F_X(x) = \begin{cases} 0 & \text{if } x < 0 \\ 1 - e^{-x/2} - 0.5xe^{-x/2} & \text{if } x \geq 0 \end{cases}$$

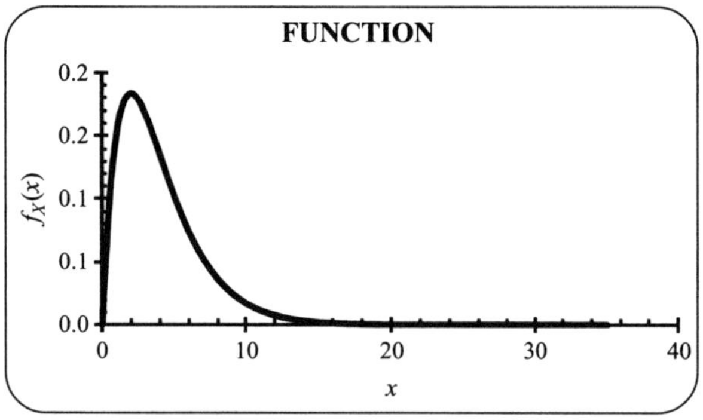


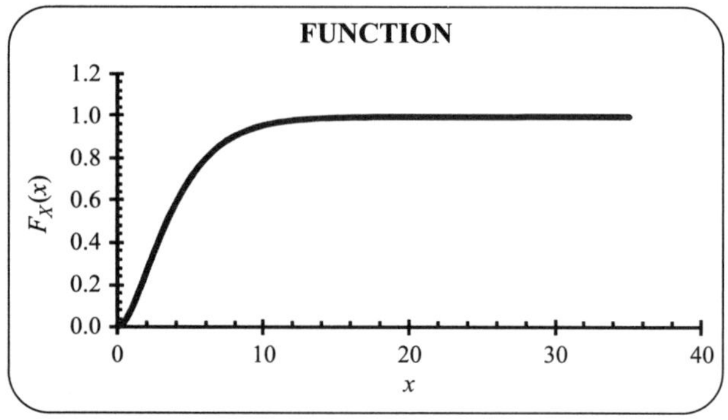


a. Shade the region of the graph of the *p.d.f.* that corresponds to $P(2 \leq X \leq 4)$.

b. Use ***Integrating.xls*** to compute $P(2 \leq X \leq 4)$.

c. Use the formula for the *c.d.f.* to compute $P(2 \leq X \leq 4)$.

## SOLUTION

a.

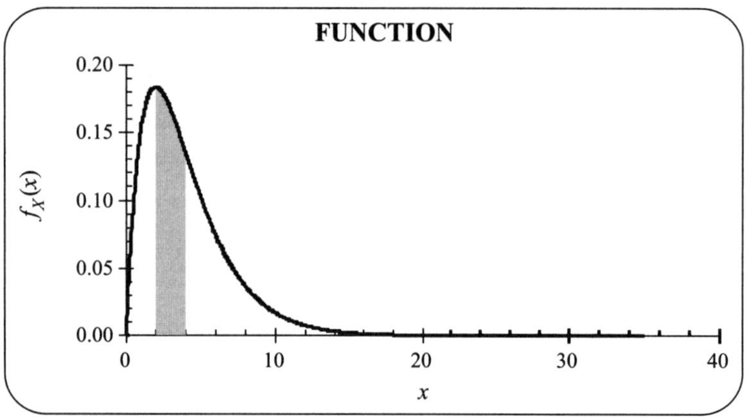


b.

$$P(2 \leq X \leq 4) = \int_2^4 0.25xe^{-x/2}\,dx = 0.3298$$

| Definition | Computation | | Plot Interval | | Integration Interval | | $\int_a^b f(x)\,dx$ |
|---|---|---|---|---|---|---|---|
| Formula for $f(x)$ | $x$ | $f(x)$ | $A$ | $B$ | $a$ | $b$ | |
| 0 | | 0 | 0 | 35 | 2 | 4 | 0.3298 |

c. $$\begin{aligned} P(2 \le X \le 4) &= P(X \le 4) - P(X < 2) \\ &= P(X \le 4) - P(X \le 2) \\ &= F_X(4) - F_X(2) \\ &= \left(1 - e^{-4/2} - 0.5(4)\left(e^{-4/2}\right)\right) - \left(1 - e^{-2/2} - 0.5(2)\left(e^{-2/2}\right)\right) \\ &\cong 0.5940 - 0.2642 \\ &= 0.3298 \end{aligned}$$

## *Uniform Random Variables*

A continuous random variable defined on an interval $[a, b]$ has a **uniform distribution** if every subinterval of $[a, b]$ having the same length has the same probability. The *p.d.f.* of a random variable $X$ that has a uniform distribution on the interval $[a, b]$ is given by

$$f_X(x) = \begin{cases} 0 & \text{if } x < a \\ \dfrac{1}{b-a} & \text{if } a \le x \le b \\ 0 & \text{if } x > b. \end{cases}$$

This function satisfies the properties that must be satisfied by any *p.d.f.*

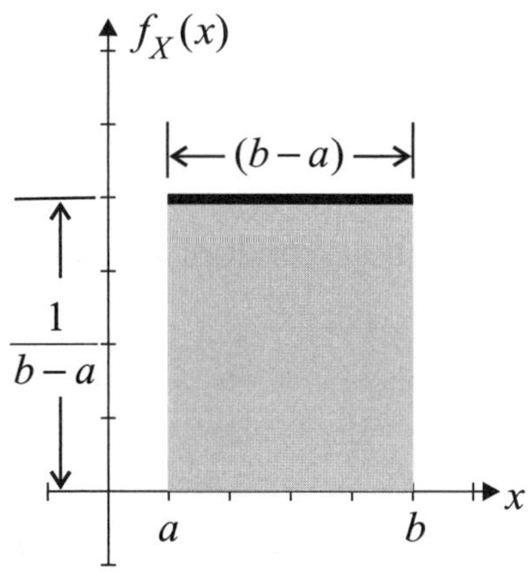


Note that:

1. $f_X(x) \ge 0$ for all $x$.
   *The graph is strictly on or above the x-axis.*

2. The total area under the graph is one.
   *Area under the curve* $= \textit{width} \cdot \textit{height}$
   $$= (b-a) \cdot \frac{1}{b-1}$$
   $$= 1$$

We can derive the *c.d.f.* of a random variable that has a uniform distribution on the interval $[a, b]$. The cumulative probability is zero for $x$ less than $a$ (the smallest possible value of $X$) and one for $x$ greater than $b$ (the largest possible value of $X$). Furthermore, the cumulative probability function is a linear function on the interval from $a$ to $b$. We must now find the equation of that line.

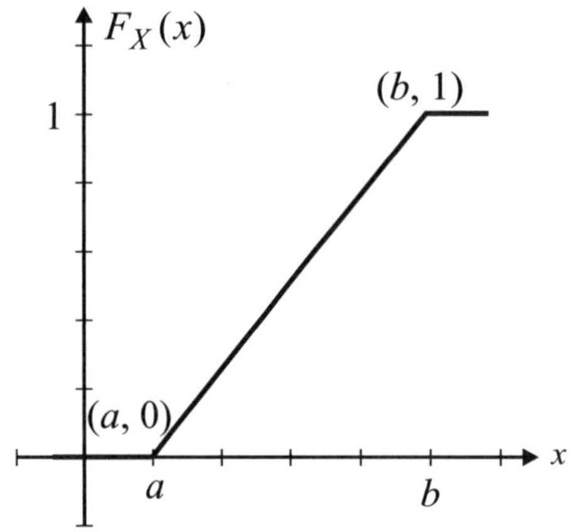


The slope of the line is given by

$$m = \frac{\Delta y}{\Delta x} = \frac{y_2 - y_1}{x_2 - x_1} = \frac{1-0}{b-a} = \frac{1}{b-a}.$$

Using the point-slope form of the equation of a line, $y - y_1 = m(x - x_1)$, we have

$$y - 0 = \frac{1}{b-a}(x-a) \text{ or } y = \frac{x-a}{b-a}.$$

The *c.d.f.* of a random variable $X$ that has a uniform distribution on the interval $[a, b]$ is given by

$$F_X(x) = \begin{cases} 0 & \text{if } x < a \\ \dfrac{x-a}{b-a} & \text{if } a \le x < b \\ 1 & \text{if } x \ge b. \end{cases}$$

The expected value of a random variable $X$ that has a uniform distribution on the interval $[a, b]$ is given by

$$E(X) = \mu_X = \frac{a+b}{2}.$$

It can also be computed using the formula for the mean of any continuous random variable.

$$E(X) = \mu_X = \int_{-\infty}^{\infty} x \cdot f_X(x)dx$$

**Example 10:** The graphs of the *p.d.f.* and *c.d.f.* of a random variable that has a uniform distribution on the interval [2, 6] are given below.

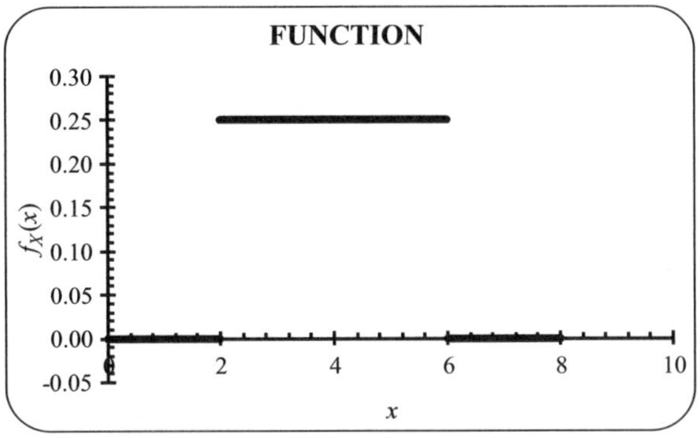


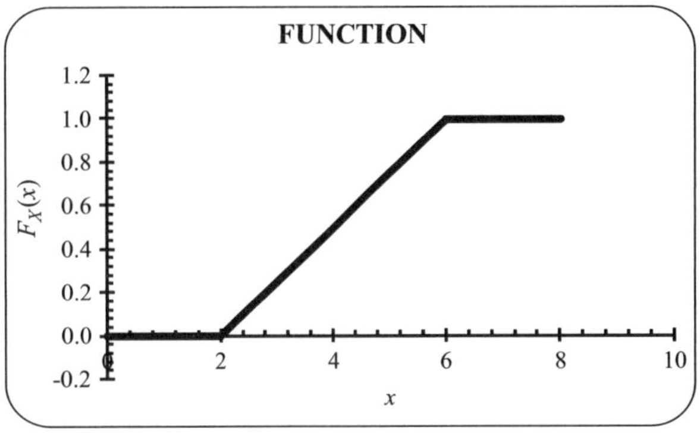


a. Find a formula for the *p.d.f.* of *X*.

b. Find a formula for the *c.d.f.* of *X*.

c. Use the formula for the *c.d.f.* to find $P(X \le 5)$.

d. Use the graph of the *p.d.f.* to find $P(3 \le X \le 5)$.

e. Use ***Integrating.xls*** to find $P(3 \le X \le 5)$.

f. Evaluate the closed-form integral that gives $P(3 \le X \le 5)$.

g. Use the graph of the *c.d.f.* to approximate $P(3 \le X \le 5)$.

h. Use the formula for the *c.d.f.* to find $P(3 \le X \le 5)$.

i. Find $P(X = 3.5)$.

j. Use the special formula for the expected value of a random variable that has a uniform distribution on the interval $[a, b]$ to find $E(X)$.

k. Use ***Integrating.xls*** to find $E(X)$.

l. Evaluate the closed-form integral that gives $E(X)$.

## SOLUTION

a. $(b-a) = (6-2) = 4$

$$f_X(x) = \begin{cases} 0 & \text{if } x < 2 \\ \dfrac{1}{4} & \text{if } 2 \le x \le 6 \\ 0 & \text{if } x > 6 \end{cases}$$

b. $$F_X(x) = \begin{cases} 0 & \text{if } x < 2 \\ \dfrac{x-2}{4} & \text{if } 2 \le x < 6 \\ 1 & \text{if } x \ge 6 \end{cases}$$

c. $P(X \le 5) = F_X(5) = \dfrac{5-2}{4} = \dfrac{3}{4} = 0.75$

d. $P(3 \le X \le 5)$ is given by the area between the graph of the *p.d.f.* and the $x$-axis over the interval [3, 5].

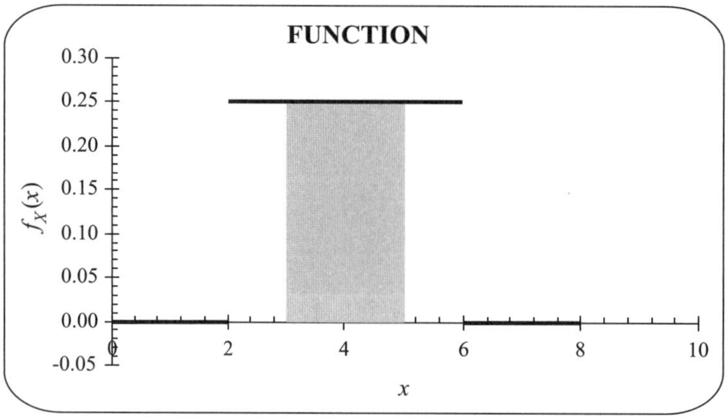


This is the area of a rectangle with base $5-3=2$ and height $\frac{1}{4}$. Therefore,

$$P(3 \le X \le 5) = 2\left(\tfrac{1}{4}\right) = 0.5.$$

e. $P(3 \le X \le 5) = \int_3^5 \frac{1}{4}\,dx = 0.5$

| Definition | Computation | | Plot Interval | | Integration Interval | | $\int_a^b f(x)\,dx$ |
|---|---|---|---|---|---|---|---|
| Formula for $f(x)$ | $x$ | $f(x)$ | $A$ | $B$ | $a$ | $b$ | |
| 0 | | 0 | 0 | 8 | 3 | 5 | 0.5000 |

f. $P(3 \le X \le 5) = \int_3^5 \frac{1}{4}\,dx = \frac{1}{4}\int_3^5 dx = \frac{1}{4}(x)\Big|_3^5 = \frac{1}{4}(5-3) = \frac{1}{4}(2) = 0.5$

g.

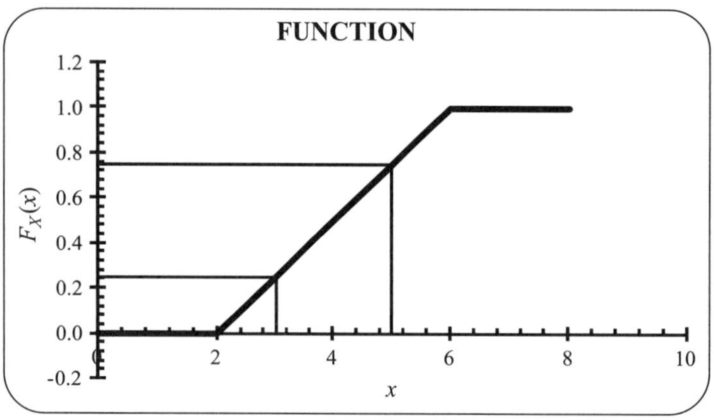


$P(3 \le X \le 5) = F_X(5) - F_X(3) \cong 0.72 - 0.23 = 0.49$

h. $P(3 \le X \le 5) = F_X(5) - F_X(3) = \frac{5-2}{4} - \frac{3-2}{4} = \frac{3}{4} - \frac{1}{4} = \frac{2}{4} = 0.5$

i. $P(X=3.5)=0$

j. $E(X)=\mu_X=\dfrac{2+6}{2}=4$

k. $E(X)=\mu_X=\displaystyle\int_2^6 x\cdot\frac{1}{4}dx=4$

| Definition | Computation | | Plot Interval | | Integration Interval | | $\int_a^b f(x)\,dx$ |
|---|---|---|---|---|---|---|---|
| Formula for $f(x)$ | $x$ | $f(x)$ | $A$ | $B$ | $a$ | $b$ | |
| 0 | | 0 | 0 | 8 | 2 | 6 | 4.0000 |

l. $E(X)=\mu_X=\displaystyle\int_2^6 x\cdot\frac{1}{4}dx=\frac{1}{4}\int_2^6 x\,dx=\frac{1}{4}\left(\frac{x^2}{2}\right)\Bigg|_2^6=\frac{1}{4}\left(\frac{(6)^2}{2}-\frac{(2)^2}{2}\right)=\frac{1}{4}(18-2)=\frac{1}{4}(16)=4$

**Example 11:** The length of time (in minutes), $T$, required to complete an exam has a uniform distribution on the interval [40, 50].

a. Plot the *p.d.f.* of $T$ on the interval [0, 60].
b. Find a formula for the *p.d.f.* of $T$.
c. Plot the *c.d.f.* of $T$ on the interval [0, 60].
d. Find a formula for the *c.d.f.* of $T$.
e. Find the probability that more than 48 minutes is required to complete the exam.
f. Find the average length of time required to complete the exam.

**SOLUTION**

a. 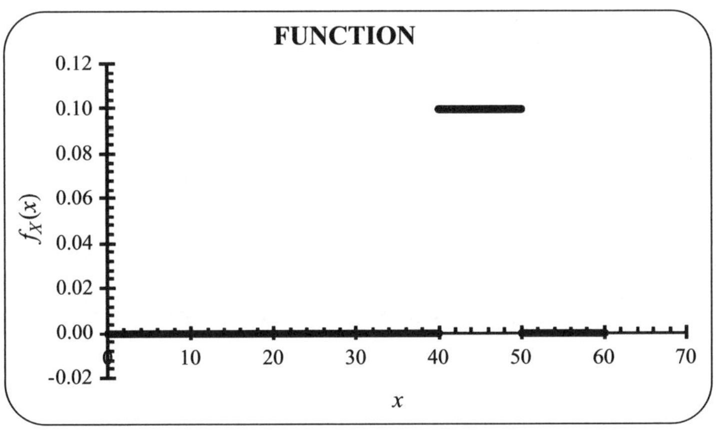


b. $f_T(t)=\begin{cases}0 & \text{if } t<40\\ \dfrac{1}{10} & \text{if } 40\le t\le 50\\ 0 & \text{if } t>50\end{cases}$

c.

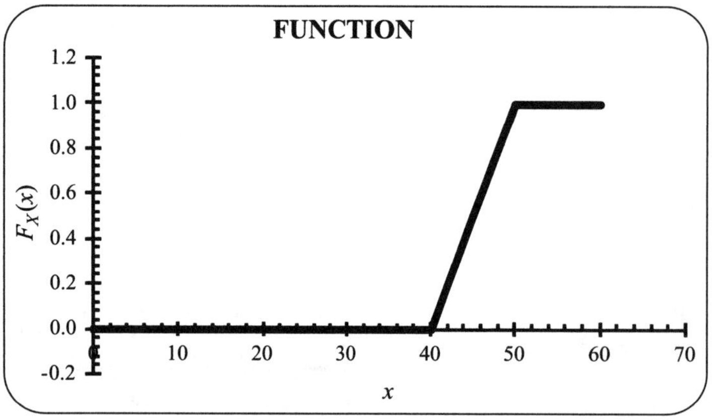


d. $F_T(t)=\begin{cases} 0 & \text{if } t<40 \\ \dfrac{t-40}{10} & \text{if } 40\le t<50 \\ 1 & \text{if } t\ge 50 \end{cases}$

e. $P(T>48)=1-P(T\le 48)=1-F_T(48)=1-\dfrac{48-40}{10}=0.20$

Note: The *p.d.f.* can also be used to answer this question.

$$P(T>48)=\text{area under curve over the interval } [48,\ 50]$$

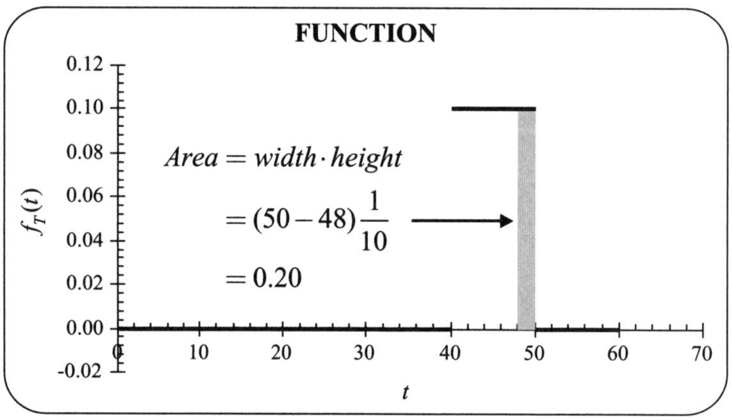


f. $E(T)=\dfrac{40+50}{2}=45$; the average length of time required to complete the exam is 45 minutes.

**Example 12:** A bus arrives at a bus stop every ten minutes. If $T$ is the waiting time (in minutes) until the next bus, then $T$ has a uniform distribution on the interval [0, 10].

a. Plot the *p.d.f.* of $T$ on the interval [0, 10].

b. Find a formula for the *p.d.f.* of $T$.

c. Use ***Integrating.xls*** to find the probability that the waiting time for the bus will be more than six minutes.

d. Use ***Integrating.xls*** to find the probability that the waiting time for the bus will be less than three minutes.

e. Use ***Integrating.xls*** to find $P(7 \le T \le 10)$.

f. Use ***Integrating.xls*** to find $P(4 \le T \le 7)$.

g. What do the answers to Parts d, e, and f have in common?

h. Find the average waiting time for the bus.

## SOLUTION

a.

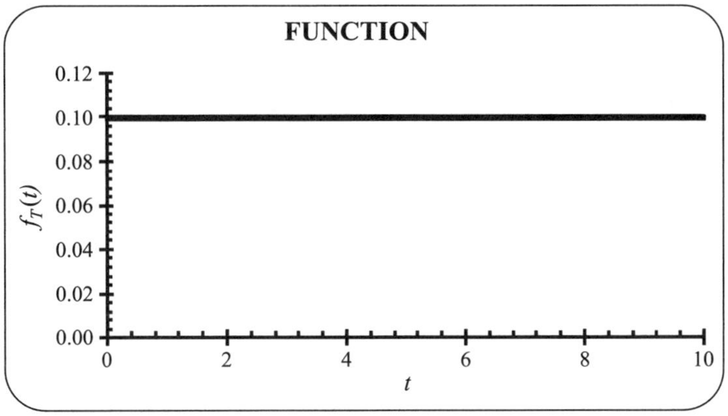


b. $f_T(t) = \begin{cases} 0 & \text{if } t < 0 \\ 0.1 & \text{if } 0 \le t \le 10 \\ 0 & \text{if } t > 10 \end{cases}$

c. $P(T > 6) = \int_6^{10} 0.1\, dt = 0.40$

| Definition | Computation | | Plot Interval | | Integration Interval | | $\int_a^b f(x)\,dx$ |
|---|---|---|---|---|---|---|---|
| Formula for $f(x)$ | $x$ | $f(x)$ | $A$ | $B$ | $a$ | $b$ | |
| 0.1 | | 0.1 | 0 | 10 | 6 | 10 | 0.4000 |

d. $P(T \le 3) = \int_0^{3} 0.1\, dt = 0.30$

| Definition | Computation | | Plot Interval | | Integration Interval | | $\int_a^b f(x)\,dx$ |
|---|---|---|---|---|---|---|---|
| Formula for $f(x)$ | $x$ | $f(x)$ | $A$ | $B$ | $a$ | $b$ | |
| 0.1 | | 0.1 | 0 | 10 | 0 | 3 | 0.3000 |

e. $P(7 \le T \le 10) = \int_7^{10} 0.1\, dt = 0.30$

| Definition | Computation | | Plot Interval | | Integration Interval | | $\int_a^b f(x)\,dx$ |
|---|---|---|---|---|---|---|---|
| Formula for $f(x)$ | $x$ | $f(x)$ | $A$ | $B$ | $a$ | $b$ | |
| 0.1 | | 0.1 | 0 | 10 | 7 | 10 | 0.3000 |

f. $P(4 \le T \le 7) = \int_4^7 0.1\, dt = 0.30$

| Definition | Computation | | Plot Interval | | Integration Interval | | $\int_a^b f(x)\,dx$ |
|---|---|---|---|---|---|---|---|
| Formula for $f(x)$ | $x$ | $f(x)$ | $A$ | $B$ | $a$ | $b$ | |
| 0.1 | | 0.1 | 0 | 10 | 4 | 7 | 0.3000 |

g. All the probabilities are equal to 0.3. This is because $T$ has a uniform distribution and all of the intervals you calculated probabilities for had a width of three ([0, 3], [7, 10], and [4, 7]).

h. $E(T) = \dfrac{0+10}{2} = 5$; the average waiting time for the bus is five minutes.

## *Exponential Random Variables*

**Exponential random variables** are used to model the length of time between consecutive occurrences of some event in a fixed unit of space or time.

The *p.d.f.* of an exponential random variable $X$ with parameter $\alpha$ is given by

$$f_X(x) = \begin{cases} 0 & \text{if } x < 0 \\ \dfrac{1}{\alpha} e^{-x/\alpha} & \text{if } x \ge 0, \end{cases}$$

and the *c.d.f.* of $X$ is given by

$$F_X(x) = \begin{cases} 0 & \text{if } x < 0 \\ 1 - e^{-x/\alpha} & \text{if } x \ge 0. \end{cases}$$

The expected value of an exponential random variable $X$ with parameter $\alpha$ is given by

$$E(X) = \mu_X = \alpha.$$

It can also be computed using the formula for the mean of any continuous random variable.

$$E(X) = \mu_X = \int_{-\infty}^{\infty} x \cdot f_X(x) dx$$

**Example 13:** Let $X$ be the exponential random variable that gives the time (in minutes) between departures of shuttle busses from the Disneyland parking lot. It is known that there are 0.25 departures per minute.

a. Find a formula for the *p.d.f.* of $X$.

b. Find a formula for the *c.d.f.* of $X$.

c. Plot the *p.d.f.* of $X$, and shade the region of the graph that corresponds to $P(2 \le X \le 4)$.

d. Use ***Integrating.xls*** to find $P(2 \leq X \leq 4)$.

e. Evaluate the closed-form integral that gives $P(2 \leq X \leq 4)$.

f. Use the formula for the *c.d.f.* to find $P(2 \leq X \leq 4)$.

g. Use the special formula for the expected value of an exponential random variable to find $E(X)$.

h. Use ***Integrating.xls*** to find $E(X)$.

## SOLUTION

a. $$\frac{1}{\alpha \text{ minutes}} = \frac{0.25 \text{ departures}}{1 \text{ minute}}$$

$$0.25\alpha = 1 \Rightarrow \alpha = \frac{1}{0.25} = 4 \text{ minutes}$$

$$f_X(x) = \begin{cases} 0 & \text{if } x < 0 \\ 0.25e^{-0.25x} & \text{if } x \geq 0 \end{cases}$$

b. $$F_X(x) = \begin{cases} 0 & \text{if } x < 0 \\ 1 - e^{-0.25x} & \text{if } x \geq 0 \end{cases}$$

c.

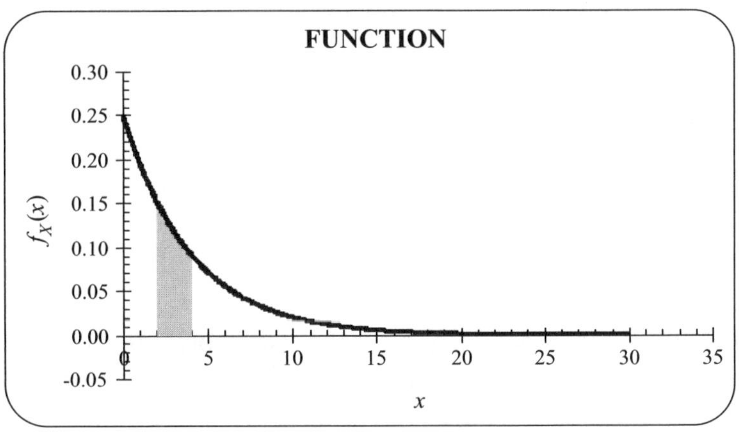


d. $$P(2 \leq X \leq 4) = \int_2^4 0.25e^{-0.25x}\,dx = 0.2387$$

| Definition | Computation | | Plot Interval | | Integration Interval | | $\int_a^b f(x)\,dx$ |
|---|---|---|---|---|---|---|---|
| Formula for $f(x)$ | $x$ | $f(x)$ | $A$ | $B$ | $a$ | $b$ | |
| 0.25 | | 0.25 | 0 | 30 | 2 | 4 | 0.2387 |

e. $$\begin{aligned} P(2 \le X \le 4) &= \int_2^4 0.25e^{-0.25x}dx \\ &= 0.25\int_2^4 e^{-0.25x}dx \\ &= 0.25\left(\frac{1}{-0.25}e^{-0.25x}\right)\Bigg|_2^4 \\ &= 0.25\left[\left(-\frac{1}{0.25}e^{-0.25(4)}\right)-\left(-\frac{1}{0.25}e^{-0.25(2)}\right)\right] \\ &\cong 0.2387 \end{aligned}$$

f. $P(2 \le X \le 4) = F_X(4) - F_X(2) = (1-e^{-0.25(4)}) - (1-e^{-0.25(2)}) \cong 0.2387$

g. $E(X) = \alpha = 4$

h. $E(X) = \int_{-\infty}^{\infty} x \cdot f_X(x)\, dx = \int_0^{\infty} x \cdot 0.25e^{-0.25x}dx = 4$

| Definition | Computation | | Plot Interval | | Integration Interval | | $\int_a^b f(x)\,dx$ |
|---|---|---|---|---|---|---|---|
| Formula for $f(x)$ | $x$ | $f(x)$ | $A$ | $B$ | $a$ | $b$ | |
| 0 | | 0 | 0 | 60 | 0 | 60 | 4.0000 |

**Example 14:** Let $X$ be the continuous random variable that gives the time (in minutes) between the arrivals/departures of planes at the Phoenix Sky Harbor Airport. There are 8.3 planes arriving or departing every ten minutes.

a. Find a formula for the *p.d.f.* of $X$.

b. Find a formula for the *c.d.f.* of $X$.

c. Compute the probability that the time between the arrivals/departures of planes at the Phoenix Sky Harbor Airport is between one and three minutes.

d. Use ***Integrating.xls*** to show that $\mu_X = 1.2$.

**SOLUTION**

a. $$\frac{1 \text{ arrival}}{\alpha \text{ minutes}} = \frac{8.3 \text{ arrivals/departures}}{10 \text{ minutes}}$$

$$8.3\alpha = 10 \Rightarrow \alpha = \frac{10}{8.3} \cong 1.2 \text{ minutes}$$

$$f_X(x) = \begin{cases} 0 & \text{if } x < 0 \\ \frac{1}{1.2}e^{-x/1.2} & \text{if } x \ge 0 \end{cases}$$

b. $F_X(x) = \begin{cases} 0 & \text{if } x < 0 \\ 1 - e^{-x/1.2} & \text{if } x \geq 0 \end{cases}$

c. $$\begin{aligned} P(1 \leq X \leq 3) &= F_X(3) - F_X(1) \\ &= \left(1 - e^{-3/1.2}\right) - \left(1 - e^{-1/1.2}\right) \\ &\cong 0.9179 - 0.5654 \\ &= 0.3525 \end{aligned}$$

d. $\mu_X = \int_0^{\infty} x \cdot \frac{1}{1.2} e^{-x/1.2} dx = 1.2$

| Definition | Computation | | Plot Interval | | Integration Interval | | $\int_a^b f(x)\,dx$ |
|---|---|---|---|---|---|---|---|
| Formula for $f(x)$ | $x$ | $f(x)$ | $A$ | $B$ | $a$ | $b$ | |
| 0 | | 0 | 0 | 30 | 0 | 30 | 1.2000 |

## *Random Samples*

Properties of a random variable can be estimated from an appropriately selected sample.

For example, the *p.m.f.* of a finite random variable can be approximated by a relative frequency histogram and the *p.d.f.* of a continuous random variable can be approximated by an adjusted relative frequency histogram. The histogram can then be used to estimate probabilities for the random variable.

In addition, the theoretical, or population, mean $\mu_X$ can be approximated by the sample mean $\overline{x}$. That is,

$$\mu_X \cong \overline{x}, \quad \text{where} \quad \overline{x} = \frac{1}{n}\sum_{i=1}^{n} x_i.$$

**Example 15:** Jack has three pennies, four nickels, two dimes, and one quarter in his pocket. Let $X$ be the denomination (in cents) of a coin selected from his pocket at random.

The *p.m.f.* of $X$ is given below.

| $x$ | 1 | 5 | 10 | 25 |
|---|---|---|---|---|
| $f_X(x)$ | 0.3 | 0.4 | 0.2 | 0.1 |

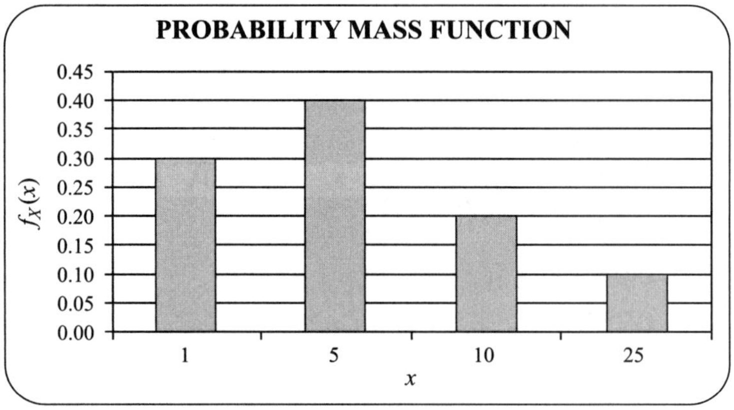


Forty observations of $X$ are described below.

| | | | | | | | |
|---|---|---|---|---|---|---|---|
| 10 | 1 | 5 | 1 | 5 | 5 | 1 | 10 |
| 1 | 1 | 5 | 25 | 10 | 1 | 5 | 1 |
| 5 | 10 | 1 | 5 | 1 | 5 | 10 | 5 |
| 5 | 1 | 5 | 10 | 5 | 5 | 1 | 10 |
| 25 | 10 | 1 | 5 | 10 | 1 | 25 | 5 |

| *Bin* | *Frequency* | Relative Frequency |
|---|---|---|
| 1 | 13 | 0.325 |
| 5 | 15 | 0.375 |
| 10 | 9 | 0.225 |
| 25 | 3 | 0.075 |

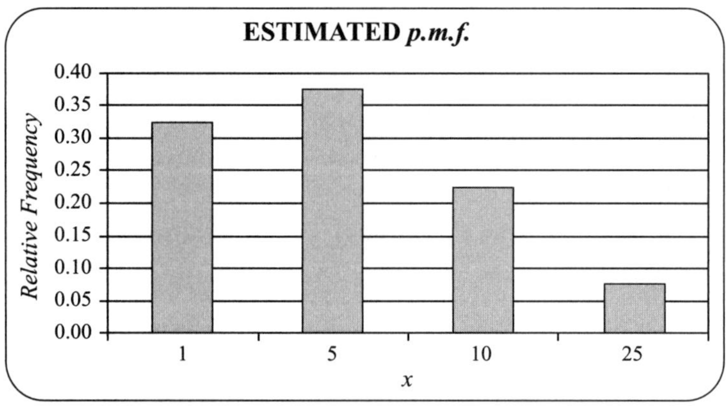


Notice that for a sample of size forty, the empirical probabilities closely approximate the theoretical probabilities. The theoretical and empirical probabilities are given in the table below.

| $x$ | 1 | 5 | 10 | 25 |
|---|---|---|---|---|
| Theoretical Probability | 0.30 | 0.40 | 0.20 | 0.10 |
| Empirical Probability | 0.325 | 0.375 | 0.225 | 0.075 |

Recall that the expected value or mean of $X$, the denomination of a coin selected at random from Jack's pocket, is given by

$$E(X) = \mu_X = 1(0.3) + 5(0.4) + 10(0.2) + 25(0.1) = 6.8 \text{ cents.}$$

In other words, Jack would expect to obtain 6.8 cents on average in a large number of repetitions of this experiment. The sample mean is given by

$$\begin{aligned}\overline{x} &= \frac{1}{n}\sum_{i=1}^{n} x_i \\ &= \frac{1}{40}(10+1+\cdots+5) \\ &= 6.325.\end{aligned}$$

This value differs from the theoretical value by only 0.475.

**Example 16**: The estimated *p.m.f.* of a finite random variable *Y* is given below.

| *Bin* | *Frequency* | Relative Frequency |
|---|---|---|
| 0 | 65 | 0.130 |
| 1 | 141 | 0.282 |
| 2 | 139 | 0.278 |
| 3 | 87 | 0.174 |
| 4 | 40 | 0.080 |
| 5 | 19 | 0.038 |
| 6 | 4 | 0.008 |
| 7 | 3 | 0.006 |
| 8 | 0 | 0.000 |
| 9 | 2 | 0.004 |

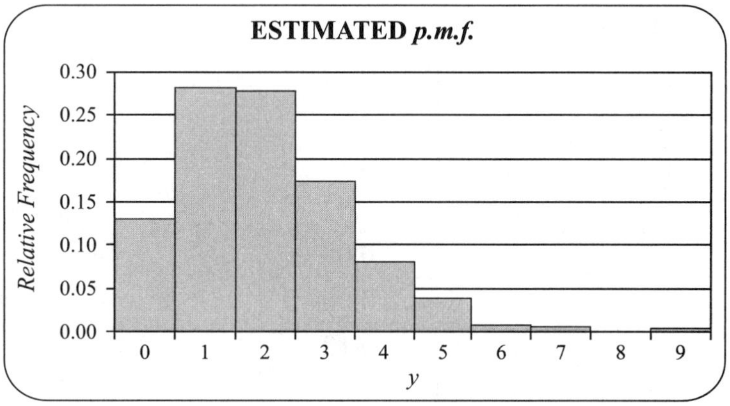


Although the theoretical distribution of *Y* is unknown, we can approximate probabilities and the expected value for *Y* from the sample. For example, the probability that *Y* is equal to 4 can be approximated by the relative frequency of 4 in the sample of 500 observations. In other words, $P(Y=4) \cong 40/500 = 0.08$. The expected value or mean of *Y* can be approximated by the mean of the sample of 500 observations, which is given by $E(Y) \cong \overline{y} = 1.996$. Since the theoretical probabilities and expected value are unknown, the probabilities and expected value approximated by the sample are only estimates of the true values. The accuracy of the estimates improves as the sample size increases.

**Example 17**: One hundred students were selected at random from those taking an exam. The length of time (in minutes) that it took each student to complete the exam was recorded. The results are described below.

| | | | | | | | | | |
|---|---|---|---|---|---|---|---|---|---|
| 43.82 | 42.45 | 43.72 | 49.52 | 40.74 | 40.05 | 47.56 | 45.29 | 49.86 | 46.72 |
| 41.01 | 40.45 | 43.56 | 40.53 | 41.98 | 49.26 | 46.27 | 47.97 | 49.26 | 47.32 |
| 45.96 | 40.32 | 49.10 | 47.05 | 40.64 | 41.00 | 41.74 | 48.06 | 49.04 | 45.85 |
| 48.99 | 41.64 | 44.66 | 48.17 | 43.58 | 42.57 | 44.05 | 42.62 | 45.45 | 41.52 |
| 48.85 | 42.20 | 44.26 | 49.73 | 44.87 | 47.76 | 45.52 | 41.78 | 45.01 | 48.92 |
| 49.58 | 40.17 | 43.04 | 44.66 | 45.11 | 46.80 | 47.12 | 48.67 | 46.75 | 43.78 |
| 40.14 | 42.85 | 49.76 | 43.00 | 43.73 | 48.09 | 45.55 | 41.15 | 44.90 | 42.00 |
| 44.07 | 43.43 | 48.07 | 47.50 | 49.86 | 47.24 | 41.81 | 40.60 | 41.46 | 42.06 |
| 48.63 | 45.54 | 49.91 | 43.51 | 40.41 | 40.85 | 49.70 | 47.62 | 40.38 | 43.34 |
| 41.39 | 43.57 | 42.56 | 47.76 | 42.31 | 41.32 | 46.87 | 47.38 | 47.96 | 43.25 |

| *Bin* | *Frequency* | Relative Frequency |
|---|---|---|
| 41 | 13 | 0.13 |
| 42 | 12 | 0.12 |
| 43 | 9 | 0.09 |
| 44 | 12 | 0.12 |
| 45 | 7 | 0.07 |
| 46 | 9 | 0.09 |
| 47 | 5 | 0.05 |
| 48 | 12 | 0.12 |
| 49 | 9 | 0.09 |
| 50 | 12 | 0.12 |

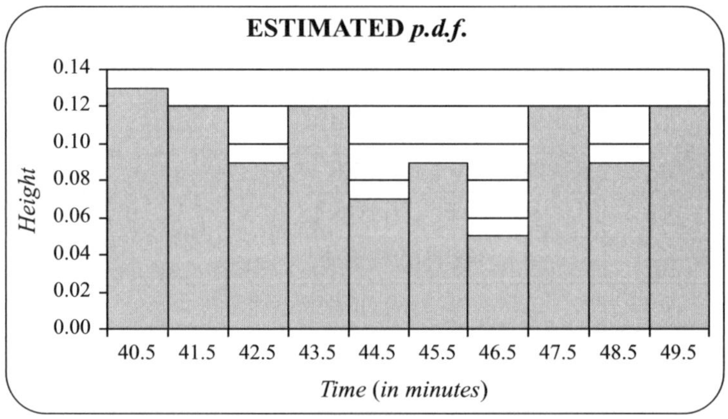


From this histogram, it appears that $T$, the time (in minutes) required for the exam, has a uniform distribution on the interval [40, 50]. The graph of the assumed *p.d.f.* of $T$ is given below.

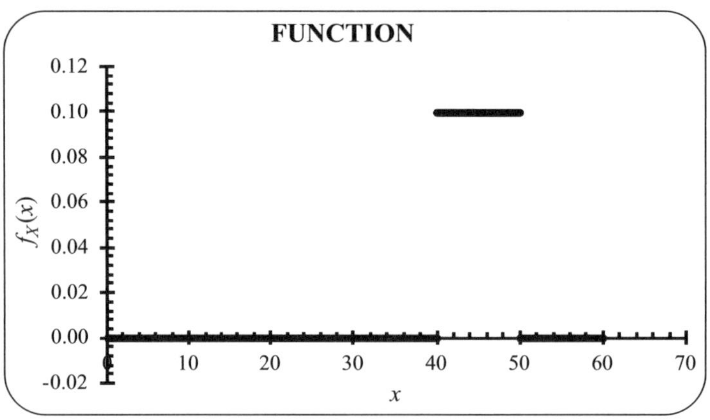


Notice that the general shape of the empirical distribution is similar to that of the theoretical distribution. A larger sample would enhance the similarities between the distributions. The assumed *p.d.f.* can now be used to estimate probabilities for $T$. For example, the probability that it will take a randomly selected student between 42.6 and 45.7 minutes to complete the exam is approximately (45.7 – 42.6)/10 = 0.31.

If $T$ has a uniform distribution on the interval [40, 50], then the expected value of $T$ is given by

$$E(T) = \mu_T = \frac{a+b}{2} = \frac{40+50}{2} = 45.$$

The sample mean is given by

$$\overline{t} = \frac{1}{n}\sum_{i=1}^{n} t_i = \frac{1}{100}(43.82 + 42.45 + \cdots + 43.25) = 44.85\,.$$

This value differs from the theoretical value by only 0.15.

Once again, it is very important to understand that the difference between the theoretical mean and the sample mean will decrease as the sample size becomes larger. Remember the saying, "the more the merrier"; in statistics, "the more data the merrier."

**Example 18:** One hundred observations of a continuous random variable $X$ are given in the table below.

| | | | | | | | | | |
|---|---|---|---|---|---|---|---|---|---|
| 1.43 | 9.01 | 2.22 | 3.35 | 3.48 | 6.92 | 4.62 | 3.30 | 0.47 | 1.31 |
| 0.35 | 7.45 | 4.14 | 13.23 | 3.49 | 4.52 | 0.60 | 0.86 | 1.79 | 5.74 |
| 7.38 | 0.17 | 1.87 | 6.42 | 5.28 | 1.96 | 1.80 | 0.20 | 0.97 | 5.96 |
| 0.75 | 2.40 | 1.51 | 0.22 | 7.22 | 8.73 | 8.81 | 0.48 | 0.64 | 1.95 |
| 0.30 | 2.04 | 2.85 | 4.38 | 1.85 | 2.14 | 0.93 | 3.12 | 0.95 | 0.03 |
| 3.66 | 3.05 | 3.14 | 5.39 | 4.28 | 6.71 | 5.17 | 4.32 | 4.59 | 13.07 |
| 0.16 | 1.73 | 0.15 | 2.75 | 8.51 | 1.46 | 12.84 | 0.72 | 1.24 | 5.05 |
| 1.14 | 2.54 | 2.52 | 1.92 | 0.51 | 4.44 | 6.09 | 1.35 | 15.85 | 2.29 |
| 0.47 | 5.99 | 3.76 | 0.47 | 3.10 | 2.13 | 6.90 | 5.57 | 1.35 | 5.24 |
| 2.11 | 0.34 | 5.05 | 2.48 | 5.99 | 4.57 | 6.43 | 6.01 | 0.81 | 3.45 |

a. Construct a histogram that would approximate the *p.d.f.* of $X$.
b. This histogram appears to approximate what type of continuous random variable?
c. Use the sample data to compute an estimate of the expected value, $E(X)$.
d. Find a formula for the assumed *p.d.f.* of $X$.
e. Find a formula for the assumed *c.d.f.* of $X$.
f. Use the formula for the assumed *c.d.f.* to compute an estimate of $P(5 \le X \le 10)$.
g. Use the formula for the assumed *c.d.f.* to compute an estimate of $P(X \ge 9)$.

**SOLUTION**

a. The adjusted relative frequency is equal to the relative frequency divided by the bin width.

| Bin | Frequency | Relative Frequency | Adjusted Relative Frequency |
|---|---|---|---|
| 2 | 39 | 0.39 | 0.195 |
| 4 | 23 | 0.23 | 0.115 |
| 6 | 20 | 0.20 | 0.100 |
| 8 | 10 | 0.10 | 0.050 |
| 10 | 4 | 0.04 | 0.020 |
| 12 | 0 | 0.00 | 0.000 |
| 14 | 3 | 0.03 | 0.015 |
| 16 | 1 | 0.01 | 0.005 |

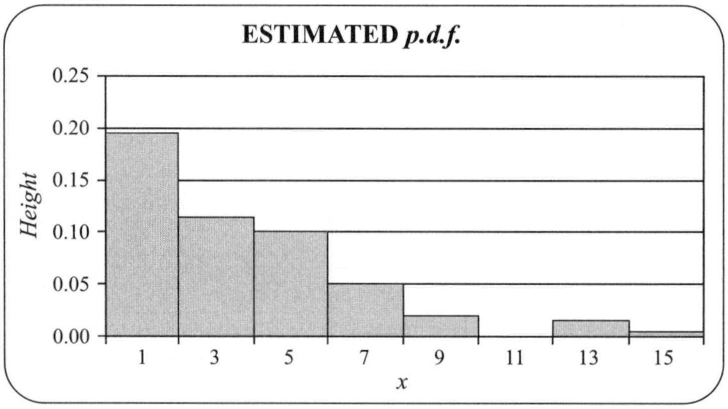


b. The shape of the histogram in Part a is similar to the shape of the graph shown below. The graph shown below is the graph of the *p.d.f.* of an exponential random variable, so it would be reasonable to assume that $X$ is an exponential random variable.

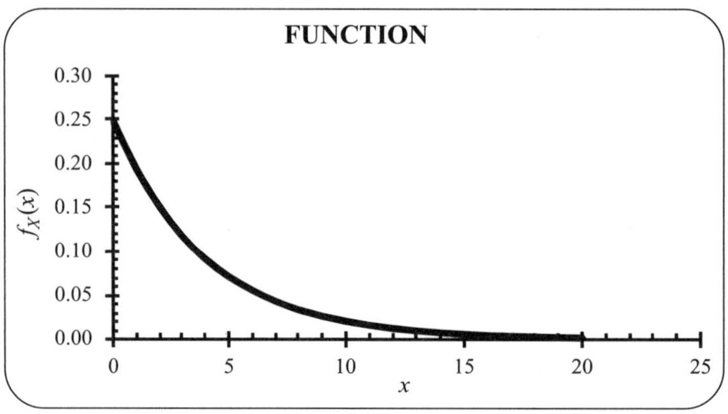


c. $E(X) \cong \overline{x} = \dfrac{1}{100}(1.43 + 0.35 + \cdots + 3.45) = 3.62$

d. $f_X(x) \cong \begin{cases} 0 & \text{if } x < 0 \\ \dfrac{1}{3.62} e^{-x/3.62} & \text{if } x \geq 0 \end{cases}$

e. $F_X(x) \cong \begin{cases} 0 & \text{if } x < 0 \\ 1 - e^{-x/3.62} & \text{if } x \geq 0 \end{cases}$

f. $$\begin{aligned} P(5 \leq X \leq 10) &= F_X(10) - F_X(5) \\ &= \left(1 - e^{-10/3.62}\right) - \left(1 - e^{-5/3.62}\right) \\ &\cong 0.9369 - 0.7487 \\ &= 0.1882 \end{aligned}$$

g. $$\begin{aligned} P(X \geq 9) &= 1 - P(X \leq 9) \\ &= 1 - \left(1 - e^{-9/3.62}\right) \\ &\cong 1 - 0.9168 \\ &= 0.0832 \end{aligned}$$

**Example 19:** Suppose a company wants to measure how accurately their machine cuts a groove in the center of a part. A random sample of parts is selected and a laser is used to measure how far the groove is off from the center. A perfect cut means that the distances from the center of the groove to each edge are equal. The figure below illustrates a perfect cut (the distances left and right of center are equal).

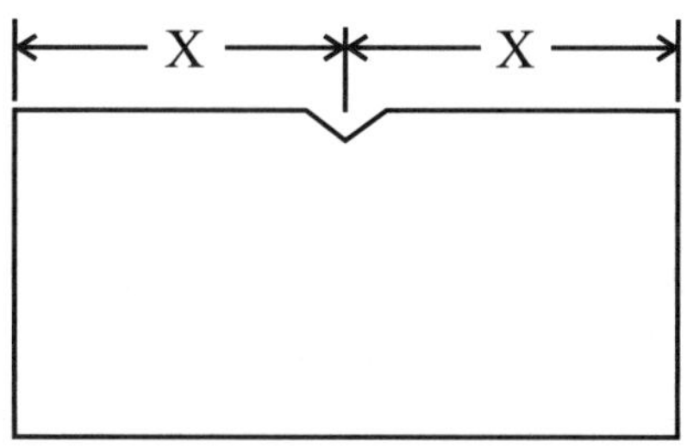


The laser-measuring device is very sophisticated — as it measures each part, it records the data. After the last observation is recorded, the device calculates a mean and standard deviation and prints out a histogram of the recorded data. The manager looks at the graphs and then looks up the specifications for the cutting tool. He sees that the machine is calibrated to have a mean accuracy of zero, $\mu_X = 0$.

Does the sample mean accurately estimate the theoretical mean? Can the type of distribution be identified from the histogram?

To answer these questions, we will look at samples of different sizes.

$n = 10 \quad \bar{x} = -0.232$

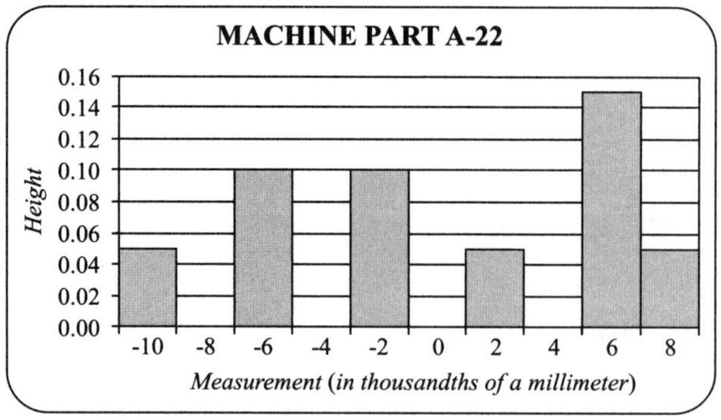


$n = 20 \quad \bar{x} = 0.984$

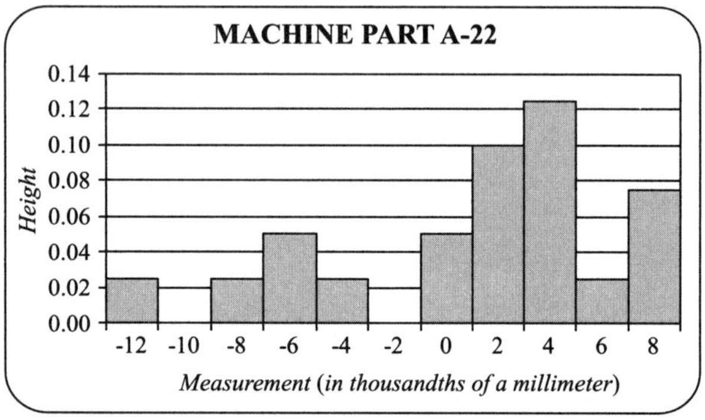


$n = 40 \quad \bar{x} = -0.477$

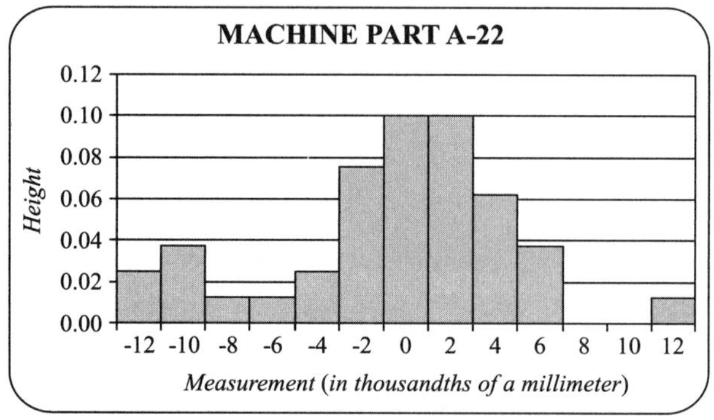


$n = 80 \quad \bar{x} = -0.515$

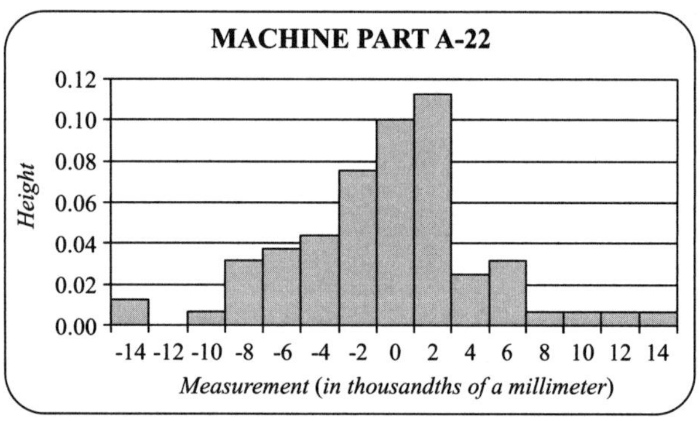


$n = 160 \quad \bar{x} = 0.051$

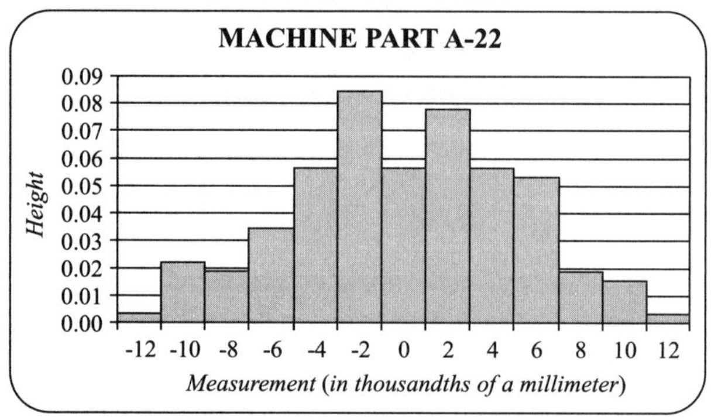


If we look at the sample statistics, we see that the sample mean closely approximates the theoretical mean ($\bar{x} \cong \mu_X$). Notice that as the sample size increases, the sample statistic more accurately estimates $\mu_X$. Also, as the sample size increases, it is possible to identify the distribution that best models the data. For this example it appears that the distribution is bell-shaped.

## *Exercises*

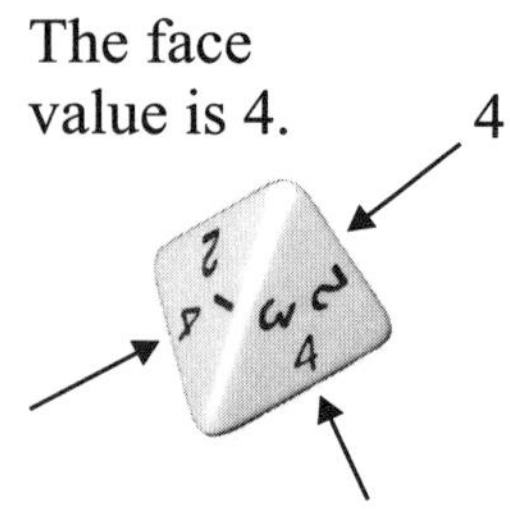


1. Two fair tetrahedral (4-sided) dice are rolled.
   a. Find the sample space.
   b. Let $E$ be the event that the sum of the numbers is even. List the outcomes in $E$.

c. Compute $P(E)$.

d. Let $F$ be the event that the sum of the numbers is even and less than 5. Compute $P(F)$.

e. Let $X$ be the finite random variable that gives the sum of the numbers on the faces of the dice. List the value of $X$ for each outcome.

f. Compute $P(X = 4)$.

g. Compute $P(X \leq 4)$.

h. Compute $P(X > 4)$.

**SOLUTION**

a. $S = \begin{Bmatrix} (1,1) & (1,2) & (1,3) & (1,4) \\ (2,1) & (2,2) & (2,3) & (2.4) \\ (3,1) & (3,2) & (3,3) & (3,4) \\ (4,1) & (4,2) & (4,3) & (4,4) \end{Bmatrix}$

b. $E = \begin{Bmatrix} (1,1) & (1,3) \\ (2,2) & (2,4) \\ (3,1) & (3,3) \\ (4,2) & (4,4) \end{Bmatrix}$

c. $P(E) = 0.5$

d. $P(F) = 0.25$

e.

| Outcome | Value of $X$ | Outcome | Value of $X$ |
|---|---|---|---|
| (1,1) | 2 | (3,1) | 4 |
| (1,2) | 3 | (3,2) | 5 |
| (1,3) | 4 | (3,3) | 6 |
| (1,4) | 5 | (3,4) | 7 |
| (2,1) | 3 | (4,1) | 5 |
| (2,2) | 4 | (4,2) | 6 |
| (2,3) | 5 | (4,3) | 7 |
| (2,4) | 6 | (4,4) | 8 |

f. $P(X = 4) = 0.1875$

g. $P(X \leq 4) = 0.375$

h. $P(X > 4) = 0.625$

2. The *p.m.f.* of $R$, the daily demand for a particular model of notebook computer, is given below.

| $r$ | 0 | 1 | 2 | 3 | 4 | 5 |
|---|---|---|---|---|---|---|
| $f_R(r)$ | 0.10 | 0.40 | 0.20 | 0.15 | 0.10 | 0.05 |

a. Find $P(R = 1)$.

b. Find $F_R(1)$.

c. Find the probability that the demand for this particular model of notebook computer is exactly three.

d. Find the probability that the demand for this particular model of notebook computer is at least three.

e. Find the probability that the demand for this particular model of notebook computer is at most three.

f. Find $\mu_R$.

**SOLUTION**

a. $P(R=1)=0.40$

b. $F_R(1)=0.50$

c. $P(R=3)=0.15$

d. $P(R\geq 3)=0.30$

e. $P(R\leq 3)=0.85$

f. $\mu_R=1.9$

3. Let $X$ be the profit (in thousands of dollars) on an investment in a bookstore. The *p.m.f.* of $X$ is given below.

| $x$ | –125 | –10 | 40 | 200 |
|---|---|---|---|---|
| $f_X(x)$ | 0.3 | 0.4 | 0.1 | 0.2 |

a. What is probability that the profit on an investment in a bookstore is at least \$20,000?
b. Find $E(X)$.
c. Give the practical interpretation of the answer found in Part b.

**SOLUTION**

a. $P(X\geq 20)=0.3$

b. $E(X)=2.5$

c. The average profit for a large number of similar investments will be approximately 2.5 thousand dollars.

4. The *c.d.f.* of a finite random variable $X$ is given below.

$$F_X(x)=\begin{cases}0 & \text{if } x<5\\ 0.2 & \text{if } 5\leq x<10\\ 0.6 & \text{if } 10\leq x<20\\ 0.9 & \text{if } 20\leq x<40\\ 1 & \text{if } x\geq 40\end{cases}$$

a. Find the values of the *p.m.f.* of $X$.

b. What value of $X$ is most likely?

c. What value of $X$ is least likely?
d. Find $P(X = 20)$.
e. Find $P(X \leq 10)$.
f. Find $E(X)$.

**SOLUTION**

a.

| $x$ | 5 | 10 | 20 | 40 |
|---|---|---|---|---|
| $f_X(x)$ | 0.2 | 0.4 | 0.3 | 0.1 |

b. 10
c. 40
d. $P(X = 20) = 0.3$
e. $P(X \leq 10) = 0.6$
f. $E(X) = 15$

5. A local computer store, DCS, was recently sold to an entrepreneur, Mr. Ron Flanders. Mr. Flanders would like to know what type of computer his customers own so that he can maintain an inventory that suits their needs. He is conducting a survey and randomly selects four customers per hour and asks if they own a PC or Mac. Past records indicate that 30% of DCS customers own a Mac. Let $X$ be the number of customers in a sample size of four who own a Mac. The *p.m.f.* of $X$ is given below.

| $x$ | $f_X(x)$ |
|---|---|
| 0 | 0.2401 |
| 1 | 0.4116 |
| 2 | 0.2646 |
| 3 | 0.0756 |
| 4 | 0.0081 |

a. Find the probability that three or more of the surveyed customers own a Mac.
b. Find the probability that none of the surveyed customers own a Mac.
c. Find the probability that only one of the surveyed customers owns a PC.
d. If the computer store expects to survey 10 groups of four customers each day, on average how many of the customers surveyed in one day are expected to own a Mac?

**SOLUTION**

a. $P(X \geq 3) = 0.0837$
b. $P(X = 0) = 0.2401$
c. $P(X = 3) = 0.0756$

d. On average, 12 of the customers surveyed in one day are expected to own a Mac.

6. An oil exploration company finds oil in approximately 5% of the test wells that it drills. Suppose that the company has plans to drill four new test wells. Then $Y$, the number of those wells that yield oil, is a binomial random variable with parameters $n=4$ and $p=0.05$.

   a. Use the ***BINOMDIST*** function in *Excel* to fill in the table.
   b. Compute $P(Y=3)$.
   c. Compute $P(Y \leq 3)$.
   d. Compute $P(Y > 3)$.
   e. Compute $\mu_Y$.

| $y$ | $f_Y(y)$ | $F_Y(y)$ |
|---|---|---|
| | | |
| | | |
| | | |
| | | |
| | | |

**SOLUTION**

a. See table to the right.

b. $P(Y=3) \cong 0.00048$

c. $P(Y \leq 3) \cong 0.99999$

d. $P(Y > 3) \cong 0.00001$

e. $\mu_Y = 0.2$

| $y$ | $f_Y(y)$ | $F_Y(y)$ |
|---|---|---|
| 0 | 0.81451 | 0.81451 |
| 1 | 0.17148 | 0.98598 |
| 2 | 0.01354 | 0.99952 |
| 3 | 0.00048 | 0.99999 |
| 4 | 0.00001 | 1.00000 |

7. Students go to a hotdog stand in groups of three. Each student chooses either a hotdog with chili or a hotdog without chili. The probability that a student chooses a chilidog is 0.75. Let $X$ be the number of students in a group of three who buy a chilidog.

   a. Fill in the table to the right.
   b. Find the probability that two or more chilidogs are sold to a group of three students.
   c. Find the probability that exactly two students don't buy chilidogs.
   d. In words, describe the meaning of $F_X(1)$ and find its value.
   e. In words, describe the meaning of $f_X(3)$ and find its value.
   f. If the hotdog stand expects 50 groups of three for lunch, how many chilidogs can they expect to sell during an average lunch period?

| $x$ | $f_X(x)$ | $F_X(x)$ |
|---|---|---|
| | 0.0156 | |
| | 0.1406 | |
| | | |
| | 0.4219 | |

**SOLUTION**

a.

| $x$ | $f_X(x)$ | $F_X(x)$ |
|---|---|---|
| 0 | 0.0156 | 0.0156 |
| 1 | 0.1406 | 0.1562 |
| 2 | 0.4219 | 0.5781 |
| 3 | 0.4219 | 1.0000 |

b. $P(X \geq 2) = 0.8438$

c. $P(X = 1) = 0.1406$

d. The probability that at most one student in a group of three buys a chilidog; $F_X(1) = 0.1562$.

e. The probability that all of the students in a group of three buy chilidogs; $f_X(3) = 0.4219$.

f. They can expect to sell 112.5 chilidogs during an average lunch period.

8. A recent Gallup poll showed that 76% of Americans have at least one credit card. Let $T$ be the number of Americans in a sample of size 20 who have at least one credit card.

a. Use the ***BINOMDIST*** function in *Excel* to compute $P(T \leq 16)$.

b. Compute $\mu_T$.

**SOLUTION**

a. $P(T \leq 16) \cong 0.7431$

b. $\mu_T = 15.2$

9. A recent survey showed that 35% of junior college students own a vehicle. Let $Y$ be the event that a junior college student owns a vehicle and let $N$ be the event that a junior college student does not own a vehicle. Let $V$ be the number of junior college students in a sample size of three who own a vehicle. Fill in the table to the right *without* using the ***BINOMDIST*** function in *Excel.*

| $v$ | $f_V(v)$ | $F_V(v)$ |
|---|---|---|
| | | |
| | | |
| | | |
| | | |

**SOLUTION**

| $v$ | $f_V(v)$ | $F_V(v)$ |
|---|---|---|
| 0 | 0.2746 | 0.2746 |
| 1 | 0.4437 | 0.7183 |
| 2 | 0.2388 | 0.9571 |
| 3 | 0.0429 | 1.0000 |

10. A recent study showed that one in four college students is age 30 or older. Let $W$ be the random variable that gives the number of college students in a sample of size ten who are age 30 or older.

a. Use the ***BINOMDIST*** function in *Excel* to find the values of the *p.m.f.* and *c.d.f.* of $W$.

b. Use the *p.m.f.* to find the probability that exactly six students in a sample of size ten are age 30 or older.

c. Use the *c.d.f.* to find the probability that at most six students in a sample of size ten are age 30 or older.

d. Use the *p.m.f.* to find the probability that at least seven students in a sample of size ten are age 30 or older.

e. Use the *c.d.f.* to answer the question in Part d.

f. Use *Excel* to find $\mu_W$.

g. Use the special formula for the expected value of a binomial random variable to find $\mu_W$.

**SOLUTION**

a.

| $w$ | $f_W(w)$ | $F_W(w)$ |
|---|---|---|
| 0 | 0.056314 | 0.056314 |
| 1 | 0.187712 | 0.244025 |
| 2 | 0.281568 | 0.525593 |
| 3 | 0.250282 | 0.775875 |
| 4 | 0.145998 | 0.921873 |
| 5 | 0.058399 | 0.980272 |
| 6 | 0.016222 | 0.996494 |
| 7 | 0.003090 | 0.999584 |
| 8 | 0.000386 | 0.999970 |
| 9 | 0.000029 | 0.999999 |
| 10 | 0.000001 | 1.000000 |

b. $P(W = 6) = 0.016222$

c. $P(W \leq 6) = 0.996494$

d. $P(W \geq 7) = 0.003506$

e. $P(W \geq 7) = 0.003506$

f. $\mu_W = 2.5$

g. $\mu_W = np = 2.5$

11. The *p.d.f.* and *c.d.f.* of a continuous random variable $X$ are plotted below.

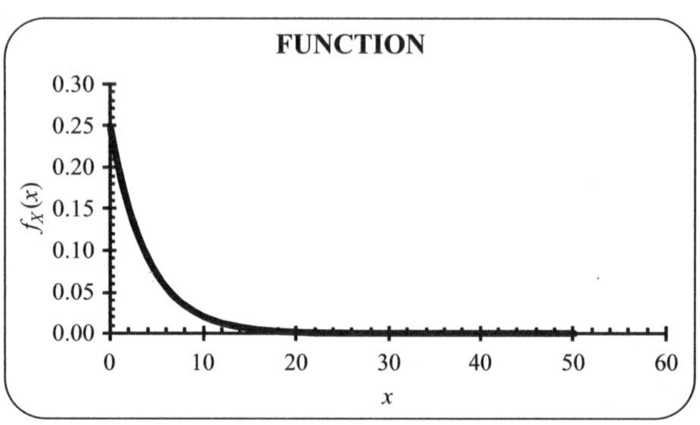


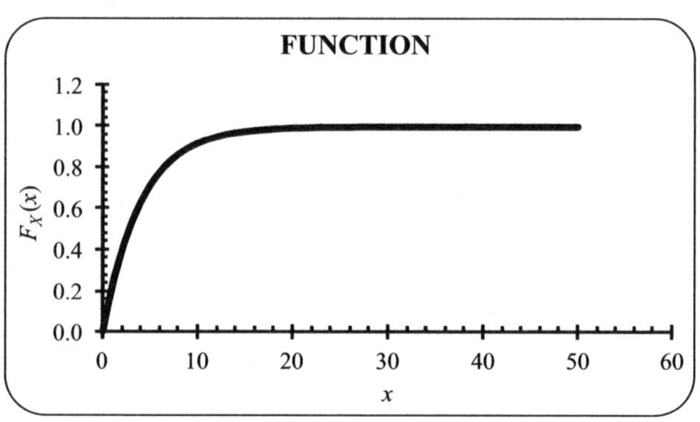

a. Shade the region of the graph of the *p.d.f.* that corresponds to $P(X \leq 10)$.
b. Use the graph of the *c.d.f.* to estimate $P(X \leq 10)$.

## SOLUTION

a.

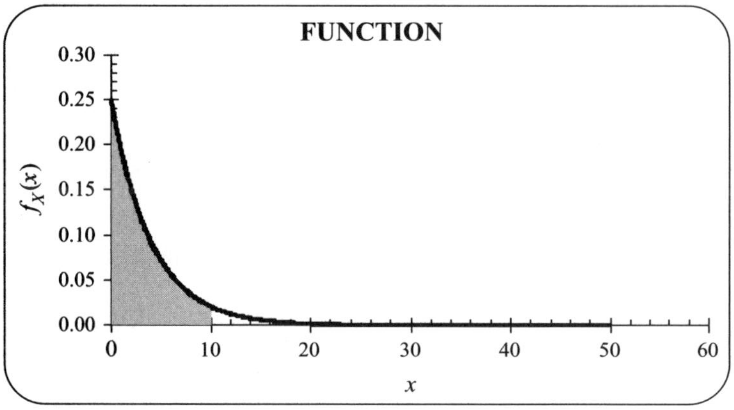


b. $P(X \leq 10) \cong 0.92$

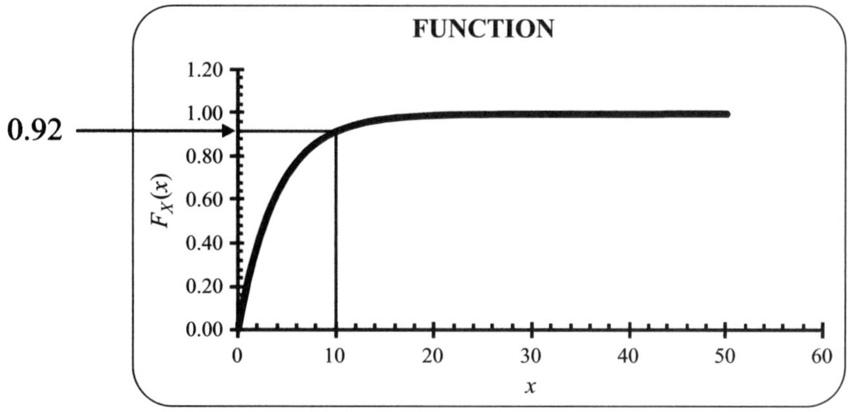


12. The formulas for and graphs of the *p.d.f.* and *c.d.f.* of a continuous random variable $X$ are given below.

$$f_X(x) = \begin{cases} 0 & \text{if } x < 0 \\ 2x & \text{if } 0 \leq x \leq 1 \\ 0 & \text{if } x > 1 \end{cases} \qquad F_X(x) = \begin{cases} 0 & \text{if } x < 0 \\ x^2 & \text{if } 0 \leq x < 1 \\ 1 & \text{if } x \geq 1 \end{cases}$$

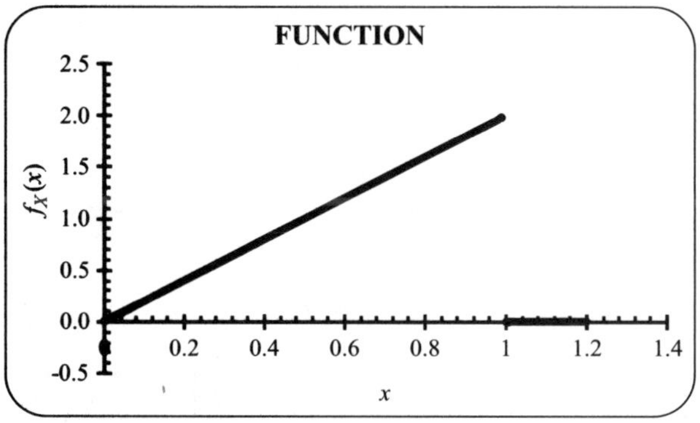


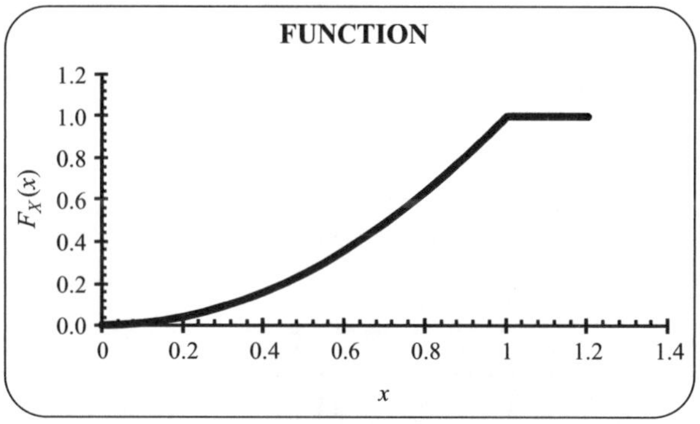

a. Use the graph of the *p.d.f.* to compute $P(0.4 \leq X \leq 0.8)$.

b. Use ***Integrating.xls*** to compute $P(0.4 \leq X \leq 0.8)$.

c. Evaluate the closed-form integral that gives $P(0.4 \leq X \leq 0.8)$.

d. Use the formula for the *c.d.f.* to compute $P(0.4 \leq X \leq 0.8)$.

e. Use ***Integrating.xls*** to compute $E(X)$.

f. Evaluate the closed-form integral that gives $E(X)$.

## SOLUTION

a. $P(0.4 \leq X \leq 0.8) = 0.48$

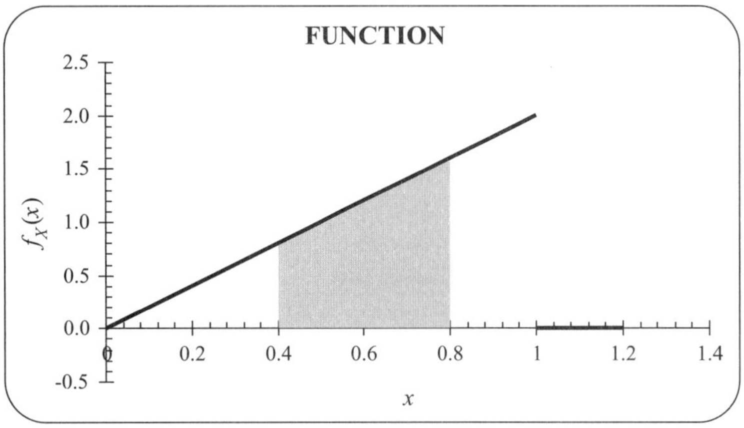


b. $P(0.4 \leq X \leq 0.8) = \int_{0.4}^{0.8} 2x\, dx = 0.48$

| Definition | Computation | | Plot Interval | | Integration Interval | | $\int_a^b f(x)\,dx$ |
|---|---|---|---|---|---|---|---|
| Formula for $f(x)$ | $x$ | $f(x)$ | $A$ | $B$ | $a$ | $b$ | |
| 0 | | 0 | 0 | 1 | 0.4 | 0.8 | 0.4800 |

c. $P(0.4 \leq X \leq 0.8) = \int_{0.4}^{0.8} 2x\, dx = 0.48$

d. $P(0.4 \leq X \leq 0.8) = F_X(0.8) - F_X(0.4) = 0.48$

e. $E(X) = \int_{0}^{1} 2x^2\, dx = 0.6667$

| Definition | Computation | | Plot Interval | | Integration Interval | | $\int_a^b f(x)\,dx$ |
|---|---|---|---|---|---|---|---|
| Formula for $f(x)$ | $x$ | $f(x)$ | $A$ | $B$ | $a$ | $b$ | |
| 0 | | 0 | 0 | 1 | 0 | 1 | 0.6667 |

f. $E(X) = \int_0^1 2x^2 \, dx = 0.6667$

13. Suppose that $X$ has a uniform distribution on the interval [0, 20].

a. Plot the *p.d.f.* of $X$ on the interval [0, 20].

b. Plot the *c.d.f.* of $X$ on the interval [0, 20].

c. Find $P(X = 10)$.

d. Find $P(X \le 10)$.

e. Find $F_X(5)$.

f. Use the graph of the *p.d.f.* to find $P(10 \le X \le 15)$.

g. Use ***Integrating.xls*** to find $P(10 \le X \le 15)$.

h. Evaluate the closed-form integral that gives $P(10 \le X \le 15)$.

i. Use the graph of the *c.d.f.* to estimate $P(10 \le X \le 15)$.

j. Use the special formula for the expected value of a random variable that has a uniform distribution on the interval $[a, b]$ to find $E(X)$.

k. Use ***Integrating.xls*** to find $E(X)$.

l. Evaluate the closed-form integral that gives $E(X)$.

## SOLUTION

a.

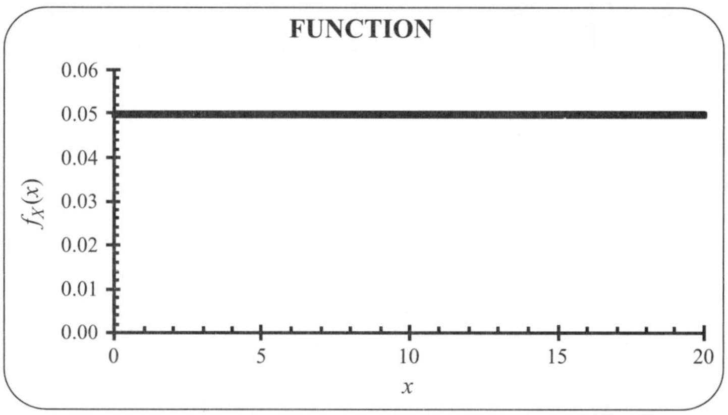


b.

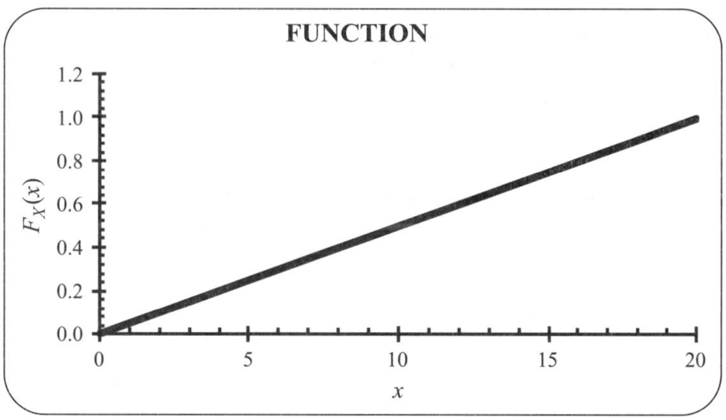

c. $P(X = 10) = 0$

d. $P(X \leq 10) = 0.50$

e. $F_X(5) = 0.25$

f. $P(10 \leq X \leq 15) = 0.25$

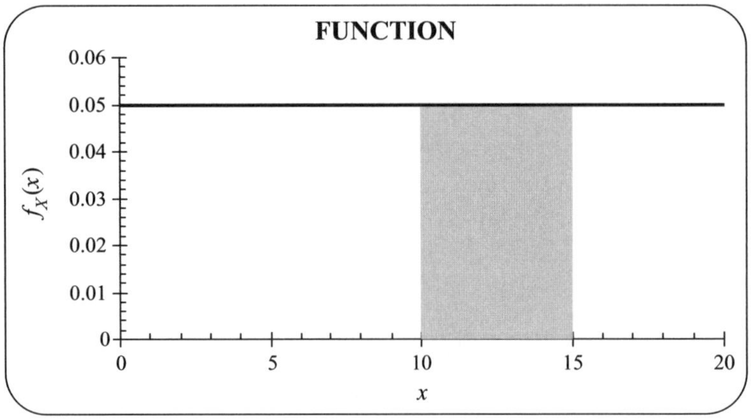


g. $P(10 \leq X \leq 15) = \int_{10}^{15} \frac{1}{20}\, dx = 0.25$

| Definition | Computation | | Plot Interval | | Integration Interval | | $\int_a^b f(x)\,dx$ |
|---|---|---|---|---|---|---|---|
| **Formula for $f(x)$** | $x$ | $f(x)$ | $A$ | $B$ | $a$ | $b$ | |
| 0.05 | | 0.05 | 0 | 20 | 10 | 15 | 0.2500 |

h. $P(10 \leq X \leq 15) = \int_{10}^{15} \frac{1}{20}\, dx = 0.25$

i. $P(10 \leq X \leq 15) = 0.25$

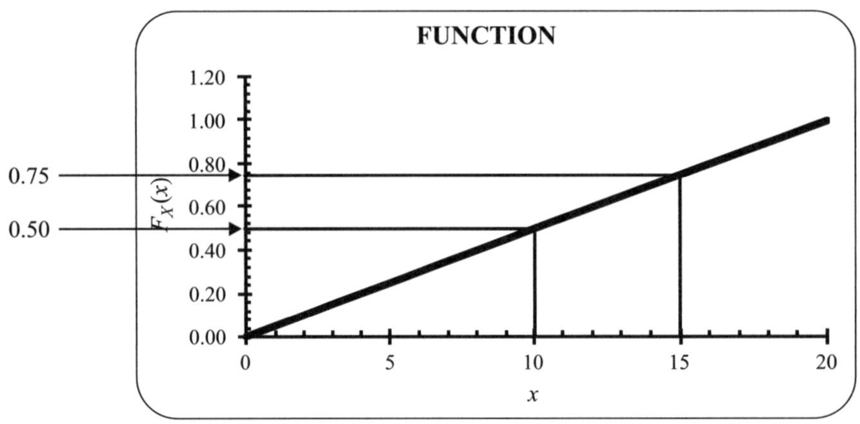


j. $E(X) = \dfrac{a+b}{2} = 10$

k. $E(X)=\int_0^{20} x\cdot\frac{1}{20}\,dx=10$

| Definition | Computation | | Plot Interval | | Integration Interval | | $\int_a^b f(x)\,dx$ |
|---|---|---|---|---|---|---|---|
| Formula for $f(x)$ | $x$ | $f(x)$ | $A$ | $B$ | $a$ | $b$ | |
| 0 | | 0 | 0 | 20 | 0 | 20 | 10.0000 |

l. $E(X)=\int_0^{20} x\cdot\frac{1}{20}\,dx=10$

14. Let $T$ be the length of time (in minutes) required for an assignment. Past experience shows that $T$ has a uniform distribution on the interval [60, 90].
   a. Find a formula for the *p.d.f.* of $T$.
   b. Find a formula for the *c.d.f.* of $T$.
   c. Use ***Integrating.xls*** to compute the probability that the time required for the assignment is between 60 and 75 minutes.
   d. Use the formula for the *c.d.f.* to compute the probability that the time required for the assignment is between 60 and 75 minutes.
   e. Compute the expected length of time required for the assignment.

**SOLUTION**

a. $$f_T(t)=\begin{cases}0 & \text{if } t<60\\ \frac{1}{30} & \text{if } 60\le t\le 90\\ 0 & \text{if } t>90\end{cases}$$

b. $$F_T(t)=\begin{cases}0 & \text{if } t<60\\ \frac{t-60}{30} & \text{if } 60\le t<90\\ 1 & \text{if } t\ge 90\end{cases}$$

c. $P(60\le T\le 75)=\int_{60}^{75}\frac{1}{30}\,dt=0.5$

| Definition | Computation | | Plot Interval | | Integration Interval | | $\int_a^b f(x)\,dx$ |
|---|---|---|---|---|---|---|---|
| Formula for $f(x)$ | $x$ | $f(x)$ | $A$ | $B$ | $a$ | $b$ | |
| 0 | | 0 | 0 | 100 | 60 | 75 | 0.5000 |

d. $P(60\le T\le 75)=0.5$

e. The expected length of time required for the assignment is 75 minutes.

15. Suppose that $X$ has a uniform distribution on the interval [4, 10].
    a. Plot the *p.d.f.* of $X$ on the interval [0, 15].
    b. Find a formula for the *p.d.f.* of $X$.
    c. Use the graph of the *p.d.f.* to find $P(6 \le X \le 8)$.
    d. Use ***Integrating.xls*** to find $P(6 \le X \le 8)$.
    e. Evaluate the closed-form integral that gives $P(6 \le X \le 8)$.
    f. Plot the *c.d.f.* of $X$ on the interval [0, 15].
    g. Find a formula for the *c.d.f.* of $X$.
    h. Use the formula for the *c.d.f.* to find $P(X \le 6)$.
    i. Find $P(X = 6)$.
    j. Use the special formula for the expected value of a random variable that has a uniform distribution on the interval $[a, b]$ to find $E(X)$.
    k. Use ***Integrating.xls*** to find $E(X)$.
    l. Evaluate the closed-form integral that gives $E(X)$.

**SOLUTION**

a.

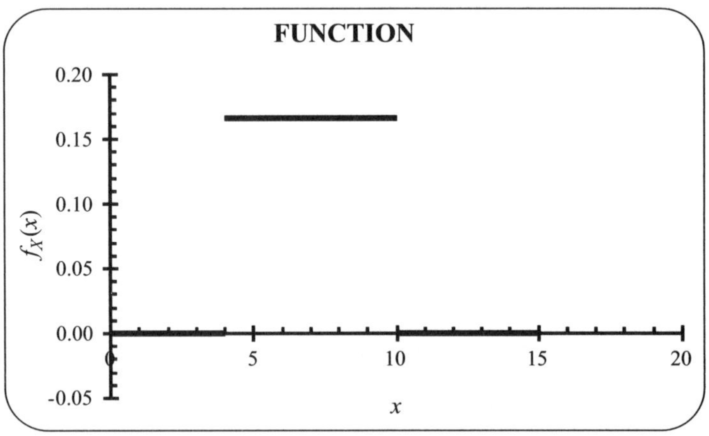


b. $$f_X(x) = \begin{cases} 0 & \text{if } x < 4 \\ \dfrac{1}{6} & \text{if } 4 \le x \le 10 \\ 0 & \text{if } x > 10 \end{cases}$$

c. $P(6 \le X \le 8) = 1/3$

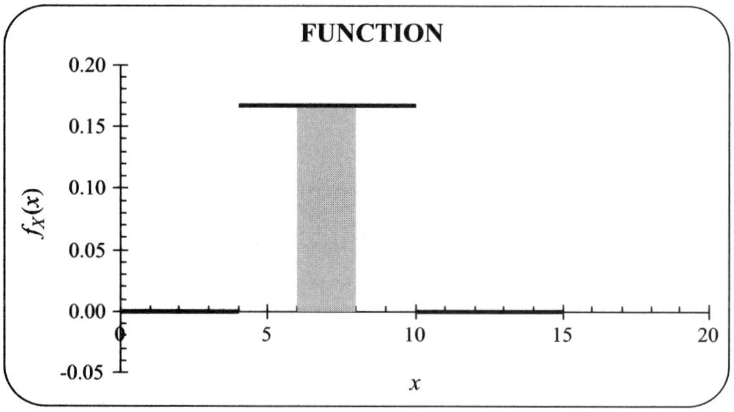


d. $P(6 \le X \le 8) = \int_6^8 \frac{1}{6}\, dx = 0.3333$

| Definition | Computation | | Plot Interval | | Integration Interval | | |
|---|---|---|---|---|---|---|---|
| Formula for $f(x)$ | $x$ | $f(x)$ | $A$ | $B$ | $a$ | $b$ | $\int_a^b f(x)\,dx$ |
| 0 | | 0 | 0 | 15 | 6 | 8 | 0.3333 |

e. $P(6 \le X \le 8) = \int_6^8 \frac{1}{6}\, dx = 0.3333$

f.

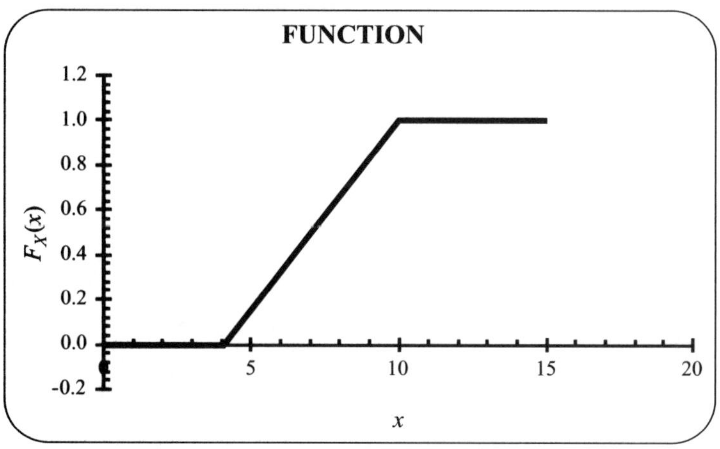


g. $F_X(x) = \begin{cases} 0 & \text{if } x < 4 \\ \dfrac{x-4}{6} & \text{if } 4 \le x < 10 \\ 1 & \text{if } x \ge 10 \end{cases}$

h. $P(X \le 6) = F_X(6) = 1/3$

i. $P(X = 6) = 0$

j. $E(X)=\dfrac{a+b}{2}=7$

k. $E(X)=\displaystyle\int_4^{10} x\cdot\frac{1}{6}\,dx=7$

| Definition | Computation | | Plot Interval | | Integration Interval | | $\int_a^b f(x)\,dx$ |
|---|---|---|---|---|---|---|---|
| Formula for $f(x)$ | $x$ | $f(x)$ | $A$ | $B$ | $a$ | $b$ | |
| 0 | | 0 | 0 | 15 | 4 | 10 | 7.0000 |

l. $E(X)=\displaystyle\int_4^{10} x\cdot\frac{1}{6}\,dx=7$

16. Let $Y$ be an exponential random variable with parameter $\alpha=20$.

a. Find a formula for the *p.d.f.* of $Y$.

b. Find a formula for the *c.d.f.* of $Y$.

c. Use ***Integrating.xls*** to find $P(0\le Y\le 20)$.

d. Evaluate the closed-form integral that gives $P(0\le Y\le 20)$.

e. Use the formula for the *c.d.f.* to find $P(0\le Y\le 20)$.

f. Use the special formula for the expected value of an exponential random variable to find $E(Y)$.

g. Use ***Integrating.xls*** to find $E(Y)$.

**SOLUTION**

a. $f_Y(y)=\begin{cases}0 & \text{if } y<0\\ \dfrac{1}{20}e^{-y/20} & \text{if } y\ge 0\end{cases}$

b. $F_Y(y)=\begin{cases}0 & \text{if } y<0\\ 1-e^{-y/20} & \text{if } y\ge 0\end{cases}$

c. $P(0\le X\le 20)=\displaystyle\int_0^{20}\frac{1}{20}e^{-x/20}dx=0.6321$

| Definition | Computation | | Plot Interval | | Integration Interval | | $\int_a^b f(x)\,dx$ |
|---|---|---|---|---|---|---|---|
| Formula for $f(x)$ | $x$ | $f(x)$ | $A$ | $B$ | $a$ | $b$ | |
| 0.05 | | 0.05 | 0 | 20 | 0 | 20 | 0.6321 |

d. $P(0\le X\le 12)=\displaystyle\int_0^{12}\frac{1}{20}e^{-x/20}dx=0.6321$

e. $P(0 \le Y \le 20) = F_Y(20) - F_Y(0) = 0.6321$

f. $E(Y) = \alpha = 20$

g. $E(Y) = \int_0^\infty y \cdot \frac{1}{20} e^{-y/20} dy = 20$

| Definition | Computation | | Plot Interval | | Integration Interval | | $\int_a^b f(x)\,dx$ |
|---|---|---|---|---|---|---|---|
| Formula for $f(x)$ | $x$ | $f(x)$ | $A$ | $B$ | $a$ | $b$ | |
| 0 | | 0 | 0 | 400 | 0 | 400 | 20.0000 |

17. On average, three customers per hour use the ATM in a local grocery store. Let $T$ be the time (in minutes) between the arrivals of consecutive customers.
   a. Find a formula for the *p.d.f.* of $T$.
   b. Use ***Integrating.xls*** to find the probability that the time between the arrivals of consecutive customers is at most 15 minutes.
   c. Evaluate the closed-form integral that gives the probability that the time between the arrivals of consecutive customers is at most 15 minutes.
   d. Find a formula for the *c.d.f.* of $T$.
   e. Use the formula for the *c.d.f.* to find the probability that the time between the arrivals of consecutive customers is between 30 and 45 minutes.
   f. Use the special formula for the expected value of an exponential random variable to find $E(T)$.
   g. Use ***Integrating.xls*** to find $E(T)$.

**SOLUTION**

a. $f_T(t) = \begin{cases} 0 & \text{if } t < 0 \\ \frac{1}{20} e^{-t/20} & \text{if } t \ge 0 \end{cases}$

b. $P(T \le 15) = \int_0^{15} \frac{1}{20} e^{-t/20} dt = 0.5276$

| Definition | Computation | | Plot Interval | | Integration Interval | | $\int_a^b f(x)\,dx$ |
|---|---|---|---|---|---|---|---|
| Formula for $f(x)$ | $x$ | $f(x)$ | $A$ | $B$ | $a$ | $b$ | |
| 0.05 | | 0.05 | 0 | 200 | 0 | 15 | 0.5276 |

c. $P(T \le 15) = \int_0^{15} \frac{1}{20} e^{-t/20} dt = 0.5276$

d. $F_T(t) = \begin{cases} 0 & \text{if } t < 0 \\ 1 - e^{-t/20} & \text{if } t \ge 0 \end{cases}$

e. $P(30 \le T \le 45) = F_T(45) - F_T(30) = 0.1177$

f. $E(T) = \alpha = 20$

g. $E(T) = \int_0^{\infty} t \cdot \frac{1}{20} e^{-t/20} dt = 20$

| Definition | Computation | | Plot Interval | | Integration Interval | | |
|---|---|---|---|---|---|---|---|
| Formula for $f(x)$ | $x$ | $f(x)$ | $A$ | $B$ | $a$ | $b$ | $\int_a^b f(x)\,dx$ |
| 0 | | 0 | 0 | 400 | 0 | 400 | 20.0000 |

18. On average, 15 patients per hour seek treatment at the emergency room of a hospital in a large metropolitan area. Let $T$ be the time (in minutes) between patient arrivals. The *p.d.f.* of $T$ is given below.

$$f_T(t) = \begin{cases} 0 & \text{if } t < 0 \\ 0.25e^{-0.25t} & \text{if } t \ge 0 \end{cases}$$

a. Find the probability that the time between patient arrivals is exactly five minutes.

b. Find a formula for the *c.d.f.* of $T$.

c. Use the formula for the *c.d.f.* to find the probability that the time between patient arrivals is at least 10 minutes.

d. Find the expected time (in minutes) between patient arrivals.

**SOLUTION**

a. $P(T = 5) = 0$

b. $F_T(t) = \begin{cases} 0 & \text{if } t < 0 \\ 1 - e^{-0.25t} & \text{if } t \ge 0 \end{cases}$

c. $P(T \ge 10) = 1 - F_T(10) \cong 0.0821$

d. The expected time between patient arrivals is 4 minutes.

19. Is

$$f_X(x) = \begin{cases} 0 & \text{if } x < 0 \\ -18x + 6 & \text{if } 0 \le x \le 1/3 \\ 0 & \text{if } x > 1/3 \end{cases}$$

a *p.d.f.*? Explain your answer. If $f_X(x)$ is a *p.d.f.*, compute $E(X)$.

**SOLUTION**

Yes, $f_X(x)$ is a *p.d.f.* because $f_X(x) \ge 0$ for all $x$ and $\int_{\text{all } x} f_X(x)\, dx = 1$. $E(X) = 0.1111$

20. Is

$$f_Z(z) = \frac{1}{\sqrt{2\pi}} e^{-0.5z^2}$$

a *p.d.f.*? Explain your answer. If $f_Z(z)$ is a *p.d.f.*, compute $E(Z)$.

**SOLUTION**

Yes, $f_Z(z)$ is a *p.d.f.* because $f_Z(z) \geq 0$ for all $z$ and $\int_{\text{all } z} f_Z(z)\, dz = 1$. $E(Z) = 0$

21. Is

$$f_X(x) = \begin{cases} 0 & \text{if } x < 0 \\ e^{-x/2} & \text{if } x \geq 0 \end{cases}$$

a *p.d.f.*? Explain your answer. If $f_X(x)$ is a *p.d.f.*, compute $E(X)$.

**SOLUTION**

No, $f_X(x)$ is not a *p.d.f.* because $\int_{\text{all } x} f_X(x)\, dx > 1$.

22. True or False? Suppose that we are given the *c.d.f.*, $F_R(r)$, for a continuous random variable $R$. We can find $P(20 \leq R \leq 80)$ by computing $\int_{20}^{80} F_R(r)\, dr$. Explain your answer.

**SOLUTION**

No! You *never* integrate the *c.d.f.* to find the probabilities. You integrate the *p.d.f.* to find probabilities.

23. Suppose that $X$ is a random variable that has a uniform distribution on the interval [0, 10]. Which is correct: $F_X(3) = 0$ or $f_X(3) = 0$?

**SOLUTION**

$f_X(3) = 0$

24. The *p.m.f.* of a finite random variable $R$ is given below.

| $r$ | 0 | 5 | 10 | 15 |
|---|---|---|---|---|
| $f_R(r)$ | 0.4 | 0.3 | 0.2 | 0.1 |

Forty observations of $R$ are given in the table below.

| | | | | | | | |
|---|---|---|---|---|---|---|---|
| 10 | 0 | 15 | 0 | 5 | 5 | 10 | 0 |
| 5 | 5 | 0 | 0 | 10 | 0 | 0 | 0 |
| 5 | 5 | 10 | 5 | 10 | 5 | 0 | 10 |
| 5 | 0 | 5 | 0 | 5 | 0 | 10 | 5 |
| 5 | 10 | 5 | 15 | 10 | 0 | 0 | 15 |

a. Find $\mu_R$.

b. Find $\overline{r}$.

**SOLUTION**

a. $\mu_R = 5$

b. $\overline{r} = 5.125$

25. The *p.m.f.* of a finite random variable *X* is given below.

| $x$ | 2 | 4 | 6 | 8 | 10 |
|---|---|---|---|---|---|
| $f_X(x)$ | 0.2 | 0.1 | 0.3 | 0.35 | 0.05 |

Nine observations of *X* are: 6, 2, 6, 8, 8, 10, 2, 6, and 8.

a. Find $\mu_X$.

b. Find $\overline{x}$.

**SOLUTION**

a. $\mu_X = 5.9$

b. $\overline{x} = 6.22$

26. Five observations of *V*, the volume (in millions of shares) of the New York Stock Exchange, are: 731, 542, 610, 689, and 558.

a. Use this information to estimate $P(V \geq 675)$.

b. Use this information to estimate $E(V)$.

**SOLUTION**

a. $P(V \geq 675) \cong 0.4$

b. $E(V) \cong \overline{v} = 626$

27. Twelve students were asked how much money they spend on textbooks per semester. The results were: \$200, \$175, \$450, \$300, \$350, \$250, \$150, \$200, \$320, \$370, \$400, and \$250.

a. Use this information to estimate the probability that a randomly selected student spends less than \$275 in a given semester.

b. Use this information to estimate the probability that a randomly selected student spends more than \$325 in a given semester.

c. Use this information to estimate the average amount of money spent on textbooks per semester.

**SOLUTION**

a. 0.50

b. 1/3

c. \$284.58

**28.** Forty observations of a continuous random variable $X$ are shown in the table below.

| | | | | | | | |
|---|---|---|---|---|---|---|---|
| 6.07 | 4.87 | 4.27 | 3.18 | 7.60 | 4.40 | 6.17 | 2.27 |
| 7.02 | 6.69 | 7.71 | 2.15 | 6.70 | 7.25 | 2.58 | 6.03 |
| 5.03 | 7.86 | 3.40 | 3.23 | 3.64 | 5.88 | 2.61 | 6.10 |
| 5.46 | 3.52 | 5.00 | 4.23 | 2.76 | 6.85 | 5.18 | 5.81 |
| 3.92 | 7.96 | 2.48 | 2.64 | 7.07 | 5.77 | 4.74 | 4.66 |

a. Construct a histogram that would approximate the *p.d.f.* of $X$.

b. This histogram would approximate what type of continuous random variable?

c. Find a formula for the assumed *p.d.f.* of $X$.

d. Find a formula for the assumed *c.d.f.* of $X$.

e. Use the formula for the assumed *c.d.f.* to compute an estimate of $P(3 \leq X \leq 7)$.

f. Use the formula for the assumed *c.d.f.* to compute an estimate of $P(X \geq 3)$.

g. Use the sample data to compute an estimate of the expected value, $E(X)$.

**SOLUTION**

a.

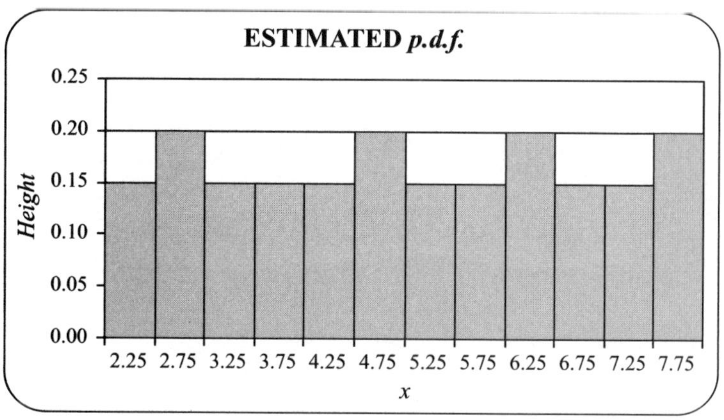


b. From this histogram, it appears that $X$ has a uniform distribution on the interval [2, 8].

c. $f_X(x) = \begin{cases} 0 & \text{if } x < 2 \\ \frac{1}{6} & \text{if } 2 \le x \le 8 \\ 0 & \text{if } x > 8 \end{cases}$

d. $F_X(x) = \begin{cases} 0 & \text{if } x < 2 \\ \frac{x-2}{6} & \text{if } 2 \le x < 8 \\ 1 & \text{if } x \ge 8 \end{cases}$

e. $P(3 \le X \le 7) = F_X(7) - F_X(3) \cong 0.667$

f. $P(X \ge 3) = 1 - F_X(3) \cong 0.833$

g. $E(X) \cong \overline{x} \cong 5.17$

# Calculus

In **Differentiation** and **Integration:** *Area Under the Curve*, two mathematical concepts were introduced — the rate of change in the value of a function and the area under the graph of a function. In *Area Under the Curve*, we said that there is a relationship between these two concepts; in fact, the derivative and the integral are, in a manner of speaking, inverses of one another. If we first integrate a given function over some interval $[a, x]$ and then differentiate the result with respect to $x$, we obtain the function that was initially given. Similarly, if we first differentiate a given function with respect to $x$ and then integrate over some interval $[a, x]$, we obtain the function that was initially given. This relationship is formally stated in ***The Fundamental Theorem of Calculus***, one of the most important theorems in the history of mathematics.

## *Probability*

**Example 1:** Let $X$ be an exponential random variable with parameter $\alpha = 12$. Use ***Differentiating.xls*** to show that the graph of the derivative of the *c.d.f.* of $X$ on the interval [0, 80] looks like the graph of the *p.d.f.* of $X$ on the same interval.

**SOLUTION**

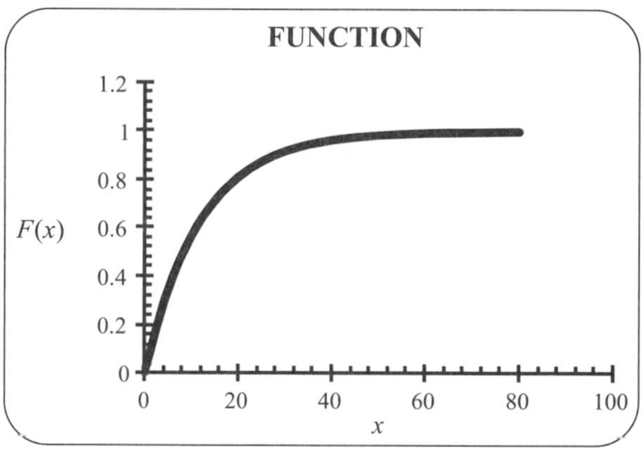


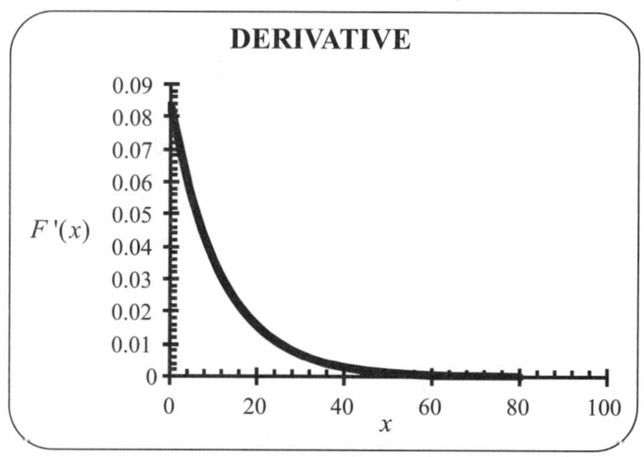


**Example 2:** Let $X$ be a random variable that has a uniform distribution on the interval [10, 50].

a. Find a formula for $F'_X(x)$, the derivative of the *c.d.f.* of $X$.

b. Find a formula for $f_X(x)$, the *p.d.f.* of $X$.

c. Compare the results from Parts a and b.

**SOLUTION**

a. The *c.d.f.* of $X$ is given by

$$F_X(x) = \begin{cases} 0 & \text{if } x < 10 \\ \dfrac{x-10}{40} & \text{if } 10 \le x < 50 \\ 1 & \text{if } x \ge 50. \end{cases}$$

Therefore,

$$F_X'(x) = \begin{cases} 0 & \text{if } x < 10 \\ \text{undefined} & \text{if } x = 10 \\ \dfrac{1}{40} & \text{if } 10 < x < 50 \\ \text{undefined} & \text{if } x = 50 \\ 0 & \text{if } x > 50. \end{cases}$$

b. The *p.d.f.* of $X$ is given by

$$f_X(x) = \begin{cases} 0 & \text{if } x < 10 \\ \dfrac{1}{40} & \text{if } 10 \le x < 50 \\ 0 & \text{if } x \ge 50. \end{cases}$$

c. For the values of $x$ at which $F_X'(x)$ is defined, the results from Parts a and b are the same.

## *Income Streams*

**Example 3:** Suppose that \$200,000 is invested at an annual rate of 1.5% compounded continuously. Then the value of the investment after $t$ years is given by $F(t) = 200,000e^{0.015t}$ and the annual income stream from this investment is given by $A(t) = 3,000e^{0.015t}$. Find $A(8)$ and show that the result is equal to 1.5% of the value of the investment after 8 years.

**SOLUTION**

$$A(8) = 3,000e^{0.015(8)} = \$3,382.49$$

$$F(8) = 200,000e^{0.015(8)} = \$225,499$$

$$\$225,499(0.015) = \$3,382.49 = A(18)$$

**Example 4:** Suppose that \$85,000 is invested at an annual rate of 5%, compounded continuously. Then the value of the investment after $t$ years is given by $F(t) = 85,000e^{0.05t}$, the total income earned during the first $t$ years is given by

$$G(t) = F(t) - 85,000 = 85,000e^{0.05t} - 85,000,$$

and the annual income stream from this investment is given by $A(t) = G'(t) = 4,250e^{0.05t}$.

Use ***Integrating.xls*** to find $\int_0^6 A(t)dt$ and show that the result is equal to $G(6)$.

**SOLUTION**

$$\int_0^6 A(t)dt = \$29,738$$

| Definition | Computation | | Plot Interval | | Integration Interval | | $\int_a^b f(x)\,dx$ |
|---|---|---|---|---|---|---|---|
| Formula for $f(x)$ | $x$ | $f(x)$ | $A$ | $B$ | $a$ | $b$ | |
| 4250 | | 4250 | 0 | 6 | 0 | 6 | 29737.9986 |

$$G(6) = 85,000e^{0.05(6)} - 85,000 = \$29,738 = \int_0^6 A(t)dt$$

**Example 5:** Suppose that \$150,000 is invested at an annual rate of 4.5%, compounded continuously. Then the annual income stream from this investment is given by $A(t) = 6,750e^{0.045t}$, the total income earned during the first $T$ years is given by

$$G(T) = \int_0^T A(t)\,dt = \int_0^T 6,750e^{0.045t}dt = 150,000e^{0.045T} - 150,000,$$

and the value of the investment after $T$ years is given by

$F(T) = 150,000 + G(T) = 150,000 + 150,000e^{0.045T} - 150,000 = 150,000e^{0.045T}$. Use ***Differentiating.xls*** to find $G'(9)$ and show that the result is equal to $A(9)$.

**SOLUTION**

$G'(9) = \$10,120.29$

**Computation**

| $x$ | $f(x)$ | $f'(x)$ |
|---|---|---|
| 9 | 74895.38 | 10120.292 |

$A(9) = 6,750e^{0.045(9)} = \$10,120.29 = G'(9)$

Since $F(T) = 150,000 + G(T)$, $F'(T) = 0 + G'(T) = G'(T)$. Therefore, $F'(9) = G'(9) = A(9)$.

## *Exercises*

1. The formulas for the *p.d.f.* and *c.d.f.* of a continuous random variable $X$ are given below.

$$f_X(x) = \begin{cases} 0 & \text{if } x < 0 \\ 4x^2e^{-2x} & \text{if } x \geq 0 \end{cases} \qquad F_X(x) = \begin{cases} 0 & \text{if } x < 0 \\ 1 - e^{-2x} - 2xe^{-2x} - 2x^2e^{-2x} & \text{if } 0 \leq x < 1 \end{cases}$$

   a. Use ***Graphing.xls*** to plot the *p.d.f.* over the interval [0, 8].
   b. Use ***Differentiating.xls*** to show that the graph of the derivative of the *c.d.f.* on the interval [0, 8] looks like the graph of the *p.d.f.* on the same interval.

**SOLUTION**

a.

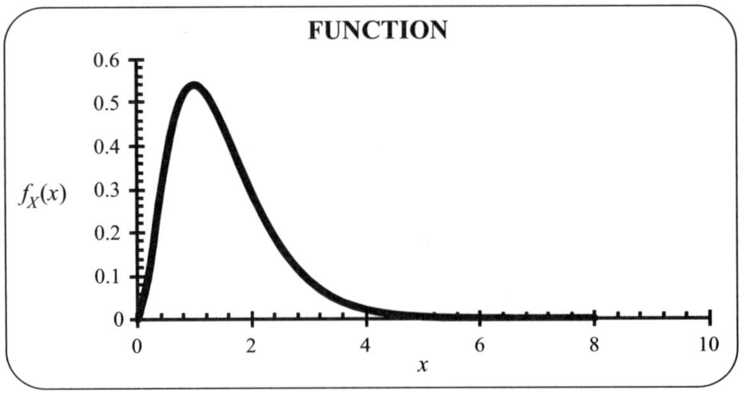


b.

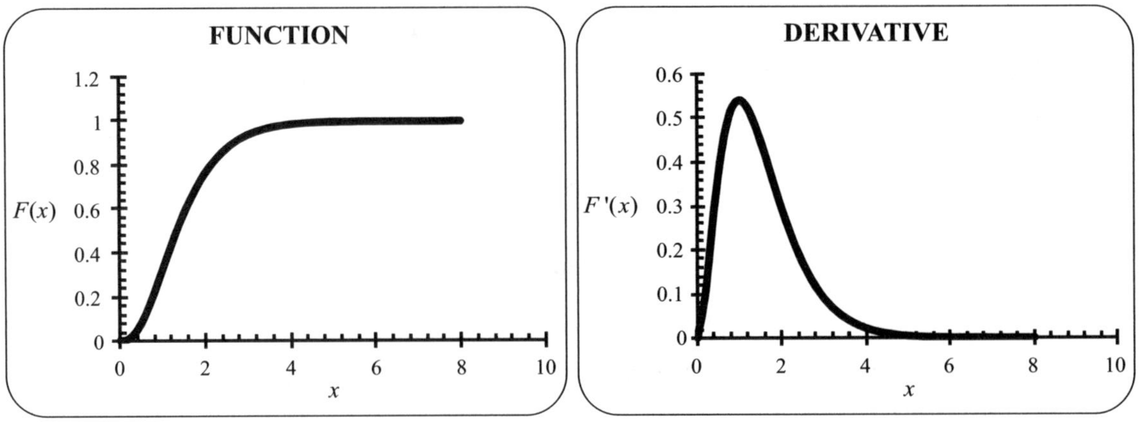


2. Let $X$ be an exponential random variable with parameter $\alpha = 4$.
   a. Find a formula for $F_X'(x)$, the derivative of the *c.d.f.* of $X$.
   b. Find a formula for $f_X(x)$, the *p.d.f.* of $X$.
   c. Compare the results from Parts a and b.

**SOLUTION**

a. $$F_X'(x) = \begin{cases} 0 & \text{if } x < 0 \\ \text{undefined} & \text{if } x = 0 \\ \frac{1}{4}e^{-x/4} & \text{if } x > 0 \end{cases}$$

b. $$f_X(x) = \begin{cases} 0 & \text{if } x < 0 \\ \frac{1}{4}e^{-x/4} & \text{if } x \geq 0 \end{cases}$$

c. For the values of $x$ at which $F_X'(x)$ is defined, the results from Parts a and b are the same.

3. Suppose that \$75,000 is invested at an annual rate of 4%, compounded continuously. Then the value of the investment after $t$ years is given by $F(t)=75,000e^{0.04t}$, the total income earned during the first $t$ years is given by

$$G(t)=F(t)-75,000=75,000e^{0.04t}-75,000,$$

and the annual income stream from this investment is given by $A(t)=G'(t)=3,000e^{0.04t}$.

Use ***Integrating.xls*** to find $\int_0^2 A(t)dt$ and show that the result is equal to $G(2)$.

**SOLUTION**

$$\int_0^2 A(t)dt=\$6,246.53=G(2)$$

| Definition | Computation | | Plot Interval | | Integration Interval | | $\int_a^b f(x)\,dx$ |
|---|---|---|---|---|---|---|---|
| Formula for $f(x)$ | $x$ | $f(x)$ | $A$ | $B$ | $a$ | $b$ | |
| 3000 | | 3000 | 0 | 2 | 0 | 2 | 6246.5301 |

4. Suppose that \$100,000 is invested at an annual rate of 2.5%, compounded continuously. Then the annual income stream from this investment is given by $A(t)=2,500e^{0.025t}$, the total income earned during the first $T$ years is given by

$$G(T)=\int_0^T A(t)\,dt=\int_0^T 2,500e^{0.025t}dt=100,000e^{0.025T}-100,000,$$

and the value of the investment after $T$ years is given by

$F(T)=100,000+G(T)=100,000+100,000e^{0.025T}-100,000=100,000e^{0.025T}$. Use ***Differentiating.xls*** to find $G'(4)$ and show that the result is equal to $A(4)$.

**SOLUTION**

$G'(4)=\$2,762.93=A(4)$

| Computation | | |
|---|---|---|
| $x$ | $f(x)$ | $f'(x)$ |
| 4 | 10517.09 | 2762.927 |

5. The Pe-lo Carbon Fiber Company manufactures and sells petroleum base carbon fiber. There are many applications for this type of carbon fiber, such as clutches and brakes in automobiles, flooring, high temperature paints, concrete, shoes, etc. The company considers its revenue as an income stream rather than a large number of discrete payments from its clients. Let $A(t)$ be the revenue rate in dollars per year where

$$A(t)=840t^6-9,990t^5+35,040t^4-40,432t^3+12,834t^2+387,700t+472,100.$$

The total revenue in dollars starting now for the next $T$ years is given by

$$G(T)=\int_0^T A(t)dt=120t^7-1{,}665t^6+7{,}008t^5-10{,}108t^4+4{,}278t^3+193{,}850t^2+472{,}100t\,.$$

a. At what rate is revenue generated at $t=2$ years?

b. Use ***Differentiating.xls*** to find $G'(2)$ and show that the result is equal to $A(2)$.

**SOLUTION**

a. $A(2)=\$1{,}270{,}100$ per year

b. $G'(2)=\$1{,}270{,}100=A(2)$

**Computation**

| $x$ | $f(x)$ | $f'(x)$ |
|---|---|---|
| 2 | 1725152 | 1270100.000 |

# Variance

## *Dispersion*

The face value is 4. 4

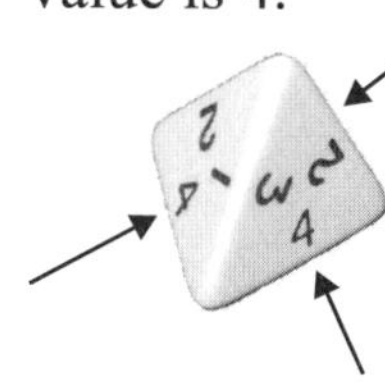

**Example 1:** Let $X$ be the number obtained on one roll of a tetrahedral die and let $Y$ be the average number obtained on two rolls of the die. The probability mass functions for $X$ and $Y$ are given below.

| $x$ | $f_X(x)$ | $x \cdot f_X(x)$ |
|---|---|---|
| 1 | 0.25 | 0.25 |
| 2 | 0.25 | 0.50 |
| 3 | 0.25 | 0.75 |
| 4 | 0.25 | 1.00 |
| | | |
| | | |
| | | |
| | | |
| $\mu_X$ | | 2.50 |

| $y$ | $f_Y(y)$ | $y \cdot f_Y(y)$ |
|---|---|---|
| 1.0 | 0.0625 | 0.06 |
| 1.5 | 0.1250 | 0.19 |
| 2.0 | 0.1875 | 0.38 |
| 2.5 | 0.2500 | 0.63 |
| 3.0 | 0.1875 | 0.56 |
| 3.5 | 0.1250 | 0.44 |
| 4.0 | 0.0625 | 0.25 |
| | | |
| $\mu_Y$ | | 2.50 |

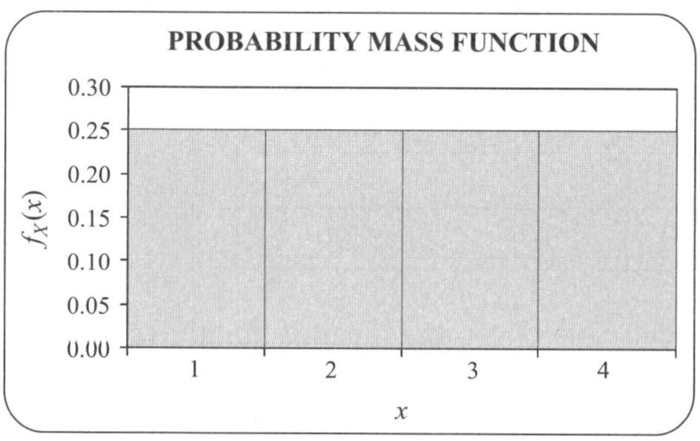


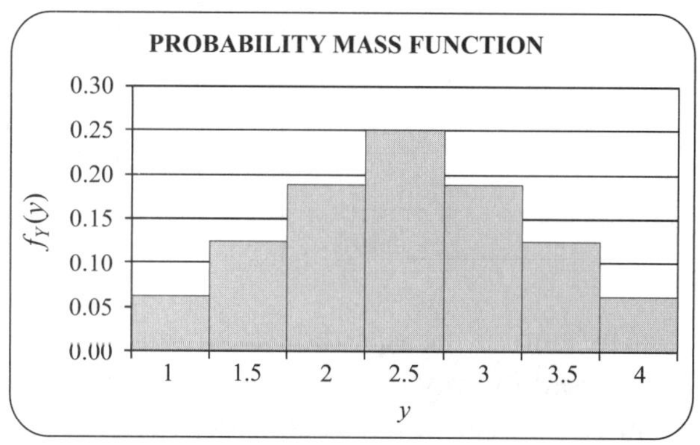


Notice that the range of values is the same for $Y$ as for $X$ and that the mean of $Y$ is equal to the mean of $X$. However, the shape of the histogram for $Y$ is noticeably different from the shape of the histogram for $X$. The possible values of $X$ are equally likely; the probability of each possible value is 0.25. However, the possible values of $Y$ are not equally likely. In particular, the probability of 2.5 is much greater than the probabilities of 1 or 4. We need a parameter that measures the dispersion of the possible values about the mean.

Consider the random variable $X$. We begin by computing the sum of the deviations of the possible values from the mean.

| $x$ | $f_X(x)$ | $x \cdot f_X(x)$ | $x-\mu_X$ |
|---|---|---|---|
| 1 | 0.25 | 0.25 | -1.5 |
| 2 | 0.25 | 0.50 | -0.5 |
| 3 | 0.25 | 0.75 | 0.5 |
| 4 | 0.25 | 1.00 | 1.5 |
| | | | |
| Sum | | $\mu_X = 2.50$ | 0 |

The sum of the deviations of the possible values from the mean is equal to 0; the positive deviations are offset by negative deviations. This result would imply that there is no dispersion in the possible values about the mean. Because there is obviously some variability about the mean, the sum of the deviations is useless as a measure of dispersion. We need to measure the deviations using a method that eliminates the sign. Although we can eliminate the sign by taking the absolute value, we will accomplish the same result by squaring the differences. The measure that is obtained by squaring the differences has some nice properties that the measure obtained by taking the absolute value does not.

| $x$ | $f_X(x)$ | $x \cdot f_X(x)$ | $x-\mu_X$ | $(x-\mu_X)^2$ |
|---|---|---|---|---|
| 1 | 0.25 | 0.25 | $-1.5$ | 2.25 |
| 2 | 0.25 | 0.50 | $-0.5$ | 0.25 |
| 3 | 0.25 | 0.75 | 0.5 | 0.25 |
| 4 | 0.25 | 1.00 | 1.5 | 2.25 |
| | | | | |
| Sum | | $\mu_X = 2.50$ | 0 | 5.00 |

The sum of the squared deviations of the possible values from the mean is equal to 5. This result would imply that there is indeed some dispersion of the values about the mean. However, we have thus far failed to consider the probabilities of the possible values. Although the possible values of this random variable are equally likely, that is not necessarily the case in general. Therefore, we will weight each squared difference by the corresponding probability.

| $x$ | $f_X(x)$ | $x \cdot f_X(x)$ | $x-\mu_X$ | $(x-\mu_X)^2$ | $(x-\mu_X)^2 \cdot f_X(x)$ |
|---|---|---|---|---|---|
| 1 | 0.25 | 0.25 | $-1.5$ | 2.25 | 0.5625 |
| 2 | 0.25 | 0.50 | $-0.5$ | 0.25 | 0.0625 |
| 3 | 0.25 | 0.75 | 0.5 | 0.25 | 0.0625 |
| 4 | 0.25 | 1.00 | 1.5 | 2.25 | 0.5625 |
| | | | | | |
| Sum | | $\mu_X = 2.50$ | 0 | 5.00 | $V(X) = 1.25$ |

The weighted sum of the squared deviations is called the **variance** of $X$ and is denoted $V(X)$. It is one useful measure of the dispersion of the possible values about the mean. We will now do the same computations for $Y$.

| $y$ | $f_Y(y)$ | $y \cdot f_Y(y)$ | $y-\mu_Y$ | $(y-\mu_Y)^2$ | $(y-\mu_Y)^2 \cdot f_Y(y)$ |
|---|---|---|---|---|---|
| 1.0 | 0.0625 | 0.0625 | −1.5 | 2.25 | 0.140625 |
| 1.5 | 0.1250 | 0.1875 | −1.0 | 1.00 | 0.125000 |
| 2.0 | 0.1875 | 0.3750 | −0.5 | 0.25 | 0.046875 |
| 2.5 | 0.2500 | 0.6250 | 0 | 0.00 | 0.000000 |
| 3.0 | 0.1875 | 0.5625 | 0.5 | 0.25 | 0.046875 |
| 3.5 | 0.1250 | 0.4375 | 1.0 | 1.00 | 0.125000 |
| 4.0 | 0.0625 | 0.2500 | 1.5 | 2.25 | 0.140625 |
| | | | | | |
| Sum | | $\mu_Y = 2.50$ | 0 | | $V(Y) = 0.625$ |

These computations show that $V(Y)$ is less than $V(X)$. In other words, the dispersion of the possible values of $Y$ about its mean is less than the dispersion of the possible values of $X$ about its mean. Although the range of possible values is the same for $Y$ as for $X$, we are much more likely to observe a value of $Y$ that is close to the mean than we are for $X$.

In general, the variance of any finite random variable $X$ is given by

$$V(X) = \sum_{\text{all } x} (x - \mu_X)^2 \cdot f_X(x).$$

The variance of any continuous random variable $X$ is given by

$$V(X) = \int_{-\infty}^{\infty} (x - \mu_X)^2 \cdot f_X(x)\, dx.$$

The variance is one measurement of dispersion. However, the square root of $V(X)$ is the preferred measure of dispersion. The square root of the variance is called the **standard deviation** and is given by $\sigma_X = \sqrt{V(X)}$. It is measured in the same units as the observations and the mean. It also has an important practical interpretation. The standard deviation, $\sigma_X$, is interpreted as the "typical amount" by which an observation of $X$ differs from the mean, $\mu_X$. In this context, "typical amount" corresponds to that which occurs with at least a 50% probability. The standard deviations of $X$ and $Y$ are given below.

$$\sigma_X = \sqrt{V(X)} = \sqrt{1.25} = 1.1180 \qquad \sigma_Y = \sqrt{V(Y)} = \sqrt{0.625} = 0.7906$$

The first result implies that the typical difference between an observed value of $X$ and its mean is 1.1180. In other words, there is at least a 50% probability that the difference between the number obtained on one roll of a tetrahedral die and the mean 2.5 is less than or equal to 1.1180. Likewise, the typical difference between an observed value of $Y$ and its mean is 0.7906. This implies that there is at least a 50% probability that the difference between the average number obtained on two rolls of the die and the mean 2.5 is less than or equal to 0.7906.

**Example 2**: The graphs of the probability mass functions of $X$ and $W$ are given below.

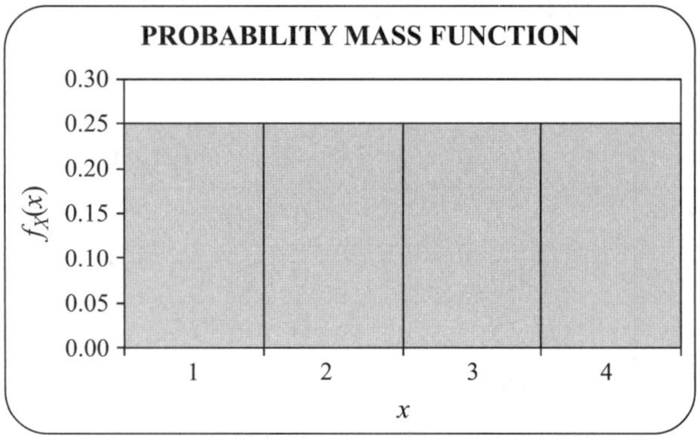


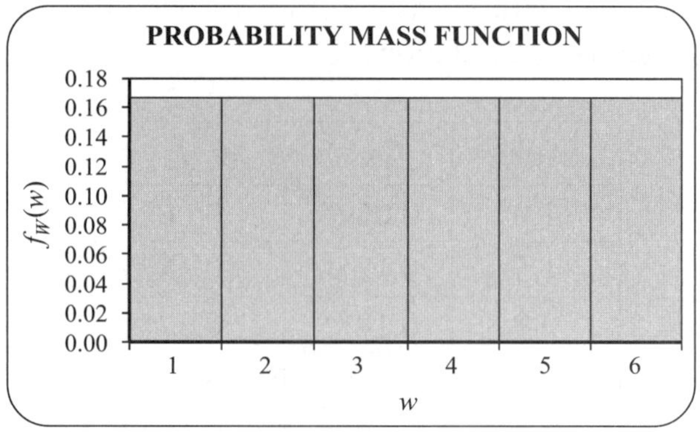


a. Find the means of $X$ and $W$.
b. For which random variable is there more dispersion about the mean?
c. Find the variance of $W$.
d. Find the standard deviation of $W$.

**SOLUTION**

a. Since the distribution of $X$ is symmetric about 2.5, the mean of $X$ is 2.5. The distribution of $W$ is symmetric about 3.5, so the mean of $W$ is 3.5.
b. We can see from the graph that the values of $W$ are more widely dispersed about $\mu_W$ than the values of $X$ are about $\mu_X$.
c.

| $w$ | $f_W(w)$ | $w \cdot f_W(w)$ | $w-\mu_W$ | $(w-\mu_W)^2$ | $(w-\mu_W)^2 \cdot f_W(w)$ |
|---|---|---|---|---|---|
| 1 | 0.166667 | 0.166667 | −2.5 | 6.25 | 1.041669 |
| 2 | 0.166667 | 0.333334 | −1.5 | 2.25 | 0.375001 |
| 3 | 0.166667 | 0.500001 | −0.5 | 0.25 | 0.041668 |
| 4 | 0.166667 | 0.666668 | 0.5 | 0.25 | 0.041668 |
| 5 | 0.166667 | 0.833335 | 1.5 | 2.25 | 0.375001 |
| 6 | 0.166667 | 1.000002 | 2.5 | 6.25 | 1.041669 |
| | | | | | |
| Sum | | $\mu_W = 3.50$ | 0 | 17.5 | $V(W) = 2.916676$ |

d. $\sigma_W = \sqrt{V(w)} = \sqrt{2.916676} = 1.707828$

**Example 3:** The *p.d.f.*, $f_X(x)$, of a continuous random variable $X$ is given below.

$$f_X(x) = \begin{cases} 0 & \text{if } x < 0 \\ 4x^2 e^{-2x} & \text{if } x \geq 0 \end{cases}$$

a. Set up, but do not evaluate, an integral that gives the mean, $\mu_X$, of $X$.
b. Use ***Integrating.xls*** to find $\mu_X$.

c. Set up, but do not evaluate, an integral that gives the variance, $V(X)$, of $X$.

d. Use ***Integrating.xls*** to find $V(X)$.

e. Find $\sigma_X$.

**SOLUTION**

a. $\mu_X = \int_{-\infty}^{\infty} x \cdot f_X(x)\,dx = \int_{0}^{\infty} x \cdot 4x^2 e^{-2x}\,dx$

b. $\mu_X = \int_{0}^{\infty} x \cdot 4x^2 e^{-2x}\,dx = 1.5$

| Definition | Computation | | Plot Interval | | Integration Interval | | $\int_a^b f(x)\,dx$ |
|---|---|---|---|---|---|---|---|
| Formula for $f(x)$ | $x$ | $f(x)$ | $A$ | $B$ | $a$ | $b$ | |
| 0 | 15 | 1.3E-09 | 0 | 15 | 0 | 15 | 1.5000 |

**Note:** The lower and upper limits of integration are 0 and ∞ respectively; however, we integrated from 0 to 15 in ***Integrating.xls***. This is because as $x$ goes to ∞, $f_X(x)$ goes to 0; as you can see in the **Computation** cells, at $x = 15$, $f_X(x) = 1.3\text{E-}09$ which is approximately equal to zero.

c. $V(X) = \int_{-\infty}^{\infty} (x-\mu_X)^2 \cdot f_X(x)\,dx = \int_{0}^{\infty} (x-1.5)^2 \cdot 4x^2 e^{-2x}\,dx$

d. $V(X) = \int_{0}^{\infty} (x-1.5)^2 \cdot 4x^2 e^{-2x}\,dx = 0.75$

| Definition | Computation | | Plot Interval | | Integration Interval | | $\int_a^b f(x)\,dx$ |
|---|---|---|---|---|---|---|---|
| Formula for $f(x)$ | $x$ | $f(x)$ | $A$ | $B$ | $a$ | $b$ | |
| 0 | | 0 | 0 | 15 | 0 | 15 | 0.7500 |

e. $\sigma_X = \sqrt{V(X)} = \sqrt{0.75} \cong 0.87$

By now you should be familiar with Bernoulli and binomial random variables, which were covered in **Distributions:** *Finite Random Variables*, and uniform and exponential random variables, which were covered in and **Distributions:** *Continuous Random Variables*. The variance can be computed for each of these random variables using the formulas below.

Bernoulli Random Variable $X$

$$V(X) = p(1-p)$$

Binomial Random Variable $X$

$$V(X) = np(1-p)$$

Uniform Random Variable $X$

$$V(X) = \frac{(b-a)^2}{12}$$

Exponential Random Variable $X$

$$V(X) = \alpha^2$$

**Example 4:** About one-third of all drivers admit to exhibiting aggressive driving behavior (tailgaiting, yelling, weaving in and out of traffic, etc.), at least some of the time. In a group of 500 randomly chosen drivers, how many would you expect to exhibit some aggressive driving behavior? What is the standard deviation?

**SOLUTION**

If we let the random variable $X$ be the number of drivers who exhibit some aggressive driving behavior, then $X$ is a binomial random variable, with the sample size $n = 500$ and the probability of success $p = 1/3$. The number of drivers in this group whom we expect to exhibit some aggressive behavior is the mean.

$$\mu_X = np = 500\left(\frac{1}{6}\right) \cong 167$$

The standard deviation is

$$\sigma_X = \sqrt{V(X)} = \sqrt{np(1-p)} = \sqrt{500\left(\frac{1}{3}\right)\left(\frac{2}{3}\right)} = \sqrt{\frac{1000}{9}} \cong 10.5409.$$

**Example 5:** The *p.d.f.* and *c.d.f.* of a random variable that has a uniform distribution on the interval [0, 6] are given below.

$$f_W(w) = \begin{cases} \frac{1}{6} & \text{if } 0 \le w \le 6 \\ 0 & \text{elsewhere} \end{cases} \qquad F_W(w) = \begin{cases} 0 & \text{if } w < 0 \\ \frac{w}{6} & \text{if } 0 \le w \le 6 \\ 1 & \text{if } w > 6 \end{cases}.$$

a. Set up, but do not evaluate, an integral that gives the mean, $\mu_W$, of $W$.
b. Use ***Integrating.xls*** to find $\mu_W$.
c. Use the special formula for the variance of a random variable that has a uniform distribution on the interval $[a, b]$ to find $V(W)$.
d. Set up, but do not evaluate, an integral that gives the variance, $V(W)$, of $W$.
e. Use ***Integrating.xls*** to find $V(W)$.
f. Find $\sigma_W$.

**SOLUTION**

a. $\mu_W = \int_0^6 x \cdot \frac{1}{6}\, dx$

b. $\mu_W = \int_0^6 x \cdot \frac{1}{6}\, dx = 3$

| Definition | Computation | | Plot Interval | | Integration Interval | | $\int_a^b f(x)\,dx$ |
|---|---|---|---|---|---|---|---|
| Formula for $f(x)$ | $x$ | $f(x)$ | $A$ | $B$ | $a$ | $b$ | |
| 1.5 | | 1.5 | 0 | 6 | 0 | 6 | 3.0000 |

**Note:** Recall that the expected value (or mean) of a random variable $X$ that has a uniform distribution on the interval $[a, b]$ can be computed using the formula

$$E(X)=\frac{a+b}{2}.$$

Since $W$ has a uniform distribution, we can use the formula above to compute $E(W)$.

$$E(W)=\frac{0+6}{2}=3$$

c. $V(W)=\dfrac{(b-a)^2}{12}=\dfrac{(6-0)^2}{12}=3$

d. $V(W)=\displaystyle\int_0^6 (w-3)^2\cdot\frac{1}{6}\,dw$

e. $V(W)=\displaystyle\int_0^6 (w-3)^2\cdot\frac{1}{6}\,dw=3$

| Definition | Computation | | Plot Interval | | Integration Interval | | $\int_a^b f(x)\,dx$ |
|---|---|---|---|---|---|---|---|
| Formula for $f(x)$ | $x$ | $f(x)$ | $A$ | $B$ | $a$ | $b$ | |
| 1.5 | | 1.5 | 0 | 6 | 0 | 6 | 3.0000 |

f. $\sigma_W=\sqrt{V(W)}=\sqrt{3}\cong 1.73$

**Example 6:** The length of time (in minutes), $T$, required to complete an exam has a uniform distribution on the interval [40, 50]. What is the standard deviation of the time required to complete the exam?

**SOLUTION**

Since $T$ has a uniform distribution,

$$\sigma_T=\sqrt{V(T)}=\sqrt{\frac{(b-a)^2}{12}}=\sqrt{\frac{(50-40)^2}{12}}=\sqrt{\frac{100}{12}}\cong 2.887.$$

The standard deviation of the time required to complete the exam is approximately 2.887 minutes.

**Example 7:** The *p.d.f.* and *c.d.f.* of $X$, an exponential random variable, are given below. Use ***Integrating.xls*** to verify that the variance, $V(X)$, of $X$ is $\alpha^2$.

$$f_X(x)=\begin{cases}0 & \text{if } x<0\\ \dfrac{1}{6}e^{-x/6} & \text{if } x\ge 0\end{cases}\qquad F_X(x)=\begin{cases}0 & \text{if } x<0\\ 1-e^{-x/6} & \text{if } x\ge 0\end{cases}$$

**SOLUTION**

$\alpha = 6$

$\mu_X = \alpha = 6$

$V(X) = \alpha^2 = 6^2 = 36$

$$V(X) = \int_0^\infty (x-6)^2 \cdot \frac{1}{6} e^{-x/6} dx = 36$$

| Definition | Computation | | Plot Interval | | Integration Interval | | $\int_a^b f(x)\,dx$ |
|---|---|---|---|---|---|---|---|
| Formula for $f(x)$ | $x$ | $f(x)$ | $A$ | $B$ | $a$ | $b$ | |
| 6 | | 6 | 0 | 150 | 0 | 150 | 36.0000 |

**Example 8:** The number of miles a certain model of car can be driven before the transmission fails is given by an exponential random variable with parameter $\alpha = 125{,}000$. What is the average number of miles that you can expect to drive this model of car before the transmission fails? What is the standard deviation?

**SOLUTION**

For an exponential random variable,

$$\mu = \alpha = 125{,}000 \text{ and } \sigma = \sqrt{V(X)} = \sqrt{\alpha^2} = \alpha = 125{,}000.$$

On average, you can expect to drive this model of car 125,000 miles before the transmission fails. The standard deviation is 125,000 miles.

We can **standardize** any random variable $X$ by subtracting its mean and dividing by its standard deviation, that is,

$$S = \frac{X - \mu_X}{\sigma_X}.$$

We call $S$ the **standardization of $X$**. The distribution of the random variable $S$ has the same shape as that of $X$, but $S$ has mean 0 and standard deviation 1. The advantages to doing this will become clearer when we get to the normal distribution, which can be used as an approximation for other distributions. Right now we want to get some practice in working with $S$.

We can find the probability that the standardized random variable $S$ is between two values $a$ and $b$ by using the relationship between $S$ and $X$.

$$\begin{aligned} P(a \le S \le b) &= P\left(a \le \frac{X - \mu_X}{\sigma_X} \le b\right) \\ &= P(a \cdot \sigma_X \le X - \mu_X \le b \cdot \sigma_X) \\ &= P(\mu_X + a \cdot \sigma_X \le X \le \mu_X + b \cdot \sigma_X) \end{aligned}$$

**Example 9:** Let $X$ be a random variable that has a uniform distribution on the interval [0, 500]. Let $S$ be the standardization of $X$. Compute the probability that $S$ falls between 0 and 1.5.

**SOLUTION**

Since $X$ has a uniform distribution, its mean is $\mu_X = (0+500)/2 = 250$ and its variance is $V(X) = (500-0)^2/12 \cong 20833.33$. The standard deviation is

$$\sigma_X = \sqrt{V(X)} \cong \sqrt{20833.33} \cong 144.3376.$$

We find the probability that the standardization of $X$ is between 0 and 1.5 as follows.

$$\begin{aligned}
P(0 \le S \le 1.5) &= P\left(0 \le \frac{X-\mu_X}{\sigma_X} \le 1.5\right) \\
&= P\left(0 \le \frac{X-250}{144.3376} \le 1.5\right) \\
&= P(0 \le X - 250 \le 216.5064) \\
&= P(250 \le X \le 466.5064) \\
&= \frac{466.5064 - 250}{500} \\
&= 0.433
\end{aligned}$$

**Example 10:** Let $X$ be an exponential random variable with parameter $\alpha = 4.5$. Let $S$ be the standardization of $X$. Find $P(S \le 0.5)$.

**SOLUTION**

Since $X$ is an exponential random variable, $\mu_X = \sigma_X = \alpha = 4.5$.

$$\begin{aligned}
P(S \le 0.5) &= P\left(\frac{X-\mu_X}{\sigma_X} \le 0.5\right) \\
&= P\left(\frac{X-4.5}{4.5} \le 0.5\right) \\
&= P(X - 4.5 \le 2.25) \\
&= P(X \le 6.75)
\end{aligned}$$

To find this last probability, we can either evaluate the integral of the *p.d.f.* of $X$,

$$\int_0^{6.75} \frac{1}{4.5} e^{-x/4.5}\, dx$$

using ***Integrating.xls***, or we can evaluate the *c.d.f.*, $F_X(6.75) = 1 - e^{-6.75/4.5}$. Either way, we get $P(S \le 0.5) = P(X \le 6.75) \cong 0.7769$.

## *Samples*

As previously discussed, we often do not know the theoretical distribution of the random variable.

Consequently, we use information from an appropriately selected sample to estimate the desired information about the random variable. For example, relative frequency histograms are used to approximate probability mass functions of finite random variables, adjusted relative frequency histograms are used to approximate probability density functions of continuous random variables, and sample means are used to approximate expected values.

The variance of a random variable can be approximated by the sample variance and the standard deviation of a random variable can be approximated by the sample standard deviation. The sample variance, $s^2$, is given by

$$s^2 = \frac{1}{n-1}\sum_{i=1}^{n}(x_i - \overline{x})^2 ,$$

and the sample standard deviation, $s$, is given by

$$s = \sqrt{s^2} .$$

**Example 11:** In **Example 15** of **Distributions:** *Random Samples*, Jack has three pennies, four nickels, two dimes and one quarter in his pocket. The random variable $X$ is the denomination of a coin selected at random from his pocket. Jack selects a single coin from his pocket 40 different times. This gives 40 observations of $X$, with the following outcomes.

| $i$ | $x_i$ | $x_i - \overline{x}$ | $(x_i - \overline{x})^2$ |
|---|---|---|---|
| 1 | 10 | 3.675 | 13.505625 |
| 2 | 1 | −5.325 | 28.355625 |
| 3 | 5 | −1.325 | 1.755625 |
| 4 | 1 | −5.325 | 28.355625 |
| 5 | 5 | −1.325 | 1.755625 |
| ⋮ | ⋮ | ⋮ | ⋮ |
| 36 | 5 | −1.325 | 1.755625 |
| 37 | 10 | 3.675 | 13.505625 |
| 38 | 1 | −5.325 | 28.355625 |
| 39 | 25 | 18.675 | 348.755625 |
| 40 | 5 | −1.325 | 1.755625 |
| | | | |
| Sum | 253 | 0 | 1562.775 |
| | $\overline{x} = 6.325$ | | $s^2 \cong 40.07$ |

The sample standard deviation was computed in *Excel* and is given by $s \cong \sqrt{40.07} \cong 6.33$. The variance of $X$, $V(X)$, is computed below.

| $x$ | $f_X(x)$ | $x \cdot f_X(x)$ | $x-\mu_X$ | $(x-\mu_X)^2$ | $(x-\mu_X)^2 \cdot f_X(x)$ |
|---|---|---|---|---|---|
| 1 | 0.3 | 0.3 | −5.8 | 33.64 | 10.092 |
| 5 | 0.4 | 2.0 | −1.8 | 3.24 | 1.296 |
| 10 | 0.2 | 2.0 | 3.2 | 10.24 | 2.048 |
| 25 | 0.1 | 2.5 | 18.2 | 331.24 | 33.124 |
| | | | | | |
| Sum | | $\mu_X = 6.8$ | | | $V(X) = 46.56$ |

The standard deviation of $X$ is given by $\sigma_X = \sqrt{V(X)} = \sqrt{46.56} \cong 6.82$. These results are summarized in the table below.

| **Theoretical** | $V(X)$ | 46.56 | $\sigma_X$ | 6.82 |
|---|---|---|---|---|
| **Sample** | $s^2$ | 40.07 | $s$ | 6.33 |

Using a sample size of 40, the sample statistics closely approximate the theoretical (population) parameters.

**Example 12:** In **Example 19** of **Distributions:** *Random Samples*, the random variable $X$ gives the error in the location of the groove cut. The table below lists the means and standard deviations for different samples of the random variable $X$.

| $n$ | $\overline{x}$ | $s$ |
|---|---|---|
| 10 | −1.52 | 6.462 |
| 20 | 0.984 | 5.684 |
| 40 | −0.477 | 5.302 |
| 80 | −0.515 | 4.909 |
| 160 | 0.051 | 5.035 |

The machine has been calibrated to have a mean accuracy of zero, $\mu_X = 0$, and a standard deviation of five, $\sigma_X = 5$. From the sample statistics in the table, we see that the sample mean and the sample standard deviation closely approximate the theoretical (population) mean and the theoretical (population) standard deviation, respectively, that is, $\overline{x} \cong \mu_X$ and $s \cong \sigma_X$.

**Example 13:** Calculate mean and standard deviation of the sample of twenty quiz scores given below.

| **Quiz Scores** | | | | | | | | | | | | | | | | | | | |
|---|---|---|---|---|---|---|---|---|---|---|---|---|---|---|---|---|---|---|---|
| 10 | 8 | 3 | 10 | 1 | 2 | 7 | 5 | 9 | 9 | 4 | 1 | 7 | 10 | 3 | 2 | 4 | 2 | 2 | 7 |

**SOLUTION**

$$\begin{aligned}\bar{x} &= \frac{1}{20}\sum_{i=1}^{20} x_i \\ &= \frac{1}{20}(10+8+3+10+1+2+7+5+9+9+4+1+7+10+3+2+4+2+2+7) \\ &= 5.3\end{aligned}$$

$$\begin{aligned}s &= \sqrt{\frac{1}{20-1}\sum_{i=1}^{20}(x_i - 5.3)^2} \\ &= \sqrt{\frac{1}{19}[(10-5.3)^2 + (8-5.3)^2 + \cdots + (7-5.3)^2]} \\ &= \sqrt{10.7474} \\ &\cong 3.278\end{aligned}$$

**Example 14:** Four observations of $X$, the cost to repair a damaged computer, are given below.

$65, $90, $110, $125

a. Find the sample mean, $\bar{x}$.
b. Find the sample variance, $s^2$.
c. Find the sample standard deviation, $s$.

**SOLUTION**

a. $\bar{x} = \dfrac{65+90+110+125}{4} = \$97.50$

b. $$\begin{aligned}s^2 &= \frac{1}{4-1}\sum_{i=1}^{4}(x_i - 97.5)^2 \\ &= \frac{1}{3}[(65-97.5)^2 + (90-97.5)^2 + (110-97.5)^2 + (125-97.5)^2 \\ &= 675\end{aligned}$$

c. $s = \sqrt{675} \cong \$25.98$

We saw in the preceding examples that the sample mean $\bar{x}$ is approximately equal to the theoretical, or population, mean $\mu_X$. Of course, in any given sample, there could be considerable difference in the specific value of $\bar{x}$ and the value of $\mu_X$. However, if we take many samples all of the same size, and take the average of the sample means of those many samples, then we *will* get a value that is very close to $\mu_X$. In symbols, $E(\bar{x}) = \mu_X$.

On the other hand, if we take many samples all of size $n$ and calculate the sample mean $\bar{x}$, will the variance of all those $\overline{x}$'s be the same as the variance of $X$? The answer to this is no! What is true is that the variance of the sample mean is equal to the theoretical (population) variance divided by the sample size. In symbols,

$$V(\overline{x}) = \frac{V(X)}{n} = \frac{{\sigma_X}^2}{n}.$$

And since the standard deviation is the square root of the variance, $\sigma_{\overline{x}} = \dfrac{\sigma_X}{\sqrt{n}}$.

**Example 15:** Suppose a random variable $X$ has mean 13.7 and standard deviation 0.097. Let $\overline{x}$ be the sample mean for random samples of size 125. What is the expected value of $\overline{x}$? What is the standard deviation of $\overline{x}$?

**SOLUTION**

$$E(\overline{x}) = \mu_X = 13.7$$

$$\sigma_{\overline{x}} = \frac{\sigma_X}{\sqrt{n}} = \frac{0.097}{\sqrt{125}} = 0.0087$$

**Example 16:** A car manufacturer wants to know the gas mileage (in miles per gallon) drivers of their new hybrid SUV are getting. The manufacturer has dealers in 30 locations who each take a sample of 50 drivers of these vehicles. The dealers report the average gas mileage found in their sample of 50. These averages (sample means) are given below.

| Values of $\overline{x}$ from 30 samples each of size 50 | | | | | | | | | |
|---|---|---|---|---|---|---|---|---|---|
| 47.715 | 46.285 | 45.871 | 44.129 | 45.289 | 44.711 | 45.837 | 44.163 | 47.751 | 46.239 |
| 45.371 | 45.046 | 44.954 | 45.981 | 44.019 | 45.681 | 44.319 | 45.963 | 44.037 | 46.648 |
| 45.352 | 47.992 | 45.008 | 47.913 | 46.087 | 47.893 | 46.107 | 45.870 | 44.130 | 46.344 |

a. What is the approximate value of the theoretical (population) mean?

b. What is the approximate value of the theoretical (population) standard deviation?

**SOLUTION**

a. Since the mean of the sample means is the theoretical (population) mean,

$$\begin{aligned}\mu_X &= E(\overline{x}) \\ &\cong \frac{1}{30}(47.715 + 46.285 + \cdots + 46.344) \\ &= 45.757.\end{aligned}$$

Here $n = 30$ since we have 30 values of $\overline{x}$.

b. We need $\sigma_{\overline{x}}$; we will use $s$ to estimate $\sigma_{\overline{x}}$. We calculate $s$ as follows.

$$
\begin{aligned}
s = \sqrt{s^2} &= \sqrt{\frac{1}{n-1}\sum_{i=1}^{n}\left(\overline{x}_i - E(\overline{x})\right)^2} \\
&= \sqrt{\frac{1}{30-1}\left((47.715-45.757)^2 + (46.285-46.757)^2 + \cdots + (46.344-45.757)^2\right)} \\
&= 1.2212.
\end{aligned}
$$

We again take $n = 30$ because we have 30 sample means.

Then we use this to find $\sigma_{\overline{x}}$. Since $\sigma_{\overline{x}} = \dfrac{\sigma_X}{\sqrt{n}}$,

$$
\begin{aligned}
\sigma_X &= \sqrt{n}\cdot\sigma_{\overline{x}} \\
&\cong \sqrt{n}\cdot s_{\overline{x}} \\
&= \sqrt{50}\cdot 1.2212 \\
&= 8.635.
\end{aligned}
$$

Here we use $n = 50$ since each sample is of size 50.

## *Exercises*

1. The *p.m.f.* of a finite random variable *Y* is given below. Compute the mean, variance, and standard deviation of *Y*.

| $y$ | $f_Y(y)$ |
|---|---|
| 1 | 0.3 |
| 5 | 0.6 |
| 10 | 0.1 |

**SOLUTION**

$\mu_Y = 4.3$; $V(Y) = 6.81$; $\sigma_Y = \sqrt{6.9} \cong 2.61$

2. Set up a table, like the one in **Example 2**, to compute the variance, $V(X)$, of the binomial random variable $X$ with parameters $n = 4$ and $p = 0.6$. Then verify your answer using the special formula for the variance of a binomial random variable.

**SOLUTION**

The ***BINOMDIST*** function in *Excel* was used to find the values of the *p.m.f.*

| $x$ | $f_X(x)$ | $x \cdot f_X(x)$ | $x-\mu_X$ | $(x-\mu_X)^2$ | $(x-\mu_X)^2 \cdot f_X(x)$ |
|---|---|---|---|---|---|
| 0 | 0.03 | 0.00 | −2.40 | 5.76 | 0.15 |
| 1 | 0.15 | 0.15 | −1.40 | 1.96 | 0.30 |
| 2 | 0.35 | 0.69 | −0.40 | 0.16 | 0.06 |
| 3 | 0.35 | 1.04 | 0.60 | 0.36 | 0.12 |
| 4 | 0.13 | 0.52 | 1.60 | 2.56 | 0.33 |
| | | | | | |
| Sum | | $\mu_X = 2.40$ | | | $V(X) = 0.96$ |

$V(X) = 4(0.6)(1-0.6) = 0.96$

3. A telemarketer's chance of convincing a randomly selected person to switch long-distance telephone carriers is about 8%. If the telemarketer calls 200 people, how many can he expect to convince to switch long-distance carriers? What is the standard deviation?

**SOLUTION**

He can expect 16 people to switch carriers, with a standard deviation of 3.8367 people.

4. Of all drivers who file a claim with their automobile insurance carrier, 12% will see an increase in rates the following year. Of 397 people who filed claims, how many can be expected to see a subsequent rise in insurance rates? What is the standard deviation?

**SOLUTION**

Of 397 people who filed claims, 48 people can be expected to see a rise in insurance rates, with a standard deviation of 6.4748 people.

5. The *p.d.f.*, $f_T(t)$, of a continuous random variable $T$ is given below.

$$f_T(t) = \begin{cases} 0 & \text{if } t \leq 1 \\ \dfrac{2}{t^3} & \text{if } t > 1 \end{cases}$$

   a. Set up, but do not evaluate, an integral that gives the mean, $\mu_T$, of $T$.
   b. Use ***Integrating.xls*** to find $\mu_T$.
   c. Set up, but do not evaluate, an integral that gives the variance, $V(T)$, of $T$.

**SOLUTION**

a. $\mu_T = \displaystyle\int_1^{\infty} t \cdot \frac{2}{t^3}\, dt$

b. $\mu_T = \int_1^{\infty} t \cdot \frac{2}{t^3}\, dt \cong 2$

| Definition | Computation | | Plot Interval | | Integration Interval | | $\int_a^b f(x)\,dx$ |
|---|---|---|---|---|---|---|---|
| Formula for $f(x)$ | $x$ | $f(x)$ | $A$ | $B$ | $a$ | $b$ | |
| #DIV/0! | | #DIV/0! | 1 | 200 | 1 | 200 | 1.9945 |

c. $V(T) = \int_1^{\infty} (t-2)^2 \cdot \frac{2}{t^3}\, dt$

6. The *p.d.f.* of a continuous random variable $X$ is given below.

$$f_X(x) = \begin{cases} \frac{1}{4} x e^{-x/2} & \text{if } x > 0 \\ 0 & \text{elsewhere} \end{cases}$$

a. Set up, but do not evaluate, an integral that gives the mean, $\mu_X$, of $X$.

b. Use ***Integrating.xls*** to find $\mu_X$.

c. Set up, but do not evaluate, an integral that gives the variance, $V(X)$, of $X$.

d. Use ***Integrating.xls*** to find $V(X)$.

e. Compute the standard deviation of $X$.

**SOLUTION**

a. $\mu_X = \int_0^{\infty} x \cdot \frac{1}{4} x e^{-x/2}\, dx$

b. $\mu_X = \int_0^{\infty} \frac{1}{4} x^2 e^{-x/2}\, dx = 4$

| Definition | Computation | | Plot Interval | | Integration Interval | | $\int_a^b f(x)\,dx$ |
|---|---|---|---|---|---|---|---|
| Formula for $f(x)$ | $x$ | $f(x)$ | $A$ | $B$ | $a$ | $b$ | |
| 0 | | 0 | 0 | 100 | 0 | 100 | 4.0000 |

c. $V(X) = \int_0^{\infty} (x-4)^2 \cdot \frac{1}{4} x e^{-x/2}\, dx$

d. $V(X) = \int_0^{\infty} (x-4)^2 \cdot \frac{1}{4} x e^{-x/2}\, dx \cong 8$

| Definition | Computation | | Plot Interval | | Integration Interval | | $\int_a^b f(x)\,dx$ |
|---|---|---|---|---|---|---|---|
| Formula for $f(x)$ | $x$ | $f(x)$ | $A$ | $B$ | $a$ | $b$ | |
| 0 | | 0 | 0 | 100 | 0 | 100 | 7.9999 |

e. $\sigma_X = \sqrt{8} \cong 2.8$

7. Let $X$ be a random variable that has a uniform distribution on the interval [0, 20]. What is $\sigma_X$?

**SOLUTION**

$\sigma_X \cong 5.774$

8. Let $T$ be the length of time (in minutes) required for an assignment. Past experience shows that $T$ has a uniform distribution on the interval [60, 90]. What is $\sigma_T$?

**SOLUTION**

$\sigma_T \cong 8.660$

9. The lifetime of a semiconductor chip has an exponential distribution with a mean of 50,000 hours. What is the standard deviation of the distribution of the lifetime of these semiconductor chips?

**SOLUTION**

$\sigma = \alpha = 50,000$

10. It is projected that the lifetime of SSL-LEDs, which are used to produce white light for illumination, will increase from a current average of 20,000 hours to an average of 100,000 hours in 10 years. What will be the standard deviation of the lifetime of SSL-LEDs ten years from now?

**SOLUTION**

$\sigma = \alpha = 100,000$

11. Information about two different random variables, $R$ and $W$, is given below.

| $R$ | $W$ |
|---|---|
| $E(R) = 80$ | $E(W) = 80$ |
| $V(R) = 15$ | $V(W) = 20$ |

a. For which random variable is there more dispersion about the mean? Explain your reasoning.

b. For which random variable is the typical difference between a value and the mean smaller? Explain your reasoning.

**SOLUTION**

a. The random variable $W$ because $V(W) > V(R)$.

b. The random variable $R$ because $\sigma_R < \sigma_W$.

12. Let $X$ be a random variable that has a uniform distribution on the interval $[0, 80]$. Let $S$ be the standardization of $X$.

   a. What is $\mu_X$?

   b. What is $\sigma_X$?

   c. What is $P(-1 \le S \le 1)$?

**SOLUTION**

   a. $\mu_X = 40$

   b. $\sigma_X \cong 23.094$

   c. $P(-1 \le S \le 1) \cong 0.5774$

13. Let $X$ be an exponential random variable with parameter $\alpha = 5$. Let $S$ be the standardization of $X$. Find $P(S \ge 1)$.

**SOLUTION**

$P(S \ge 1) = 0.1353$

14. The charitable contributions on six randomly selected federal tax returns are given below.

$495, $2550, $951, $734, $552, $1394

   a. Find the sample mean, $\bar{x}$.

   b. Find the sample variance, $s^2$.

   c. Find the sample standard deviation, $s$.

**SOLUTION**

   a. $\bar{x} \cong 1{,}112.67$

   b. $s^2 \cong 602{,}091.87$

   c. $s \cong 775.95$

15. The *p.m.f.* of a finite random variable $R$ is given below.

| $r$ | 0 | 5 | 10 | 15 |
|---|---|---|---|---|
| $f_R(r)$ | 0.4 | 0.3 | 0.2 | 0.1 |

The experiment is conducted four times and the observed values of $R$ are, respectively, 5, 10, 0, and 10.

a. Find $\mu_R$.

b. Find $\sigma_R$.

c. Find $\overline{r}$.

d. Find $s$.

**SOLUTION**

a. $\mu_R = 5$

b. $\sigma_R = 5$

c. $\overline{r} = 6.25$

d. $s \cong 4.79$

16. Let $X$ be an exponential random variable. One hundred observations of $X$ are given in the table below.

| | | | | | | | | | |
|---|---|---|---|---|---|---|---|---|---|
| 5.95 | 9.80 | 2.09 | 11.55 | 0.30 | 7.36 | 5.08 | 2.80 | 0.33 | 9.52 |
| 12.84 | 15.59 | 1.48 | 6.55 | 0.82 | 7.18 | 1.59 | 8.20 | 5.38 | 5.15 |
| 3.25 | 1.02 | 5.93 | 3.15 | 6.13 | 11.41 | 2.73 | 3.79 | 9.18 | 3.66 |
| 3.01 | 16.28 | 2.44 | 13.29 | 5.14 | 6.54 | 14.56 | 18.76 | 0.02 | 2.69 |
| 11.54 | 7.83 | 1.14 | 4.53 | 5.98 | 33.45 | 10.79 | 8.03 | 32.31 | 9.20 |
| 3.66 | 17.58 | 2.72 | 14.17 | 8.27 | 5.11 | 7.75 | 1.98 | 2.92 | 18.20 |
| 20.25 | 2.14 | 7.26 | 1.30 | 2.97 | 0.52 | 23.17 | 2.10 | 6.66 | 23.73 |
| 2.01 | 4.96 | 0.15 | 2.40 | 2.67 | 15.46 | 10.03 | 3.07 | 16.41 | 2.57 |
| 8.63 | 9.51 | 0.64 | 0.10 | 1.31 | 2.05 | 2.74 | 7.85 | 4.46 | 2.51 |
| 3.26 | 1.89 | 4.78 | 6.93 | 6.83 | 5.68 | 3.47 | 2.34 | 0.95 | 10.70 |

a. Use *Excel* to compute an estimate of the expected value of $X$.

b. Use *Excel* to compute an estimate of the variance of $X$.

c. Use *Excel* to compute an estimate of the standard deviation of $X$.

d. Use the results from above to estimate $\alpha$.

e. Using the estimate of $\alpha$, find a formula for the *p.d.f.* of $X$.

f. Using the estimate of $\alpha$, find a formula for the *c.d.f.* of $X$.

**SOLUTION**

a. $E(X) \cong \overline{x} \cong 6.96$

b. $V(X) \cong s^2 \cong 42.82$

c. $\sigma_X \cong s \cong 6.54$

d. $\alpha = E(X) \cong 6.96$

e. $f_X(x) = \begin{cases} 0 & \text{if } x < 0 \\ \dfrac{1}{6.96} e^{-x/6.96} & \text{if } x \geq 0 \end{cases}$

f. $F_X(x) = \begin{cases} 0 & \text{if } x < 0 \\ 1 - e^{-x/6.96} & \text{if } x \geq 0 \end{cases}$

17. If $W$ is an exponential random variable and we have only a sample of size 50 observations, then which of the following statements is/are true?
   a. The sample mean can be used to estimate the variance.
   b. The sample standard deviation, $s$, can be used to estimate the standard deviation, $\sigma_X$.
   c. The expected value, $E(X)$, can be approximated by $\overline{x}$.
   d. The variance, $V(X)$, can be approximated by $s$.
   e. The sample variance, $s^2$, can be used to approximate $V(X)$.

**SOLUTION**

Only b, c, and e are true.

18. Suppose a random variable $X$ has mean 1436 and variance 201.64. Let $\overline{x}$ be the sample mean for a random sample of size 75. What is the expected value of $\overline{x}$? What is the standard deviation of $\overline{x}$?

**SOLUTION**

$E(\overline{x}) = 1436$; $\sigma_{\overline{x}} = 1.6397$

# The Sample Mean

Recall from **Distributions:** *Random Samples*, that the sample mean, $\overline{x}$, is used to estimate the mean, $\mu_X$, of the corresponding random variable.

**Example 1**: Two different samples of size $n = 100$ observations each of *X*, an exponential random variable with parameter $\alpha = 4$, are given below.

| **SAMPLE 1** ($n = 100$) | | | | | | | | | |
|---|---|---|---|---|---|---|---|---|---|
| 3.08 | 8.26 | 5.76 | 1.31 | 3.85 | 9.52 | 0.06 | 1.34 | 1.34 | 4.96 |
| 5.99 | 2.09 | 0.60 | 0.63 | 1.97 | 0.62 | 9.38 | 2.08 | 4.66 | 2.89 |
| 4.17 | 1.85 | 4.99 | 5.58 | 0.60 | 6.27 | 4.05 | 7.44 | 13.54 | 1.71 |
| 9.27 | 2.83 | 2.75 | 0.60 | 15.30 | 6.32 | 5.84 | 1.09 | 0.87 | 2.34 |
| 0.48 | 10.32 | 1.95 | 0.57 | 2.30 | 1.19 | 4.34 | 0.91 | 0.71 | 18.37 |
| 5.01 | 4.58 | 1.02 | 0.14 | 1.03 | 3.55 | 7.63 | 0.27 | 6.93 | 1.83 |
| 1.99 | 4.99 | 0.68 | 6.90 | 0.23 | 2.86 | 5.46 | 1.94 | 4.56 | 1.76 |
| 12.41 | 2.29 | 2.25 | 1.31 | 3.18 | 1.32 | 4.43 | 0.30 | 2.66 | 3.45 |
| 1.40 | 1.44 | 0.26 | 1.20 | 8.33 | 7.01 | 3.16 | 4.29 | 1.83 | 5.68 |
| 3.45 | 0.01 | 0.74 | 7.21 | 8.52 | 11.35 | 1.56 | 5.35 | 1.72 | 1.14 |

| **SAMPLE 2** ($n = 100$) | | | | | | | | | |
|---|---|---|---|---|---|---|---|---|---|
| 0.18 | 14.79 | 0.08 | 2.89 | 5.30 | 9.12 | 8.53 | 7.92 | 8.18 | 2.08 |
| 2.39 | 0.30 | 5.71 | 7.88 | 3.94 | 5.04 | 2.36 | 4.53 | 6.49 | 5.19 |
| 1.62 | 12.93 | 2.79 | 3.72 | 1.06 | 1.19 | 0.72 | 4.52 | 3.89 | 5.12 |
| 1.70 | 0.82 | 2.56 | 1.03 | 1.73 | 4.75 | 11.13 | 3.46 | 0.80 | 0.52 |
| 9.14 | 1.11 | 4.06 | 6.68 | 5.69 | 0.78 | 3.26 | 4.21 | 7.23 | 1.50 |
| 3.96 | 5.68 | 1.83 | 0.10 | 1.93 | 5.77 | 2.40 | 0.14 | 5.09 | 2.78 |
| 2.36 | 1.36 | 7.97 | 10.49 | 0.69 | 0.61 | 1.51 | 2.36 | 0.45 | 0.27 |
| 4.99 | 1.93 | 0.52 | 5.76 | 1.81 | 0.03 | 3.37 | 0.04 | 2.63 | 0.86 |
| 1.21 | 22.37 | 16.82 | 3.10 | 0.73 | 0.36 | 2.28 | 2.71 | 0.54 | 3.60 |
| 5.83 | 1.54 | 0.11 | 0.99 | 2.39 | 1.03 | 0.98 | 0.38 | 0.47 | 4.73 |

Compute the mean of each sample.

**SOLUTION**

The sample means were computed in *Excel*; the mean of Sample 1 is $\overline{x}_1 = 3.82$ and the mean of Sample 2 is $\overline{x}_2 = 3.64$.

Although both samples are from the same distribution, the sample means are different. In fact, if more samples of the same size were selected, the means of those samples would vary as well. *Vary* is the key word here. Since the means vary from sample to sample, the sample mean is a random variable.

Now that we have identified the sample mean, $\overline{x}$, as another random variable, we are interested in finding its *p.m.f.* and computing its mean and standard deviation.

**Example 2**: Let $X$ be random variable that gives the number obtained on one roll of a fair eight-sided die, and let $\overline{x}$ be the random variable that gives the mean number obtained on two rolls of the die.

a. Find the values of and plot the *p.m.f.* of $X$.
b. Compute the mean and standard deviation of $X$.
c. Find the values of and plot the *p.m.f.* of $\overline{x}$.
d. Compute the mean and standard deviation of $\overline{x}$.

**SOLUTION**

a. The possible values of $X$ are 1, 2, …, 8. Since the die is fair, we will assume that each of these values is equally likely. The values of and graph of the *p.m.f.* of $X$ are given below.

| $x$ | $f_X(x)$ |
|---|---|
| 1 | $1/8 = 0.125$ |
| 2 | $1/8 = 0.125$ |
| 3 | $1/8 = 0.125$ |
| 4 | $1/8 = 0.125$ |
| 5 | $1/8 = 0.125$ |
| 6 | $1/8 = 0.125$ |
| 7 | $1/8 = 0.125$ |
| 8 | $1/8 = 0.125$ |

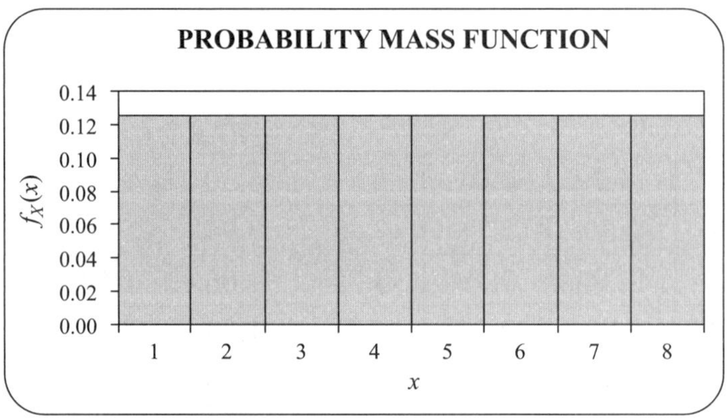


Notice that all of the bars in the graph of the *p.m.f.* have the same height and that the graph is symmetric about 4.5.

b. The mean and standard deviation of $X$ are computed below.

$$\mu_X = \sum_{\text{all } x} x \cdot f_X(x) = 1(0.125) + 2(0.125) + \cdots + 8(0.125) = 4.5$$

$$\begin{aligned} V(X) &= \sum_{\text{all } x} (x - \mu_X)^2 \cdot f_X(x) \\ &= (1-4.5)^2(0.125) + (2-4.5)^2(0.125) + \cdots + (8-4.5)^2(0.125) \\ &= 5.25 \end{aligned}$$

$$\sigma_X = \sqrt{V(X)} = \sqrt{5.25} \cong 2.2913$$

c. The sample space $S$ for the experiment of rolling a fair eight-sided die twice and the corresponding sample means are given below.

$$S = \begin{Bmatrix} (1,1) & (1,2) & (1,3) & (1,4) & (1,5) & (1,6) & (1,7) & (1,8) \\ (2,1) & (2,2) & (2,3) & (2,4) & (2,5) & (2,6) & (2,7) & (2,8) \\ (3,1) & (3,2) & (3,3) & (3,4) & (3,5) & (3,6) & (3,7) & (3,8) \\ (4,1) & (4,2) & (4,3) & (4,4) & (4,5) & (4,6) & (4,7) & (4,8) \\ (5,1) & (5,2) & (5,3) & (5,4) & (5,5) & (5,6) & (5,7) & (5,8) \\ (6,1) & (6,2) & (6,3) & (6,4) & (6,5) & (6,6) & (6,7) & (6,8) \\ (7,1) & (7,2) & (7,3) & (7,4) & (7,5) & (7,6) & (7,7) & (7,8) \\ (8,1) & (8,2) & (8,3) & (8,4) & (8,5) & (8,6) & (8,7) & (8,8) \end{Bmatrix}$$

| **Mean number obtained on two rolls of an eight-sided die** | | | | | | | |
|---|---|---|---|---|---|---|---|
| 1 | 1.5 | 2 | 2.5 | 3 | 3.5 | 4 | 4.5 |
| 1.5 | 2 | 2.5 | 3 | 3.5 | 4 | 4.5 | 5 |
| 2 | 2.5 | 3 | 3.5 | 4 | 4.5 | 5 | 5.5 |
| 2.5 | 3 | 3.5 | 4 | 4.5 | 5 | 5.5 | 6 |
| 3 | 3.5 | 4 | 4.5 | 5 | 5.5 | 6 | 6.5 |
| 3.5 | 4 | 4.5 | 5 | 5.5 | 6 | 6.5 | 7 |
| 4 | 4.5 | 5 | 5.5 | 6 | 6.5 | 7 | 7.5 |
| 4.5 | 5 | 5.5 | 6 | 6.5 | 7 | 7.5 | 8 |

Notice that there 64 means ($\overline{x}_1$, $\overline{x}_2, \ldots,$ $\overline{x}_{64}$). These means are the possible values of $\overline{x}$. Since the die is fair, we will assume that each of the outcomes is equally likely. Therefore, $f_{\overline{X}}(1) = P(\overline{x} = 1) = 1/64 = 0.015625$, $f_{\overline{X}}(1.5) = P(\overline{x} = 1.5) = 2/64 = 0.031250$, …, $f_{\overline{X}}(8) = P(\overline{x} = 8) = 1/64 = 0.015625$. The values of and graph of the *p.m.f.* of $\overline{x}$ are given below.

| $\bar{x}$ | $f_{\bar{X}}(\bar{x})$ |
|---|---|
| 1.0 | $1/64 = 0.015625$ |
| 1.5 | $2/64 = 0.031250$ |
| 2.0 | $3/64 = 0.046875$ |
| 2.5 | $4/64 = 0.062500$ |
| 3.0 | $5/64 = 0.078125$ |
| 3.5 | $6/64 = 0.093750$ |
| 4.0 | $7/64 = 0.109375$ |
| 4.5 | $8/64 = 0.125000$ |
| 5.0 | $7/64 = 0.109375$ |
| 5.5 | $6/64 = 0.093750$ |
| 6.0 | $5/64 = 0.078125$ |
| 6.5 | $4/64 = 0.062500$ |
| 7.0 | $3/64 = 0.046875$ |
| 7.5 | $2/64 = 0.031250$ |
| 8.0 | $1/64 = 0.015625$ |

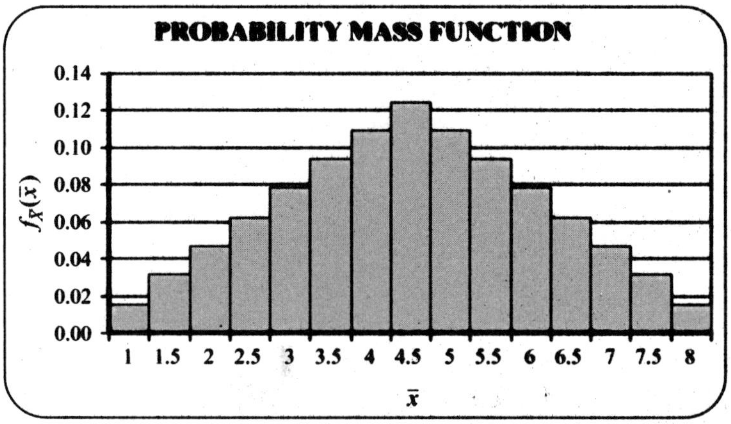


Recall that all of the bars in the graph of the *p.m.f.* of $X$ have the same height and that the graph is symmetric about 4.5. The graph of the *p.m.f.* of $\bar{x}$ is also symmetric about 4.5; however, the bars in this graph do not all have the same height. In particular, this graph has a distinct peak over the value 4.5 and tapers off as the values of $\bar{x}$ become larger or smaller.

d. The mean and standard deviation of $\bar{x}$ are computed below.

$$\begin{aligned}\mu_{\bar{x}} &= \sum_{\text{all } \bar{x}} \bar{x} \cdot f_{\bar{X}}(\bar{x}) \\ &= 1(0.015625) + 1.5(0.031250) + \cdots + 8(0.015625) \\ &= 4.5\end{aligned}$$

$$\begin{aligned}V(\bar{x}) &= \sum_{\text{all } \bar{x}} (\bar{x} - \mu_{\bar{x}})^2 \cdot f_{\bar{X}}(\bar{x}) \\ &= (1-4.5)^2(0.015625) + (1.5-4.5)^2(0.031250) + \cdots + (8-4.5)^2(0.015625) \\ &= 2.625\end{aligned}$$

$$\sigma_{\bar{x}} = \sqrt{V(\bar{x})} = \sqrt{2.625} \cong 1.6202$$

Notice that the mean of the random variable $\bar{x}$ is equal to the mean of the random variable $X$ and the standard deviation of the random variable $\bar{x}$ is equal to the standard deviation of the random variable $X$ divided by the square root of the sample size.

$$\begin{aligned}\sigma_{\overline{x}} &= \sqrt{2.625} \\ &= \sqrt{\frac{5.25}{2}} \\ &= \frac{\sqrt{5.25}}{\sqrt{2}} \\ &= \frac{\sigma_X}{\sqrt{n}}\end{aligned}$$

To summarize this important example, the graph of the *p.m.f.* of $\overline{x}$, the mean number obtained on two rolls of a fair eight-sided die, was compared to the graph of the *p.m.f.* of *X*, the number obtained on one roll of the die. Although both graphs are symmetric about 4.5, a distinct difference between the shapes of the graphs was noted. In addition, the relationship between the means of the random variables and the standard deviations of the random variables was discussed. These relationships are summarized below.

$$\mu_{\overline{x}} = \mu_X \qquad \sigma_{\overline{x}} = \frac{\sigma_X}{\sqrt{n}}$$

**Example 3:** Let *X* be a random variable that can assume only the values 0 and 1, with $P(X=0)=0.55$ and $P(X=1)=0.45$. Let $\overline{x}$ be the sample mean for a sample of size $n=10$ observations of *X*.

a. Find the values of the *p.m.f.* of $\overline{x}$.

b. Compute the mean and standard deviation of $\overline{x}$.

**SOLUTION**

a. The sum of the *X*'s, $\sum_{i=1}^{10} X_i$, is a binomial random variable with $n=10$. Therefore, the possible values of $\sum_{i=1}^{10} X_i$ are 0, 1,…,10. We can use ***BINOMDIST*** to compute the corresponding values of the *p.m.f.* Since $\overline{x} = \frac{1}{10}\sum_{i=1}^{10} X_i$, the possible values of $\overline{x}$ are 0, 0.1,…,1. Furthermore, $f_{\overline{X}}(0) = P(\overline{x}=0) = P\left(\sum_{i=1}^{10} X_i = 0\right)$,

$$f_{\overline{X}}(0.1) = P(\overline{x}=0.1) = P\left(\sum_{i=1}^{10} X_i = 1\right), \ldots, f_{\overline{X}}(1) = P(\overline{x}=1) = P\left(\sum_{i=1}^{10} X_i = 10\right)$$

Therefore, we have the following.

| $\sum_{i=1}^{10} X_i$ | $\overline{x}$ | $f_{\overline{X}}(\overline{x})$ |
|---|---|---|
| 0 | 0 | 0.002533 |
| 1 | 0.1 | 0.020724 |
| 2 | 0.2 | 0.076303 |
| 3 | 0.3 | 0.166478 |
| 4 | 0.4 | 0.238367 |
| 5 | 0.5 | 0.234033 |
| 6 | 0.6 | 0.159568 |
| 7 | 0.7 | 0.074603 |
| 8 | 0.8 | 0.022890 |
| 9 | 0.9 | 0.004162 |
| 10 | 1 | 0.000341 |

b. The mean and standard deviation of $\overline{x}$ are computed below.

$$\begin{aligned}\mu_{\overline{x}} &= \sum_{\text{all } \overline{x}} \overline{x} \cdot f_{\overline{X}}(\overline{x}) \\ &= 0(0.002533) + 0.1(0.020724) + \cdots + 1(0.000341) \\ &= 0.45\end{aligned}$$

$$\begin{aligned}V(\overline{x}) &= \sum_{\text{all } \overline{x}} (\overline{x} - \mu_{\overline{x}})^2 \cdot f_{\overline{X}}(\overline{x}) \\ &= (0-0.45)^2(0.002533) + (0.1-0.45)^2(0.020724) + \cdots + (1-0.45)^2(0.000341) \\ &= 0.02475\end{aligned}$$

$$\sigma_{\overline{x}} = \sqrt{V(\overline{x})} = \sqrt{0.02475} \cong 0.157321$$

Notice that $\mu_{\overline{x}} = \mu_X$ and $\sigma_{\overline{x}} = \dfrac{\sigma_X}{\sqrt{n}}$.

**Example 4:** Suppose that the Macnut Company grows macadamia nuts, packages the nuts, and sells them to various suppliers. They sell many forms of the macadamia nuts, including plain, salted, honey roasted, and chocolate covered. Each flavor comes in packages that contain 8, 16, or 32 ounces of macadamia nuts. Since the Macnut Company distributes its macadamia nuts all over the world, they package thousands of tons of nuts every year. In order to monitor the manufacturing process, the Macnut Company measures the weights of 50 randomly selected packages of each flavor four times per month. For this example, we will consider only samples of the 32-ounce bags of salted macadamia nuts. Use the given information to estimate the mean and standard deviation of the sample mean, and create a histogram that approximates the graph of the *p.d.f.* of the sample mean.

## SOLUTION

Note that the Macnut Company takes a sample of 32-oz. packages of salted macadamia four times a month giving a total of 48 samples per year. Consequently, there are 48 different sample means ($m_1, m_2, \ldots, m_{48}$). Each sample mean, $m_i$, is an observation of the random variable $\bar{x}$. (Note that, in this example, $m_i$ is used to denote the mean of a single sample.) The Macnut Company weighs the 50 packages in each sample and computes the average weight of the packages in the sample. The formula given below is used to compute the sample means.

$$m_i = \frac{1}{50}\sum_{j=1}^{50} x_{i,j}\,,\ i = 1,2,\ldots,48\,;\ j = 1,2,\ldots,50$$

In this formula, $i$ corresponds to a particular sample and the $x_{i,j}$ 's are the observations within that sample. For example, $m_{30}$ corresponds to the mean of the 30th sample and $x_{30,15}$ corresponds to the 15th observation in the 30th sample.

The table below shows the means (in ounces) of the 48 yearly samples.

| Monthly Sample | Jan. | Feb. | March | April | May | June | July | Aug. | Sept. | Oct. | Nov. | Dec. |
|---|---|---|---|---|---|---|---|---|---|---|---|---|
| 1 | 31.79 | 32.21 | 32.36 | 32.07 | 31.93 | 31.80 | 31.98 | 31.68 | 31.87 | 31.64 | 32.26 | 31.92 |
| 2 | 32.24 | 32.28 | 32.05 | 32.13 | 32.36 | 31.54 | 32.18 | 31.79 | 31.50 | 32.11 | 31.48 | 31.78 |
| 3 | 31.60 | 32.10 | 31.82 | 31.91 | 32.25 | 31.88 | 32.16 | 32.07 | 31.91 | 32.59 | 32.38 | 32.55 |
| 4 | 32.11 | 31.93 | 31.72 | 31.83 | 31.92 | 31.36 | 31.92 | 31.98 | 31.74 | 32.56 | 31.91 | 31.74 |

Now we estimate the mean, $\mu_{\bar{x}}$, of $\bar{x}$. What we are doing is calculating the mean of the 48 means shown above.

$$\begin{aligned}\mu_{\bar{x}} &\cong \bar{m} \\ &= \frac{1}{48}\sum_{i=1}^{48} m_i \\ &= \frac{1}{48}(31.79 + 32.24 + \cdots + 31.74) \\ &= 31.98\end{aligned}$$

The mean weight of all 32-oz. packages of salted macadamia nuts should be 32 ounces. Notice that our estimate of the mean weight is extremely close to the desired value.

Next we estimate the standard deviation, $\sigma_{\bar{x}}$, of $\bar{x}$. If we knew the theoretical standard deviation of $X$, we could use the relationship $\sigma_{\bar{x}} = \sigma_X/\sqrt{n}$ to find the exact value of $\sigma_{\bar{x}}$. Since the value of $\sigma_X$ is unknown, we will use $s$, the sample standard deviation of the 48 means, to estimate $\sigma_{\bar{x}}$.

$$\begin{aligned} V(\overline{x}) &\cong s^2 \\ &= \frac{1}{48-1}\sum_{i=1}^{48}(m_i - \overline{m})^2 \\ &= \frac{1}{47}[(31.79-31.98)^2 + (132.24-31.98)^2 + \cdots + (31.74-31.98)^2] \\ &= 0.2861 \end{aligned}$$

$$\begin{aligned} \sigma_{\overline{x}} &\cong s \\ &= \sqrt{0.2861} \\ &\cong 0.5349 \end{aligned}$$

Finally, we approximate the graph of the *p.d.f.* of $\overline{x}$. We use the histogram function in *Excel* to sort the 48 means into 11 bins and then create a histogram that approximates the *p.d.f.* of $\overline{x}$. The output is shown below.

| *Bin* | *Frequency* | Relative Frequency | Adjusted Relative Frequency |
|---|---|---|---|
| 31.3 | 0 | 0.00000 | 0.00000 |
| 31.45 | 1 | 0.02083 | 0.13889 |
| 31.6 | 4 | 0.08333 | 0.55556 |
| 31.75 | 5 | 0.10417 | 0.69444 |
| 31.9 | 8 | 0.16667 | 1.11111 |
| 32.05 | 11 | 0.22917 | 1.52778 |
| 32.2 | 8 | 0.16667 | 1.11111 |
| 32.35 | 5 | 0.10417 | 0.69444 |
| 32.5 | 3 | 0.06250 | 0.41667 |
| 32.65 | 3 | 0.06250 | 0.41667 |
| 32.8 | 0 | 0.00000 | 0.00000 |

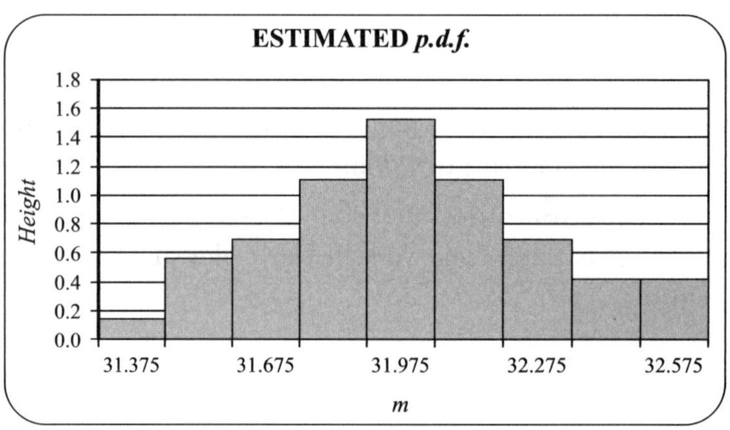

Notice that the histogram resembles a bell curve. This type of curve is associated with a special family of continuous random variables, which will be discussed in the next section.

**Example 5:** Let $S_2$ be the standardized sample mean for a sample of size $n = 2$ observations of the random variable in **Example 2**.

a. Find the values of the *p.m.f.* of $S_2$.

b. Compute the mean and standard deviation of $S_2$.

**SOLUTION**

a. The values of $S_2$ are given by

$$s_2 = \frac{\overline{x} - \mu_{\overline{x}}}{\sigma_{\overline{x}}} = \frac{\overline{x} - 4.5}{\sqrt{2.625}}$$

and the corresponding values of the *p.m.f.* are given by

$$f_{S_2}(s_2) = f_{\overline{X}}(\overline{x}).$$

Therefore, we have the following.

| $s_2$ | $f_{S_2}(s_2)$ |
|---|---|
| -2.160247 | 0.015625 |
| -1.851640 | 0.031250 |
| -1.543033 | 0.046875 |
| -1.234427 | 0.062500 |
| -0.925820 | 0.078125 |
| -0.617213 | 0.093750 |
| -0.308607 | 0.109375 |
| 0 | 0.125000 |
| 0.308607 | 0.109375 |
| 0.617213 | 0.093750 |
| 0.925820 | 0.078125 |
| 1.234427 | 0.062500 |
| 1.543033 | 0.046875 |
| 1.851640 | 0.031250 |
| 2.160247 | 0.015625 |

b. The mean and standard deviation of $S_2$ are computed below.

$$\mu_{S_2} = \sum_{\text{all } s_2} s_2 \cdot f_{S_2}(s_2)$$
$$= (-2.16025)(0.015625) + (-1.85164)(0.031250) + \cdots + 2.160247(0.015625)$$
$$= 0$$
$$V(S_2) = \sum_{\text{all } s_2} (s_2 - \mu_{S_2})^2 \cdot f_{S_2}(s_2)$$
$$= (-2.16025 - 0)^2 (0.015625) + (-1.85164 - 0)^2 (0.031250) + \cdots + (2.160247 - 0)^2 (0.015625)$$
$$= 1$$
$$\sigma_{S_2} = \sqrt{V(S_2)} = \sqrt{1} = 1$$

**Example 6:** Let $S_{10}$ be the standardized sample mean for a sample of size $n = 10$ observations of the random variable in **Example 3**.

a. Find the values of the *p.m.f.* of $S_{10}$.

b. Compute the mean and standard deviation of $S_{10}$.

**SOLUTION**

a. The values of $S_{10}$ are given by

$$s_{10} = \frac{\overline{x} - \mu_{\overline{X}}}{\sigma_{\overline{X}}} = \frac{\overline{x} - 0.45}{\sqrt{0.02475}}$$

and the corresponding values of the *p.m.f.* are given by

$$f_{S_{10}}(s_{10}) = f_{\overline{X}}(\overline{x}).$$

Therefore, we have the following.

| $s_{10}$ | $f_{S_{10}}(s_{10})$ |
|---|---|
| -2.860388 | 0.002533 |
| -2.224746 | 0.020724 |
| -1.589104 | 0.076303 |
| -0.953463 | 0.166478 |
| -0.317821 | 0.238367 |
| 0.317821 | 0.234033 |
| 0.953463 | 0.159568 |
| 1.589104 | 0.074603 |
| 2.224746 | 0.022890 |
| 2.860388 | 0.004162 |
| 3.496029 | 0.000341 |

b. The mean, variance, and standard deviation of $S_{10}$ are computed below.

$$\mu_{S_{10}} = \sum_{\text{all } s_{10}} s_{10} \cdot f_{S_{10}}(s_{10})$$
$$= (-2.860388)(0.002533) + (-2.224746)(0.020724) + \cdots + 3.496029(0.000341)$$
$$= 0$$
$$V(S_{10}) = \sum_{\text{all } s_{10}} (s_{10} - \mu_{S_{10}})^2 \cdot f_{S_{10}}(s_{10})$$
$$= (-2.860388 - 0)^2 (0.002533) + (-2.224746 - 0)^2 (0.020724) + \cdots + (3.496029 - 0)^2 (0.000341)$$
$$= 1$$
$$\sigma_{S_{10}} = \sqrt{V(S_{10})} = \sqrt{1} = 1$$

If $X$ is any random variable, then the standardized value of $\overline{x}$ is given by

$$\frac{\overline{x} - \mu_{\overline{x}}}{\sigma_{\overline{x}}} = \frac{\overline{x} - \mu_X}{\sigma_X / \sqrt{n}}.$$

This transformation yields a random variable, which has a mean of 0 and standard deviation of 1. If $n$ is large, then that random variable has a bell-shaped density curve that is centered about zero with inflection points at $(-1, f(-1))$ and $(1, f(1))$. Generally a sample size of $n \geq 30$ is large enough.

The graph of that bell-shaped density curve is given below.

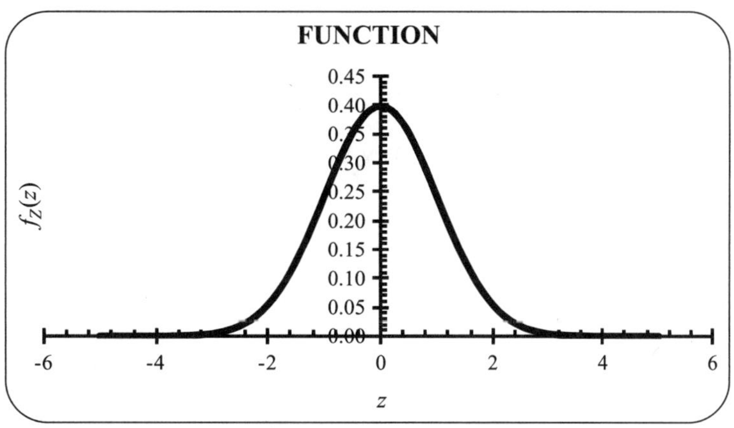


This is the graph of the *p.d.f.* of $Z$, the standard normal random variable, where the density function of $Z$ is given by

$$f_Z(z) = \frac{1}{\sqrt{2\pi}} e^{-0.5z^2}.$$

## *Exercises*

1. Four observations of a random variable $W$ with $\mu_W = 10$ and $\sigma_W = 4$ are given below.

8, 12, 6, 19

a. Compute the sample mean, $\overline{w}$.

b. Compute the sample standard deviation, $s$.

c. Compute $\mu_{\overline{w}}$.

d. Compute $\sigma_{\overline{w}}$.

**SOLUTION**

a. $\overline{w} = 11.25$

b. $s \cong 5.74$

c. $\mu_{\overline{w}} = 10$

d. $\sigma_{\overline{w}} = 2$

2. Approximately 12% of the students taking *Mathematics For Business Decisions Part 2: Calculus and Optimization* plan to major in Accounting, approximately 8% plan to major in Finance, approximately 12% plan to major in Marketing, approximately 28% plan to major in MIS, and the remainder plan to major in some other area. Let the random variable $W$ equal 1 if a randomly selected student plans to major in Accounting and 0 if the student does not plan to major in Accounting.

   a. Compute $\mu_W$.

   b. Compute $\sigma_W$.

   c. Compute $\mu_{\overline{w}}$ for samples of 25 randomly selected students.

   d. Compute $\sigma_{\overline{w}}$ for samples of 25 randomly selected students.

**SOLUTION**

a. $\mu_W = 0.12$

b. $\sigma_W \cong 0.325$

c. $\mu_{\overline{w}} = 0.12$

d. $\sigma_{\overline{w}} \cong 0.065$

3. The *p.m.f.* of a finite random variable $X$ is given below.

| $x$ | 0 | 5 | 10 | 15 |
|---|---|---|---|---|
| $f_X(x)$ | 0.4 | 0.3 | 0.2 | 0.1 |

   Let $\overline{x}$ be the sample mean for a sample of size $n = 2$ observations of $X$.

   a. Compute $\mu_X$.

   b. Compute $\sigma_X$

   c. Find the *p.m.f.* of $\overline{x}$.

d. Compute $\mu_{\overline{x}}$.

e. Compute $\sigma_{\overline{x}}$.

**SOLUTION**

a. $\mu_X = 5$

b. $\sigma_X = 5$

c.

| $\overline{x}$ | $f_{\overline{X}}(\overline{x})$ |
|---|---|
| 0 | 0.16 |
| 2.5 | 0.24 |
| 5 | 0.25 |
| 7.5 | 0.20 |
| 10 | 0.10 |
| 12.5 | 0.04 |
| 15 | 0.01 |

d. $\mu_{\overline{x}} = 5$

e. $\sigma_{\overline{x}} \cong 3.5355$

4. Let $X$ be a random variable that can assume only the values 0 and 1, with $P(X = 0) = 0.45$ and $P(X = 1) = 0.55$. Let $\overline{x}$ be the sample mean for a sample of size $n = 10$ observations of $X$.

   a. Find the *p.m.f.* of $\overline{x}$.

   b. Compute $\mu_{\overline{x}}$.

   c. Compute $\sigma_{\overline{x}}$.

**SOLUTION**

a.

| $\overline{x}$ | $f_{\overline{X}}(\overline{x})$ |
|---|---|
| 0 | 0.000341 |
| 0.1 | 0.004162 |
| 0.2 | 0.022890 |
| 0.3 | 0.074603 |
| 0.4 | 0.159568 |
| 0.5 | 0.234033 |
| 0.6 | 0.238367 |
| 0.7 | 0.166478 |
| 0.8 | 0.076303 |
| 0.9 | 0.020724 |
| 1 | 0.002533 |

b. $\mu_{\overline{x}} = 0.55$

c. $\sigma_{\overline{x}} \cong 0.157321$

5. Let $S_2$ be the standardized sample mean for a sample of size $n = 2$ observations of the random variable in **Exercise 3**.

   a. Find the *p.m.f.* of $S_2$.

   b. Compute $\mu_{S_2}$.

   c. Compute $\sigma_{S_2}$.

**SOLUTION**

a.

| $s_2$ | $f_{S_2}(s_2)$ |
|---|---|
| -1.414214 | 0.16 |
| -0.707107 | 0.24 |
| 0 | 0.25 |
| 0.707107 | 0.20 |
| 1.414214 | 0.10 |
| 2.121320 | 0.04 |
| 2.828427 | 0.01 |

b. $\mu_{S_2} = 0$

c. $\sigma_{S_2} = 1$

6. Let $S_{10}$ be the standardized sample mean for a sample of size $n = 10$ observations of the random variable in **Exercise 4**.

   a. Find the *p.m.f.* of $S_{10}$.

   b. Compute $\mu_{S_{10}}$.

   c. Compute $\sigma_{S_{10}}$.

**SOLUTION**

a.

| $s_{10}$ | $f_{S_{10}}(s_{10})$ |
|---|---|
| -3.496029 | 0.000341 |
| -2.860388 | 0.004162 |
| -2.224746 | 0.022890 |
| -1.589104 | 0.074603 |
| -0.953463 | 0.159568 |
| -0.317821 | 0.234033 |
| 0.317821 | 0.238367 |
| 0.953463 | 0.166478 |
| 1.589104 | 0.076303 |
| 2.224746 | 0.020724 |
| 2.860388 | 0.002533 |

b. $\mu_{S_{10}} = 0$

c. $\sigma_{S_{10}} = 1$

7. Explain the difference between $\mu_X$ and $\mu_{\overline{x}}$ and give the formulas that are used to compute them.

**SOLUTION**

$\mu_X$ is the mean of the random variable *X*.
$\mu_{\overline{x}}$ is the mean of the random variable $\overline{x}$; equivalently, $\mu_{\overline{x}}$ is the mean of the sample means.

The mean of a finite random variable *X* is given by

$$\mu_X = \sum_{\text{all } x} x \cdot f_X(x).$$

The mean of a continuous random variable *X* is given by

$$\mu_X = \int_{-\infty}^{\infty} x \cdot f_X(x)\, dx.$$

The mean of the random variable $\overline{x}$ is given by $\mu_{\overline{x}} = \mu_X$.
Although $\mu_X$ and $\mu_{\overline{x}}$ have different meanings, they have the same value.

8. Explain the difference between $\mu_X$ and $\overline{x}$ and give the formulas that are used to compute them.

**SOLUTION**

$\mu_X$ is the mean of the random variable *X*.

$\overline{x}$ is the mean of a sample.

The mean of a finite random variable *X* is given by $\mu_X = \sum_{\text{all } x} x \cdot f_X(x)$.

The mean of a continuous random variable *X* is given by $\mu_X = \int_{-\infty}^{\infty} x \cdot f_X(x)\, dx$.

The mean of a sample is given by $\overline{x} = \frac{1}{n}\sum_{i=1}^{n} x_i$.

9. Explain the difference between $E(X)$ and $\mu_X$.

**SOLUTION**

There is no difference. The expected value of a random variable *X* can be expressed as $E(X)$ or $\mu_X$.

10. Explain the difference between $\sigma_X$ and $s$ and give the formulas that are used to compute them.

**SOLUTION**

$\sigma_X$ is the standard deviation of the random variable $X$.

$s$ is the standard deviation of a sample of size $n$.

The standard deviation of a finite random variable $X$ is given by

$$\sigma_X = \sqrt{\sum_{\text{all } x} (x - \mu_X)^2 \cdot f_X(x)} .$$

The standard deviation of a continuous random variable $X$ is given by

$$\sigma_X = \sqrt{\int_{-\infty}^{\infty} (x - \mu_X)^2 \cdot f_X(x)\, dx} .$$

The standard deviation of a sample is given by

$$s = \sqrt{\frac{1}{n-1} \sum_{i=1}^{n} (x - \bar{x})^2} .$$

11. Explain the difference between $\sigma_X$ and $\sigma_{\bar{x}}$ and give the formulas that are used to compute them

**SOLUTION**

$\sigma_X$ is the standard deviation of the random variable $X$.

$\sigma_{\bar{x}}$ is the standard deviation of the random variable $\bar{x}$.

The standard deviation of a finite random variable $X$ is given by

$$\sigma_X = \sqrt{\sum_{\text{all } x} (x - \mu_X)^2 \cdot f_X(x)}.$$

The standard deviation of a continuous random variable $X$ is given by

$$\sigma_X = \sqrt{\int_{-\infty}^{\infty} (x - \mu_X)^2 \cdot f_X(x)\, dx}.$$

The standard deviation of the random variable $\bar{x}$ is given by $\sigma_{\bar{x}} = \sigma_X / \sqrt{n}$.

# Normal Distributions

## *Standard Normal*

The standard normal distribution can be used to derive the formula for a confidence interval for the mean $\mu_X$ of a random variable *X*. We will illustrate the derivation of an 80% confidence interval. First we find the value $z_0$ such that $P(-z_0 \le Z \le z_0) = 0.80$. This is equivalent to finding the value $z_0$ such that $P(0 \le Z \le z_0) = 0.80/2 = 0.40$.

The probability that *Z* is between 0 and $z_0$ is given by

$$\int_0^{z_0} \frac{1}{\sqrt{2\pi}} e^{-0.5z^2}\, dz = 0.40.$$

Therefore, in order to find $z_0$, we use ***Integrating.xls*** to find the value *b* such that

$$\int_0^{b} \frac{1}{\sqrt{2\pi}} e^{-0.5z^2}\, dz = 0.40.$$

| Definition | Computation | | Plot Interval | | Integration Interval | | $\int_a^b f(x)\,dx$ |
|---|---|---|---|---|---|---|---|
| Formula for $f(x)$ | $x$ | $f(x)$ | $A$ | $B$ | $a$ | $b$ | |
| 0.39894 | | 0.39894 | -5 | 5 | 0 | 1.28155 | 0.4000 |

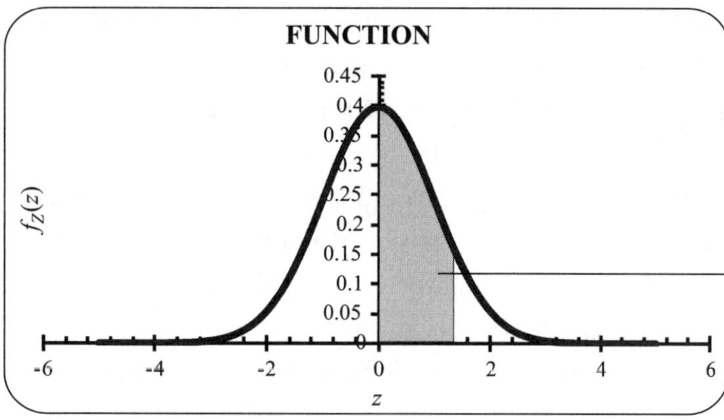


After substituting $(\bar{x} - \mu_{\bar{x}})/\sigma_{\bar{x}}$ for *Z*, we have

$$P\left(-1.282 \le \frac{\bar{x} - \mu_{\bar{x}}}{\sigma_{\bar{x}}} \le 1.282\right) = 0.80.$$

We use the algebraic manipulations given below to rewrite the expression inside of the parentheses.

$$P\left(-1.282 \le \frac{\bar{x} - \mu_{\bar{x}}}{\sigma_{\bar{x}}} \le 1.282\right) = 0.80$$

Multiply each part of the inequality by $\sigma_{\bar{x}}$.

$$P(-1.282\sigma_{\bar{x}} \le \bar{x} - \mu_{\bar{x}} \le 1.282\sigma_{\bar{x}}) = 0.80$$

Subtract $\bar{x}$ from each part of the inequality.

$$P(-\bar{x} - 1.282\sigma_{\bar{x}} \le -\mu_{\bar{x}} \le -\bar{x} + 1.282\sigma_{\bar{x}}) = 0.80$$

Multiply each part of the inequality by –1 and reverse the direction of the inequalities.

$P(\overline{x}+1.282\sigma_{\overline{x}} \geq \mu_{\overline{x}} \geq \overline{x}-1.282\sigma_{\overline{x}})=0.80$ Apply $b \geq x \geq a$ is equivalent to $a \leq x \leq b$.

$P(\overline{x}-1.282\sigma_{\overline{x}} \leq \mu_{\overline{x}} \leq \overline{x}+1.282\sigma_{\overline{x}})=0.80$ Substitute $\mu_{\overline{x}}=\mu_X$ and $\sigma_{\overline{x}}=\frac{\sigma_X}{\sqrt{n}}$.

$$P\left(\overline{x}-1.282\frac{\sigma_X}{\sqrt{n}} \leq \mu_X \leq \overline{x}+1.282\frac{\sigma_X}{\sqrt{n}}\right)=0.80$$

If $\sigma_X$ is not known, then the sample standard deviation $s$ is used to approximate $\sigma_X$.
An 80% confidence interval for $\mu_X$ is given by

$$\left(\overline{x}-1.282\frac{\sigma_X}{\sqrt{n}},\ \overline{x}+1.282\frac{\sigma_X}{\sqrt{n}}\right).$$

Approximately 80% of all such intervals will contain $\mu_X$. Alternatively, we are 80% confident that the interval

$$\left(\overline{x}-1.282\frac{\sigma_X}{\sqrt{n}},\ \overline{x}+1.282\frac{\sigma_X}{\sqrt{n}}\right)$$

contains $\mu_X$.

**Example 1:** The Macnut Company, from **Example 4** of **The Sample Mean**, wants to minimize the number of ounces that each package goes over the 8, 16, or 32 ounces. Also, they want to give the customer the most value for their money so they want to minimize the number of ounces that each package goes under the 8, 16, or 32 ounces. In other words, the company wants the weight of each package to be as close as possible to the stated weight. It is too time consuming and expensive to test the weight of every package; therefore, samples of size 50 are selected and a confidence interval for the actual weight is computed from each sample. For this example we will use the observations in the first of 48 samples to construct an 80% confidence interval for the true mean weight of 32-oz. packages of salted macademia nuts. Those observations are given below.

| | | | | | | | | | |
|---|---|---|---|---|---|---|---|---|---|
| 31.40 | 35.47 | 30.62 | 27.76 | 31.35 | 30.97 | 35.32 | 31.83 | 29.11 | 32.06 |
| 29.44 | 27.63 | 28.62 | 30.86 | 31.26 | 35.94 | 28.78 | 30.95 | 30.31 | 31.35 |
| 32.49 | 31.53 | 28.31 | 31.19 | 34.69 | 33.73 | 33.08 | 33.35 | 28.96 | 36.39 |
| 34.55 | 34.19 | 30.04 | 32.27 | 31.83 | 36.75 | 33.80 | 31.24 | 31.27 | 28.52 |
| 34.40 | 29.83 | 30.45 | 31.27 | 31.63 | 30.69 | 35.84 | 33.52 | 31.94 | 30.53 |

**SOLUTION**

First we compute the mean and standard deviation of the sample.

$$\overline{x} = \frac{1}{n}\sum_{i=1}^{n} x_i = \frac{1}{50}(31.40 + 29.44 + \cdots + 30.53) \cong 31.79$$

$$\begin{aligned} s^2 &= \frac{1}{n-1}\sum_{i-1}^{n}(x_i - \overline{x}) \\ &= \frac{1}{49}[(31.40-31.79)^2 + (29.44-31.79)^2 + \cdots + (30.53-31.79)^2] \\ &\cong 5.41 \end{aligned}$$

$$s = \sqrt{s^2} = \sqrt{5.41} \cong 2.33$$

Recall that $P(-1.282 \le Z \le 1.282) = 0.80$. We now have all the information that is needed to construct the confidence interval.

An 80% confidence interval for the true mean weight of 32-oz. packages of salted macadamia nuts is

$$\left(31.79 - 1.282 \cdot \frac{2.33}{\sqrt{50}},\ 31.79 + 1.282 \cdot \frac{2.33}{\sqrt{50}}\right) = (31.368, 32.212)\cdot$$

We are 80% confident that the interval $(31.368, 32.2123)$ contains the true mean weight of 32-oz. packages of salted macadamia nuts. Since the stated weight of the packages (32 ounces) is contained within this interval, the data do not indicate that the packages are being improperly filled.

**Example 2:** Using the information in **Example 1**, construct a 90% confidence interval for the true mean weight of 32-oz. packages of salted macadamia nuts.

**SOLUTION**

First we find the value $z_0$ such that $P(-z_0 \le Z \le z_0) = 0.90$. This is equivalent to finding the value $z_0$ such that $P(0 \le Z \le z_0) = 0.90/2 = 0.45$.

| Definition | Computation | | Plot Interval | | Integration Interval | | $\int_a^b f(x)\,dx$ |
|---|---|---|---|---|---|---|---|
| Formula for $f(x)$ | $x$ | $f(x)$ | $A$ | $B$ | $a$ | $b$ | |
| 0.39894 | | 0.39894 | -5 | 5 | 0 | 1.64486 | 0.4500 |

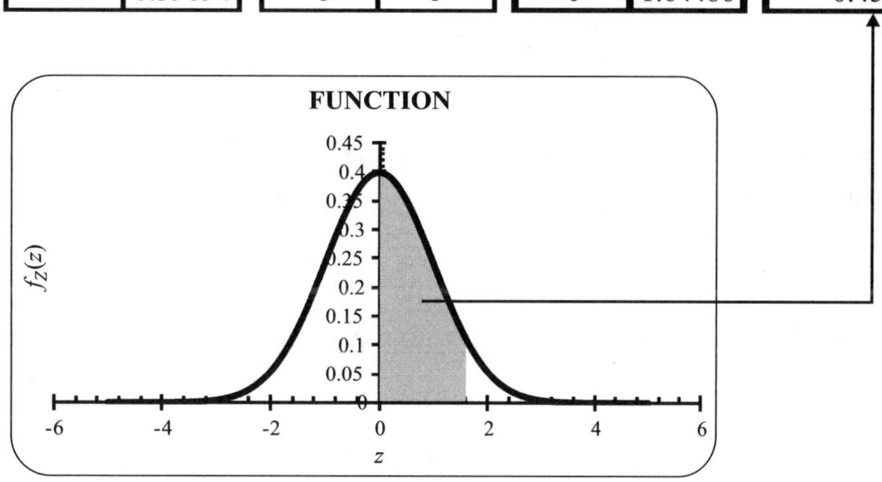

Now that we have found $z_0$, a 90% confidence interval for $\mu_X$ is given by

$$\left(\bar{x} - 1.645\frac{\sigma_X}{\sqrt{n}}, \bar{x} + 1.645\frac{\sigma_X}{\sqrt{n}}\right).$$

Using $\bar{x} = 31.79$, $\sigma_X \cong s = 2.33$, and $n = 50$, a 90% confidence interval for the true mean weight of 32 oz. packages of salted macadamia nuts is

$$\left(31.79 - 1.645 \cdot \frac{2.33}{\sqrt{50}},\ 31.79 + 1.645 \cdot \frac{2.33}{\sqrt{50}}\right) = (31.248,\ 32.332).$$

We are 90% confident that the interval (31.248,32.332) contains the true mean weight of 32-oz. packages of salted macadamia nuts. Once again the stated weight of the packages (32 ounces) is contained within this interval. The data do not indicate that the packages are being improperly filled. Diagrams of this confidence interval and the 80% confidence interval are given below.

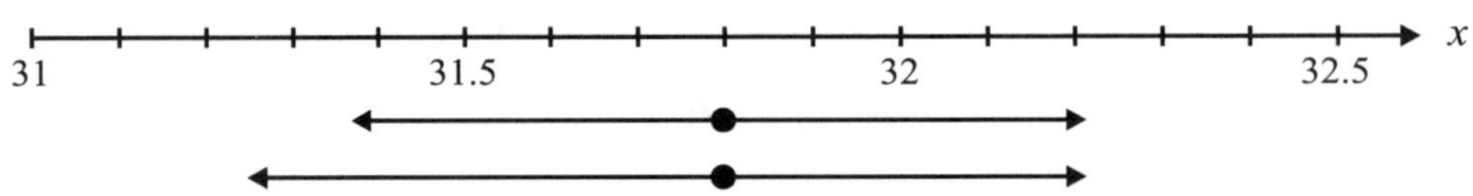


Note that the 90% confidence interval computed in **Example 2** is wider than the 80% confidence interval computed in **Example 1**. A 95% confidence interval would be even wider.

**Example 3:** Which of the two functions plotted below is the *p.d.f.* of the standard normal random variable?

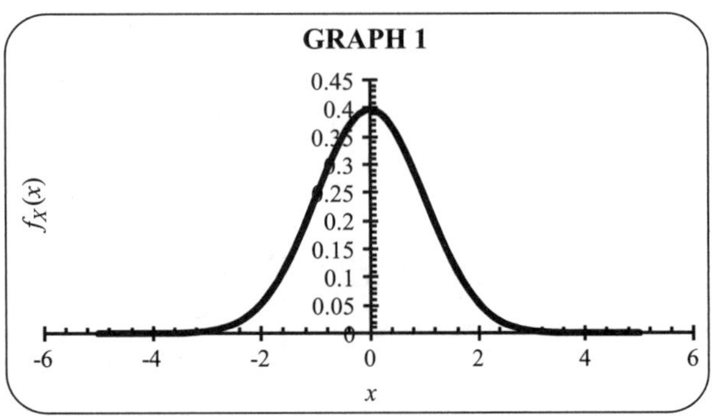


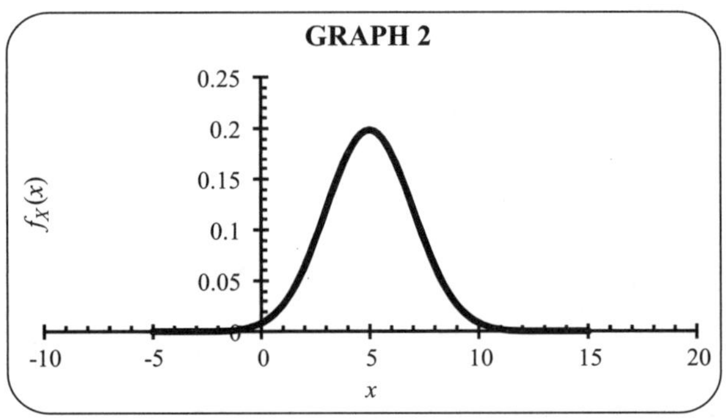


**SOLUTION**

The random variable whose *p.d.f.* is shown in Graph 1 has a mean of zero and appears to have a standard deviation of one.

## *General Normal*

The *p.d.f.* of a normal random variable *X* with mean $\mu_X$ and standard deviation $\sigma_X$ is given by

$$f_X(x) = \frac{1}{\sigma_X\sqrt{2\pi}} e^{-0.5\left(\frac{x-\mu_X}{\sigma_X}\right)^2}.$$

**Example 4:** A company that produces snack foods uses a machine to package 454-gram bags of pretzels. However, the variability inherent in any machine causes the actual weights of the bags to vary. Let $W$ be the net weight of a randomly selected bag of pretzels. Then $W$ is a normal random variable with mean $\mu_W = 454$ grams and standard deviation $\sigma_W = 7.8$ grams.

a. Find a formula for the *p.d.f.* of $W$.

b. Set up, but do not evaluate, an integral that gives the probability that a randomly selected bag of pretzels weighs between 450 and 460 grams.

c. Use ***Integrating.xls*** to evaluate the integral in Part b.

**SOLUTION**

a. $f_W(w) = \dfrac{1}{7.8\sqrt{2\pi}} e^{-0.5\left(\frac{w-454}{7.8}\right)^2}$

b. $P(450 \le W \le 460) = \displaystyle\int_{450}^{460} \frac{1}{7.8\sqrt{2\pi}} e^{-0.5\left(\frac{w-454}{7.8}\right)^2} dw$

c. $P(450 \le W \le 460) = \displaystyle\int_{450}^{460} \frac{1}{7.8\sqrt{2\pi}} e^{-0.5\left(\frac{w-454}{7.8}\right)^2} dw = 0.4751$

| Definition | Computation | | Plot Interval | | Integration Interval | | $\int_a^b f(x)\,dx$ |
|---|---|---|---|---|---|---|---|
| Formula for $f(x)$ | $x$ | $f(x)$ | $A$ | $B$ | $a$ | $b$ | |
| 0 | | 0 | 415 | 493 | 450 | 460 | 0.4751 |

Formula for $f(x)$: =1/(7.8*SQRT(2*PI()))*EXP(-0.5*((x-454)/7.8)^2)

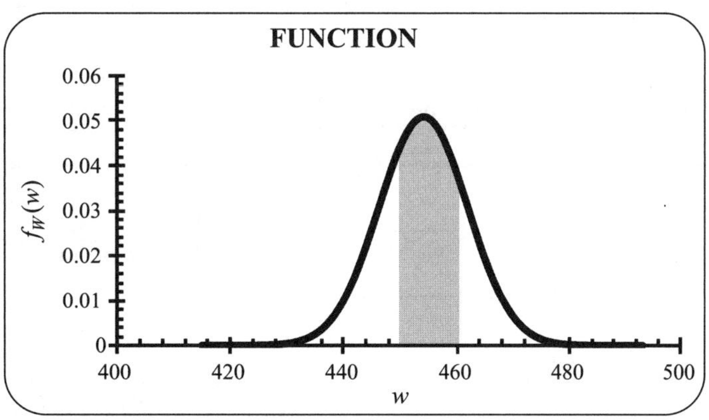


**Example 5:** The *p.d.f.* of a normal random variable $X$ is given below.

$$f_X(x) = \frac{1}{5\sqrt{2\pi}} e^{-0.5\left(\frac{x-10}{5}\right)^2}$$

a. Use ***Integrating.xls*** to compute $P(7 \le X \le 12)$.

b. Use ***Integrating.xls*** to compute the mean, $\mu_X$, of $X$.

c. Use ***Integrating.xls*** to compute the variance, $V(X)$, of $X$.

d. Compute the standard deviation, $\sigma_X$, of $X$.

**SOLUTION**

a. $$P(7 \le X \le 12) = \int_{7}^{12} \frac{1}{5\sqrt{2\pi}} e^{-0.5\left(\frac{x-10}{5}\right)} dx = 0.3812$$

| Definition | Computation | | Plot Interval | | Integration Interval | | $\int_a^b f(x)\,dx$ |
|---|---|---|---|---|---|---|---|
| **Formula for *f*(*x*)** | ***x*** | ***f*(*x*)** | ***A*** | ***B*** | ***a*** | ***b*** | |
| 0.0108 | | 0.0108 | -15 | 35 | 7 | 12 | 0.3812 |

Formula for $f(x)$: =1/(5*SQRT(2*PI()))*EXP(-0.5*((x-10)/5)^2)

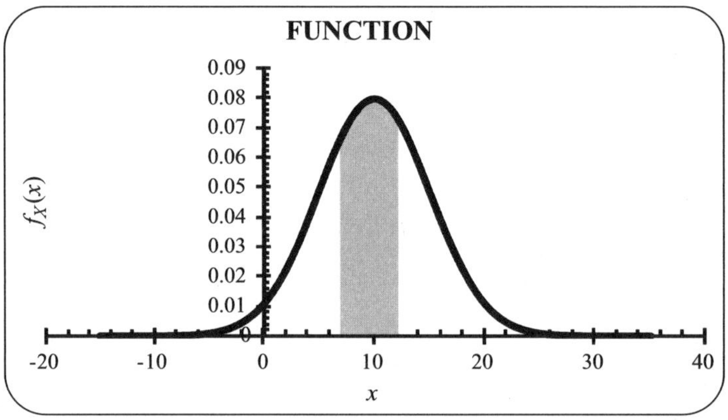


b. $$\mu_X = \int_{-\infty}^{\infty} x \cdot f_X(x)\, dx = \int_{-\infty}^{\infty} x \cdot \frac{1}{5\sqrt{2\pi}} e^{-0.5\left(\frac{x-10}{5}\right)^2} dx = 10$$

| Definition | Computation | | Plot Interval | | Integration Interval | | $\int_a^b f(x)\,dx$ |
|---|---|---|---|---|---|---|---|
| **Formula for *f*(*x*)** | ***x*** | ***f*(*x*)** | ***A*** | ***B*** | ***a*** | ***b*** | |
| 0 | | 0 | -15 | 35 | -15 | 35 | 10.0000 |

Formula for $f(x)$: =x*1/(5*SQRT(2*PI()))*EXP(-0.5*((x-10)/5)^2)

c. $$V(X) = \int_{-\infty}^{\infty} (x-\mu_X)^2 \cdot f_X(x)\, dx = \int_{-\infty}^{\infty} (x-10)^2 \cdot \frac{1}{5\sqrt{2\pi}} e^{-0.5\left(\frac{x-10}{5}\right)^2} dx = 25$$

| Definition | Computation | | Plot Interval | | Integration Interval | | $\int_a^b f(x)\,dx$ |
|---|---|---|---|---|---|---|---|
| **Formula for *f*(*x*)** | ***x*** | ***f*(*x*)** | ***A*** | ***B*** | ***a*** | ***b*** | |
| 1.07982 | | 1.07982 | -15 | 35 | -15 | 35 | 24.9996 |

Formula for $f(x)$: =(x-10)^2*1/(5*SQRT(2*PI()))*EXP(-0.5*((x-10)/5)^2)

d. $\sigma_X = \sqrt{V(X)} = \sqrt{25} = 5$

**Example 6:** Find the indicated values.

a. $f_Z(z) = \frac{1}{\sqrt{2\pi}} e^{-0.5z^2}$ $\mu_Z =$ $\sigma_Z =$ $V(Z) =$

b. $f_T(t) = \frac{1}{3\sqrt{2\pi}} e^{-0.5\left(\frac{t-8}{3}\right)^2}$ $\mu_T =$ $\sigma_T =$ $V(T) =$

**SOLUTION**

a. $f_Z(z)$ is the *p.d.f.* of the standard normal random variable *Z*; therefore, $\mu_Z = 0$, $\sigma_Z = 1$, and $V(Z) = 1$.

b. $f_T(t)$ is the *p.d.f.* of a normal random variable *T* with $\mu_T = 8$, $\sigma_T = 3$, and $V(T) = \sigma_T^2 = 9$.

## *Normal in Excel*

Up to this point we have been computing probabilities for normal random variables using ***Integrating.xls***. Probabilities for normal random variables can also be computed using the ***NORMDIST*** function in *Excel*. In fact, it is easier to compute probabilities using ***NORMDIST***. This is because entering the formula for the *p.d.f.* of a normal random variable in ***Integrating.xls*** is much more difficult than entering the arguments of the ***NORMDIST*** function.

**Example 7:** Let *T* be the random variable that gives the length of time required to complete an algebra exam. Assume that *T* has a normal distribution with mean $\mu_T = 66$ minutes and standard deviation $\sigma_T = 3$ minutes.

a. Find a formula for the *p.d.f.* of *T*

b. Use ***Integrating.xls*** to compute $P(T \leq 70)$.

c. Use the ***NORMDIST*** function to compute $P(T \leq 70)$.

**SOLUTION**

a. $f_T(t) = \frac{1}{3\sqrt{2\pi}} e^{-0.5\left(\frac{t-66}{3}\right)^2}$

b. $P(T \leq 70) = \int_{-\infty}^{70} \frac{1}{3\sqrt{2\pi}} e^{-0.5\left(\frac{t-66}{3}\right)^2} dt = 0.9088$

| Definition | Computation | | Plot Interval | | Integration Interval | | $\int_a^b f(x)\,dx$ |
|---|---|---|---|---|---|---|---|
| Formula for $f(x)$ | $x$ | $f(x)$ | $A$ | $B$ | $a$ | $b$ | |
| 1E-106 | | 1E-106 | 51 | 81 | 51 | 70 | 0.9088 |

Formula for $f(x)$: =1/(3*SQRT(2*PI()))*EXP(-0.5*((x-66)/3)^2)

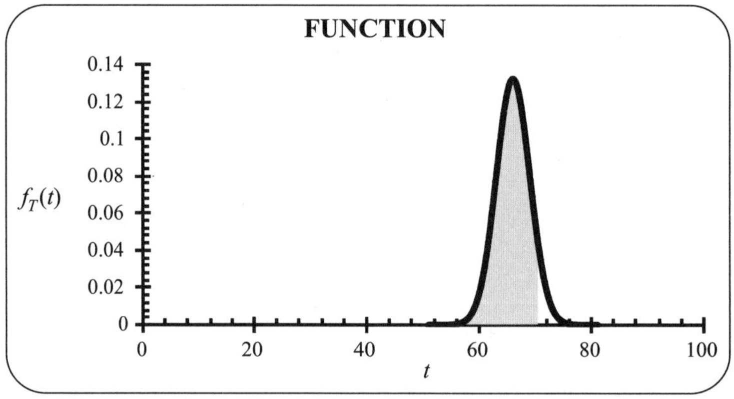


c. Insert the ***NORMDIST*** function, enter the information shown below, and click **OK**.

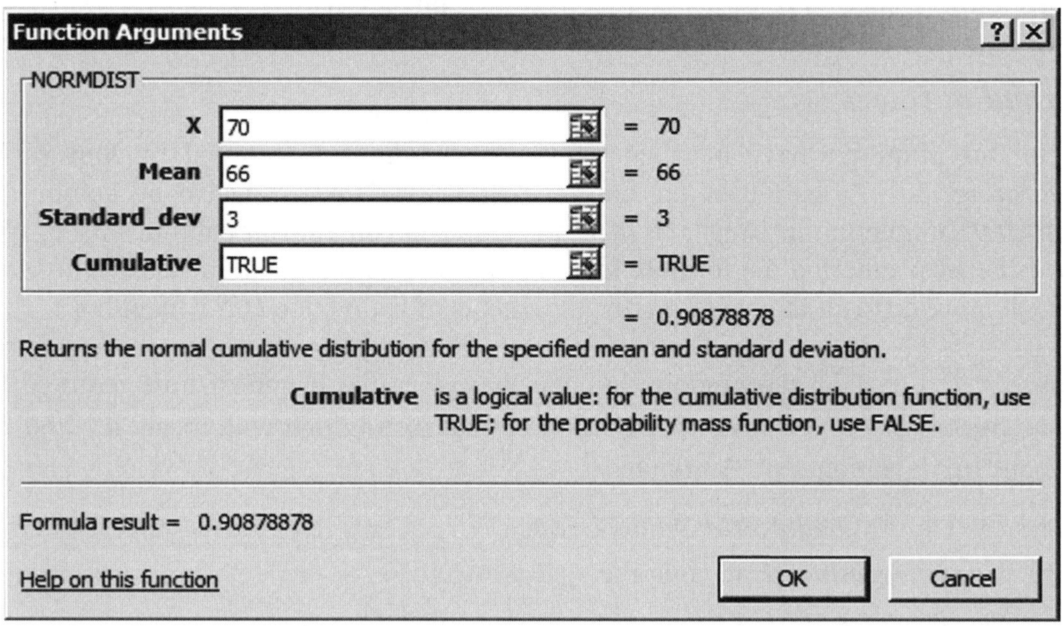


Alternatively, we could have entered the formula =NORMDIST(70,66,3,TRUE) directly. $P(T \le 70) = 0.9088$. This is the same as the result found in Part b. As you can see, entering the arguments of the ***NORMDIST*** function is much less complicated than entering the formula for the *p.d.f.* of *X*.

**Example 8:** A study published in the Accounting Review showed that the mean annual salary and bonus compensation to CEO's is \$634,961 with a standard deviation of \$124,690. Let *X* be the random variable that gives the annual compensation of a randomly selected CEO.

a. Find a formula for the *p.d.f.* of *X*.

b. Use ***Integrating.xls*** to compute the probability that the compensation to a randomly selected CEO is at least \$750,000.

c. Use ***Integrating.xls*** to compute the probability that the compensation to a randomly selected CEO is between \$500,000 and \$750,000.

d. Use the ***NORMDIST*** function to compute the probability that the compensation to a randomly selected CEO is between \$500,000 and \$750,000.

## SOLUTION

a. $$f_X(x) = \frac{1}{124{,}690\sqrt{2\pi}} e^{-0.5\left(\frac{x-634{,}961}{124{,}690}\right)^2}$$

b. $$P(X \geq 750{,}000) = \int_{750{,}000}^{\infty} \frac{1}{124{,}690\sqrt{2\pi}} e^{-0.5\left(\frac{x-634{,}961}{124{,}690}\right)^2} dx = 0.1781$$

| Definition | Computation | | Plot Interval | | Integration Interval | | $\int_a^b f(x)\,dx$ |
|---|---|---|---|---|---|---|---|
| **Formula for $f(x)$** | ***x*** | ***f*(*x*)** | ***A*** | ***B*** | ***a*** | ***b*** | |
| 7.5E-12 | | 7.5E-12 | 11511 | 1258411 | 750000 | 1258411 | 0.1781 |

Formula for $f(x)$: =1/(124690*SQRT(2*PI()))*EXP(-0.5*((x-634961)/124690)^2)

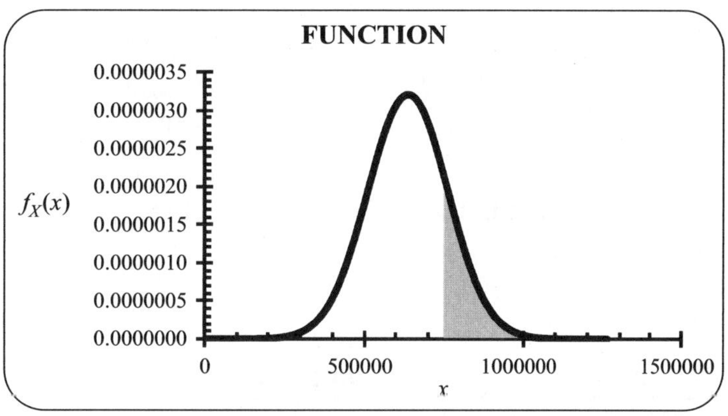


c. $$P(500{,}000 \leq X \leq 750{,}000) = \int_{500{,}000}^{750{,}000} \frac{1}{124{,}690\sqrt{2\pi}} e^{-0.5\left(\frac{x-634{,}961}{124{,}690}\right)^2} dx = 0.6823$$

| Definition | Computation | | Plot Interval | | Integration Interval | | $\int_a^b f(x)\,dx$ |
|---|---|---|---|---|---|---|---|
| **Formula for $f(x)$** | ***x*** | ***f*(*x*)** | ***A*** | ***B*** | ***a*** | ***b*** | |
| 7.5E-12 | | 7.5E-12 | 11511 | 1258411 | 500000 | 750000 | 0.6823 |

Formula for $f(x)$: =1/(124690*SQRT(2*PI()))*EXP(-0.5*((x-634961)/124690)^2)

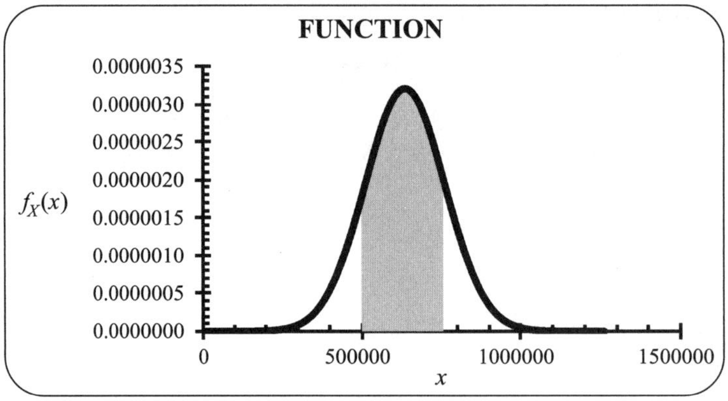


d. $P(500{,}000 \leq X \leq 750{,}000)$
$= \text{NORMDIST}(750000{,}634961{,}124690{,}\text{TRUE}) - \text{NORMDIST}(500000{,}634961{,}124690{,}\text{TRUE})$
$= 0.8218922 - 0.1395436$
$\cong 0.6823$

The ***NORMDIST*** function can also be used in ***Graphing.xls*** to plot the *p.d.f.* or *c.d.f.* for a normal random variable. As was previously mentioned, it is easier to enter the arguments of the ***NORMDIST*** function than the formula for $f_X(x)$ or $F_X(x)$.

**Example 9**: Let *X* be a normal random variable with *p.d.f.* given by

$$f_X(x) = \frac{1}{5000} e^{-0.5\left(\frac{x-45000}{5000}\right)^2}.$$

Use the ***NORMDIST*** function in ***Graphing.xls*** to plot the *p.d.f.* and *c.d.f.* of *X* over the interval [20000, 70000].

**SOLUTION**

**Definition**

| **Formula for $f(x)$** |
|---|
| 2.05595E-22 |

**Computation**

| $x$ | $f(x)$ |
|---|---|
| | 2.06E-22 |

**Plot Interval**

| $a$ | $b$ |
|---|---|
| 20000 | 70000 |

Formula for $f(x)$: =NORMDIST(x,45000,5000,FALSE)

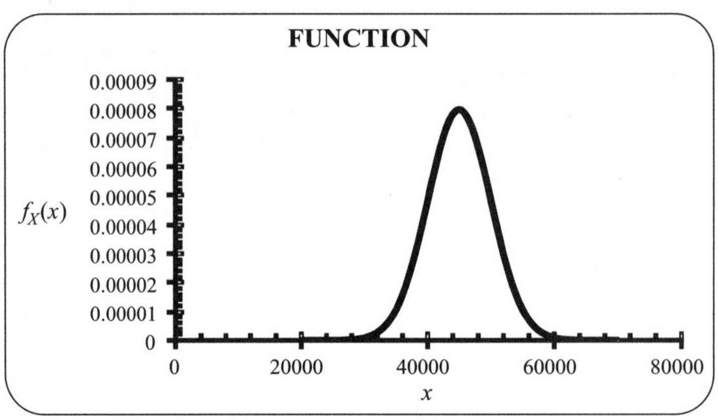

| Definition | Computation | | Plot Interval | |
|---|---|---|---|---|
| Formula for $f(x)$ | $x$ | $f(x)$ | $a$ | $b$ |
| 1.12859E-19 | | 1.13E-19 | 20000 | 70000 |

Formula for $f(x)$: =NORMDIST(x,45000,5000,TRUE)

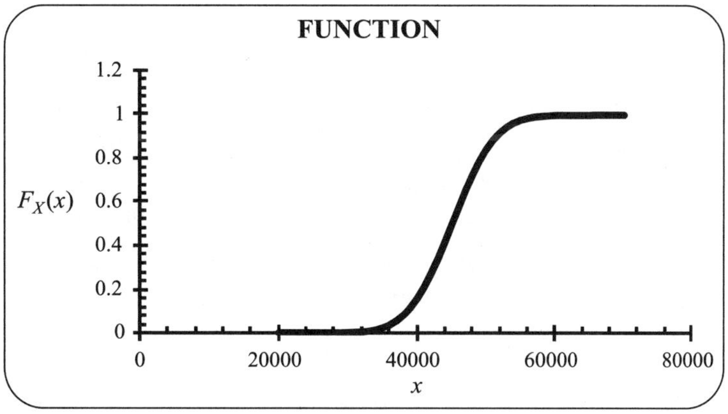


## *Exercises*

1. Let $Z$ be the standard normal random variable.
   a. Use ***Integrating.xls*** to compute $P(0.3 \le Z \le 1.5)$.
   b. Use ***NORMDIST*** to compute $P(0.3 \le Z \le 1.5)$.

### SOLUTION

a. $P(0.3 \le Z \le 1.5) = \int_{0.3}^{1.5} \frac{1}{\sqrt{2\pi}} e^{-0.5z^2} dz = 0.3153$

| Definition | Computation | | Plot Interval | | Integration Interval | | $\int_a^b f(x)\,dx$ |
|---|---|---|---|---|---|---|---|
| Formula for $f(x)$ | $x$ | $f(x)$ | $A$ | $B$ | $a$ | $b$ | |
| 0.39894 | | 0.39894 | -5 | 5 | 0.3 | 1.5 | 0.3153 |

b. $P(0.3 \le Z \le 1.5) = F_Z(1.5) - F_Z(0.3)$
$=$ NORMDIST(1.5,0,1,TRUE) – NORMDIST(0.3,0,1,TRUE)
$= 0.3153$

2. Thirty-six observations of $X$, the lifetime of a flashlight battery, are given below.

| | | | | | |
|---|---|---|---|---|---|
| 38 | 33 | 41 | 48 | 52 | 43 |
| 37 | 38 | 41 | 38 | 39 | 37 |
| 41 | 35 | 32 | 40 | 42 | 43 |
| 38 | 51 | 43 | 37 | 37 | 61 |
| 44 | 41 | 38 | 44 | 40 | 43 |
| 47 | 31 | 33 | 36 | 41 | 38 |

a. Find $\overline{x}$ and $s$.

b. Construct a 90% confidence interval for $\mu_X$.

**SOLUTION**

a. $\overline{x} = 40.6$; $s \cong 2.4$

b. (39.9, 41.3)

3. A hotel chain is opening a new upscale restaurant and is in desperate need of experienced waitresses. To attract quality waitresses, the management team wants to advertise an accurate estimate of the average salary (including tips) that waitresses can expect to earn per week. A survey of 64 randomly selected waitresses from the hotel chain's other restaurants found that, on average, waitresses earn \$715 per week in tips with a standard deviation of \$80. Let $W$ be the random variable that gives the amount that a waitress earns per week in tips.

   a. Construct a 98% confidence interval for $\mu_W$.

   b. Can the management team claim, at the 98% level, that the average amount that waitresses can expect to earn per week in tips is greater than \$700?

**SOLUTION**

a. (691.75, 738.25)

b. No

4. The mean and standard deviation of 100 observations of $C$, the amount of cola in a 16 oz. can, are 15.99 oz. and 0.15 oz., respectively.

   a. Find the value $b$ such that $P(Z \leq b) = 0.96$.

   b. Use the given information and the result from Part a to construct a confidence interval for $\mu_C$.

   c. What is the confidence level of the confidence interval constructed in Part b?

**SOLUTION**

a. $b = 1.751$

b. (15.96, 16.02)

c. 92%

5. Since 1945, the annual rate of return on the Standard & Poors's 500 Index has had a normal distribution with a mean of 12% and a standard deviation of 16.5%. Let $R$ be the rate of return on the Standard & Poor's 500 Index in a randomly selected year.

   a. Find a formula for the *p.d.f.* of $R$.

   b. Use ***Integrating.xls*** to compute the probability that the rate of return on the Standard & Poor's 500 Index in a randomly selected year is between 10% and 20%.

   c. Use ***NORMDIST*** to compute the probability that the rate of return on the Standard & Poor's 500 Index in a randomly selected year is between 10% and 20%.

## SOLUTION

a. $f_R(r) = \dfrac{1}{16.5\sqrt{2\pi}} e^{-0.5\left(\frac{r-12}{16.5}\right)^2}$

b. $P(10 \le R \le 20) = \displaystyle\int_{10}^{20} \frac{1}{16.5\sqrt{2\pi}} e^{-0.5\left(\frac{r-12}{16.5}\right)^2} dr = 0.2343$

| Definition | Computation | | Plot Interval | | Integration Interval | | $\int_a^b f(x)\,dx$ |
|---|---|---|---|---|---|---|---|
| Formula for $f(x)$ | $x$ | $f(x)$ | $A$ | $B$ | $a$ | $b$ | |
| 0.01856 | | 0.01856 | -70.5 | 94.5 | 10 | 20 | 0.2343 |

c. $P(10 \le R \le 20) = 0.2343$

6. Let $X$ be the random variable that gives the yearly mileage driven for an automobile that is at least five years old and registerd in Tucson, Arizona. A study conducted by the Arizona Department of Transportation found that $X$ has a normal distribution with a mean of 12,127.4 miles and a standard deviation of 1,050.6 miles.
   a. Find a formula for the *p.d.f.* of $X$.
   b. Use ***Integrating.xls*** to compute the probability that an automobile that is at least five years old and registered in Tucson, Arizona is driven over 14,000 miles in one year.
   c. Use ***NORMDIST*** to compute the probability that an automobile that is at least five years old and registered in Tucson, Arizona is driven over 14,000 miles.

## SOLUTION

a. $f_X(x) = \dfrac{1}{1{,}050.6\sqrt{2\pi}} e^{-0.5\left(\frac{x-12{,}127.4}{1{,}050.6}\right)^2}$

b. $P(X > 14{,}000) = \displaystyle\int_{14{,}000}^{\infty} \frac{1}{1{,}050.6\sqrt{2\pi}} e^{-0.5\left(\frac{x-12{,}127.4}{1{,}050.6}\right)^2} dx = 0.0373$

| Definition | Computation | | Plot Interval | | Integration Interval | | $\int_a^b f(x)\,dx$ |
|---|---|---|---|---|---|---|---|
| Formula for $f(x)$ | $x$ | $f(x)$ | $A$ | $B$ | $a$ | $b$ | |
| 4.4E-33 | | 4.4E-33 | 6874.4 | 17380.4 | 14000 | 17380.4 | 0.0373 |

c. $P(X > 14{,}000) = 0.0373$

7. The price, $P$, of electronic textbooks is an approximately normal random variable with mean $\mu_P = \$400$ and standard deviation $\sigma_P = \$150$.
   a. Use ***NORMDIST*** in ***Graphing.xls*** to plot the *p.d.f.* and *c.d.f.* of $P$ over the interval [–350, 1150].
   b. Use ***NORMDIST*** to compute the probability that the price of a randomly selected electronic textbook is between \$250 and \$550.

## SOLUTION

a.

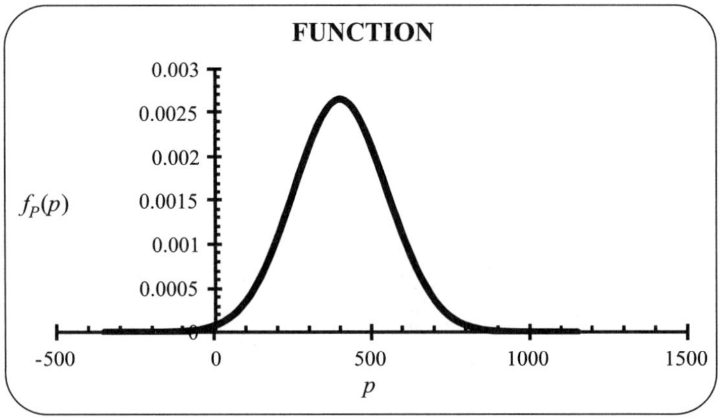


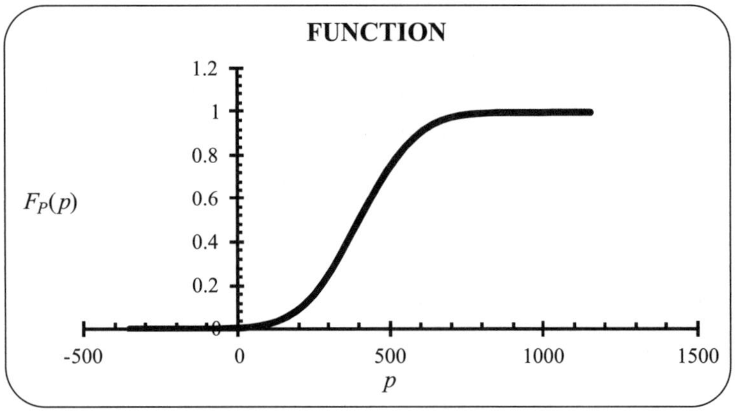


b. $P(250 \leq P \leq 550) = 0.6827$

8. The Portland Cement Company has developed a fast-drying indoor grout for use in showers. The grout needs to be reasonably dry before it is sealed. Studies found that the drying time has a normal distribution with a mean of 20 hours and a standard deviation of 4 hours. Let $T$ be the random variable that gives the the drying time for the fast-drying grout.
   a. Use ***NORMDIST*** in ***Graphing.xls*** to plot the *p.d.f.* and *c.d.f.* of $T$ over the interval [0, 40].
   b. Use ***NORMDIST*** to compute $P(18 \leq T \leq 24)$.

## SOLUTION

a.

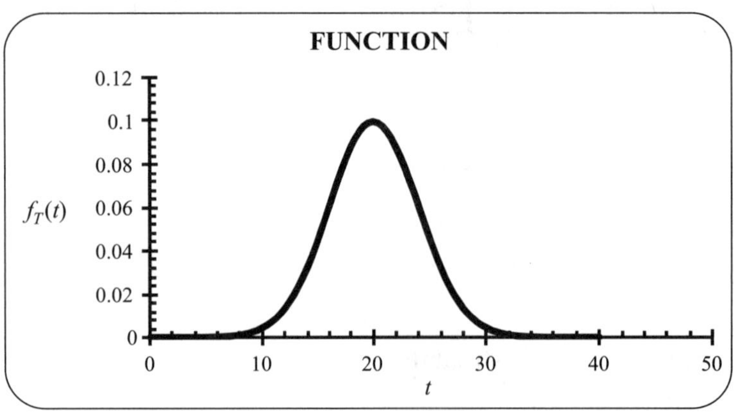

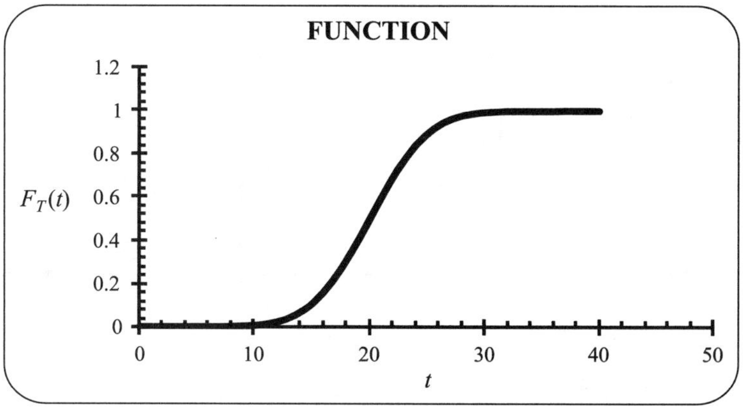


b. $P(18 \le T \le 24) = 0.5328$

9. Identify the graph that corresponds to each set of parameters.

   a. $\mu_X = 15$ and $\sigma_X = 5$

   b. $\mu_X = 0$ and $\sigma_X = 1$

   c. $\mu_X = 12$ and $\sigma_X = 3$

   d. $\mu_X = 0$ and $\sigma_X = 5$

   e. $\mu_X = 3$ and $\sigma_X = 6$

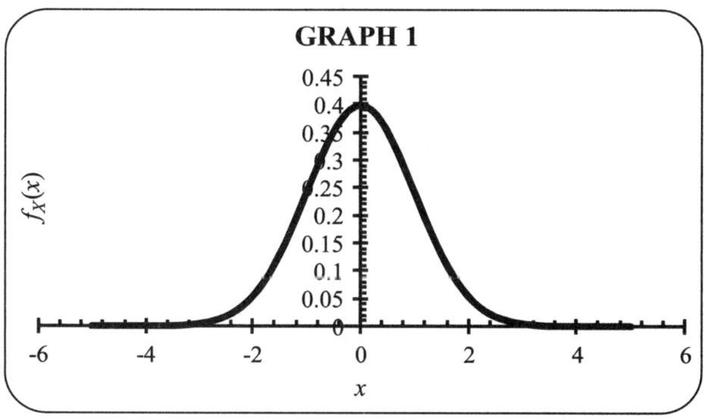


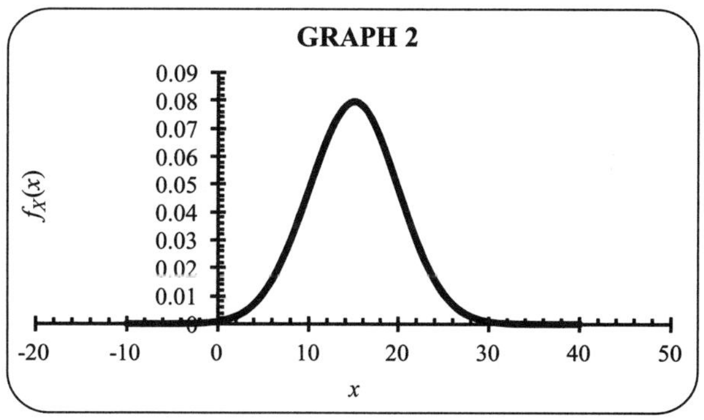


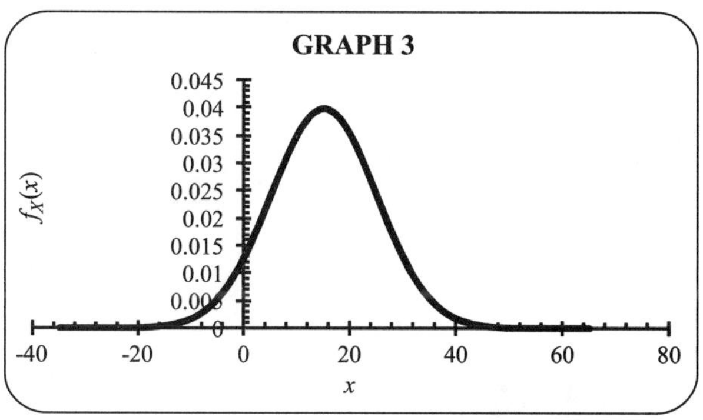


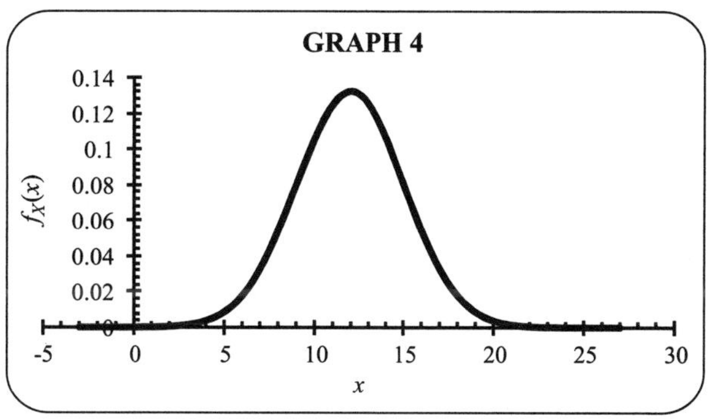

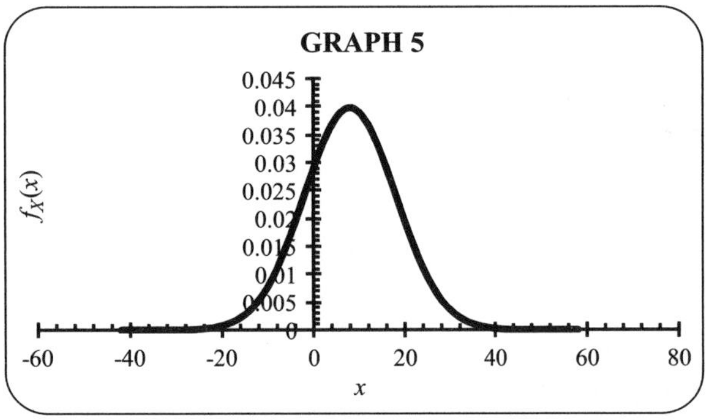


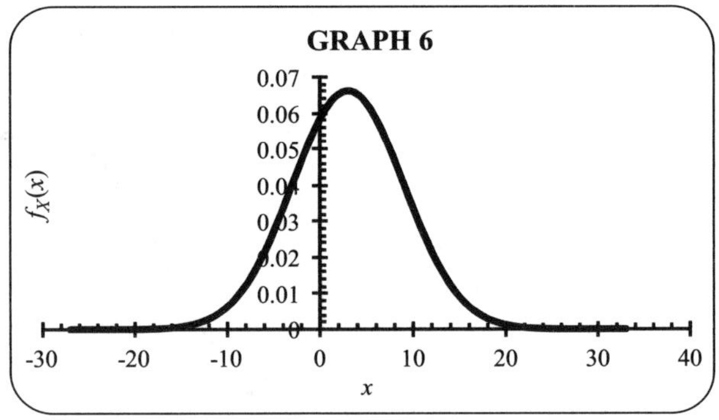


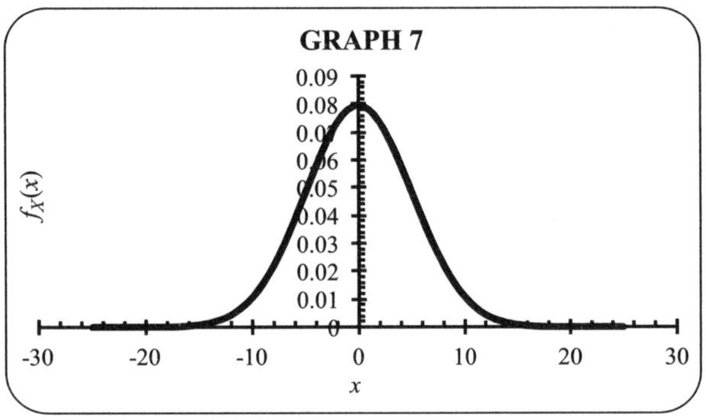


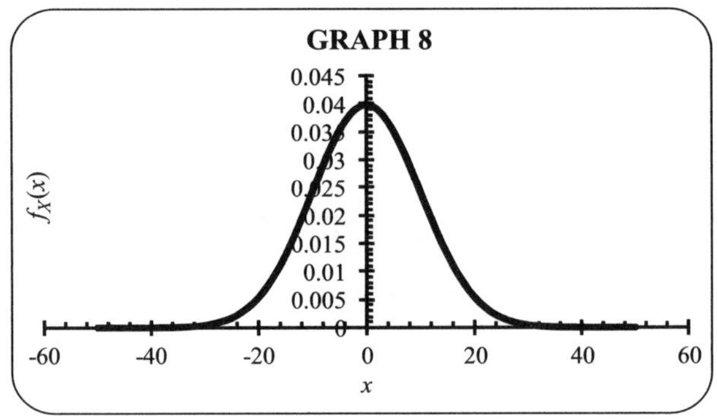


## SOLUTION

a. Graph 2

b. Graph 1

c. Graph 4

d. Graph 7

e. Graph 6

# Simulating Normal Random Variables

Two functions in *Excel* can be used to simulate normal random variables: ***Random Number Generation*** is used to generate fixed random samples, numbers that do not change in value when the workbook is recalculated. The ***NORMINV*** function is used to generate changeable random samples, numbers that do change in value when the workbook is recalculated.

**Example 1:** The net weight, $T$, of a randomly selected bag of pretzels has a normal distribution with mean $\mu_T = 454$ grams and $\sigma_T = 7.8$ grams. Use **Random Number Generation** to generate a sample of size $n = 50$ observations of $T$.

**SOLUTION**

To open ***Random Number Generation*** in Microsoft *Excel* 2003, pull down the **Tools** menu and select **Data Analysis**. Then click **Random Number Generation** and click **OK**. If you do not see **Data Analysis**, click the menu expansion arrows. If **Data Analysis** does not appear on the menu, you will need to load it. Pull down the **Tools** menu and select **Add-Ins**. Check the **Analysis TookPak Add-In** check box and click **OK**. (You may be prompted to insert the *Excel* installation disk.)

Click the menu expansion if you do not see Data Analysis. If it does not appear on the menu, follow the instructions in the paragraph above.

To open ***Random Number Generation*** in Microsoft *Excel* 2007, click the **Data** tab, then click **Data Analysis** and select **Random Number Generation**.

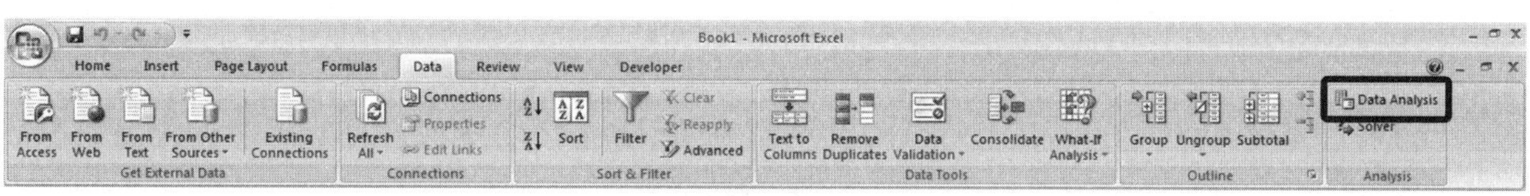


If **Data Analysis** does not appear on the **Data** tab, you will need to load it. Click the **Microsoft Office Button** and then click **Excel Options**. Click **Add-Ins**, select **Excel Add-Ins** (in the Manage box), and click **Go**.

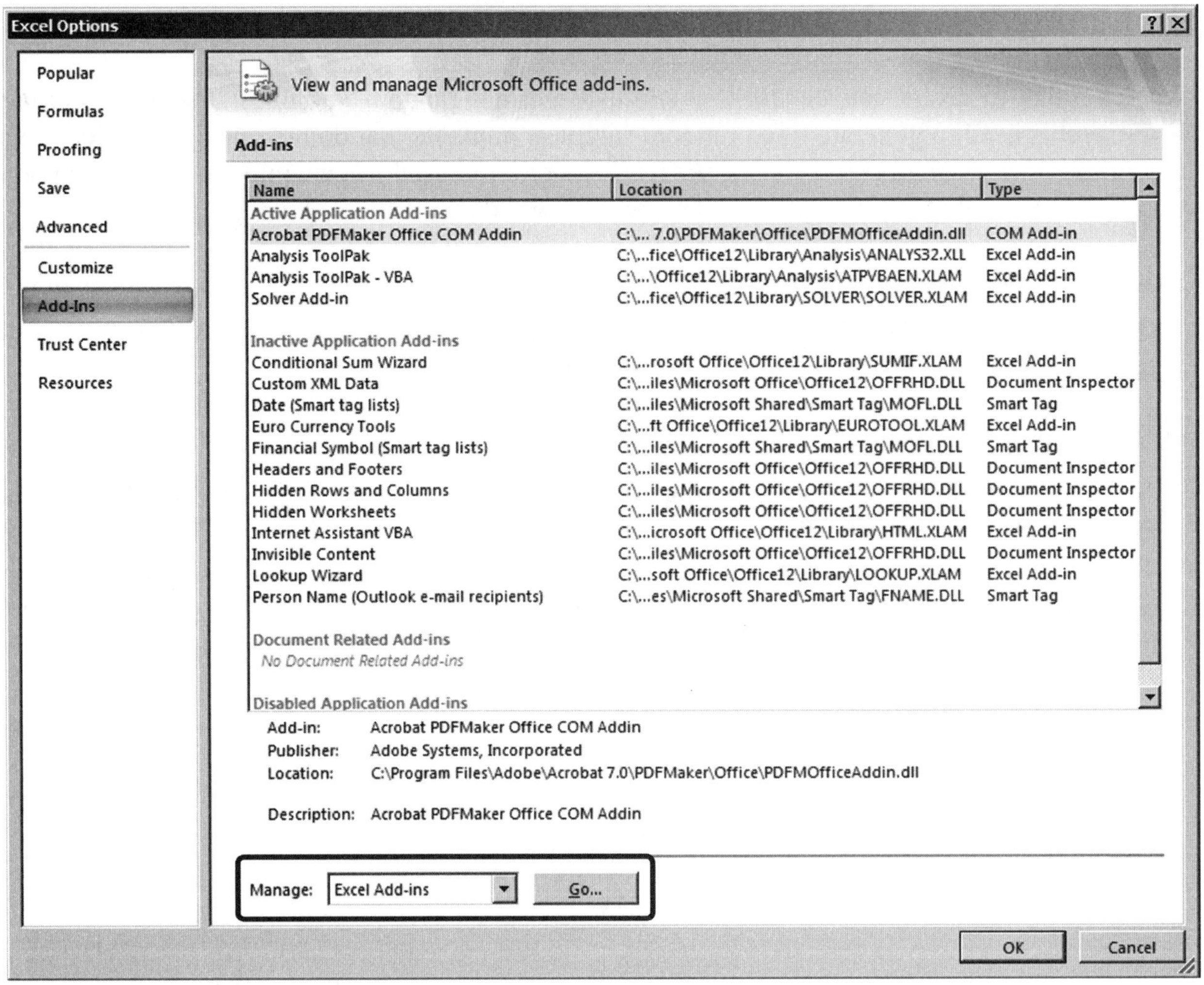


Check the **Analysis ToolPak Add-in** check box and click **OK**. (You may be prompted to insert the *Excel* installation disk.)

Enter the value 1 in the **Number of Variables** box and the value 50 in the **Number of Random Numbers** box. Click the down arrow at the right of the **Distribution** box and select **Normal**. Enter the value 454 in the **Mean** box and the value 7.8 in the **Standard Deviation** box. Click the **Output Range** option button, then click in the **Option Range** box and select *Cell A2*. Then click **OK**. The sample will be displayed in *Cells A2:A51*.

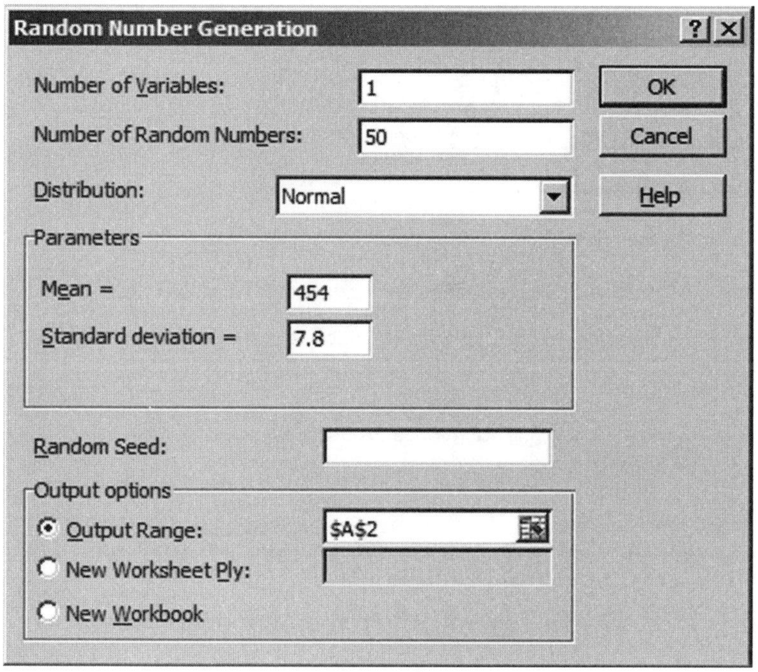


| Sample |
|---|
| 457.1218 |
| 437.6878 |
| 454.8703 |
| 447.6585 |
| 458.8819 |
| ⋮ |
| 460.0834 |
| 449.8678 |
| 468.5666 |
| 464.3476 |
| 464.7898 |

**Example 2:** Let $X$ be a normal random variable with mean $\mu_X = 10$ and $\sigma_X = 5$. Use the ***NORMINV*** function to generate a sample of size $n = 20$ observations of $X$.

## SOLUTION

The *c.d.f.* of $X$ is a one-to-one function from the real numbers to the interval [0, 1]. Thus, the inverse exists and is a function from the interval [0, 1] to the real numbers. In other words, the inverse *c.d.f.* is a function from the values of the *c.d.f.* to the possible values of $X$. The ***RAND*** is used to randomly generate a value of the *c.d.f.* Then the ***NORMINV*** function is used to find the corresponding value of $X$.

Graphically, this is equivalent to finding the value on the $x$-axis of the graph of $F_X(x)$ that corresponds to a randomly selected value on the $F_X$ -axis.

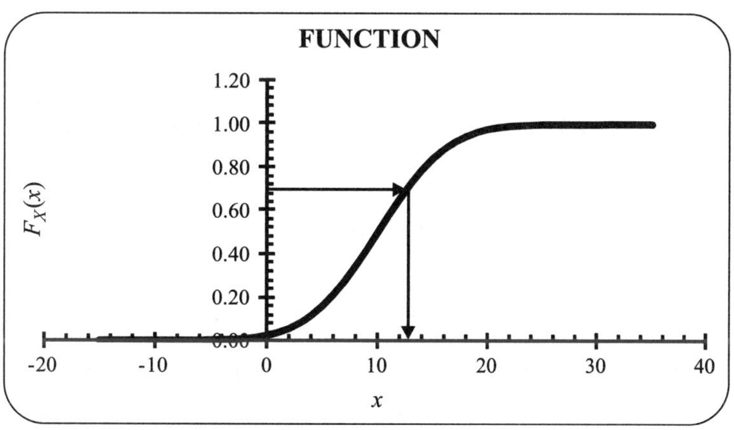


Insert the ***NORMINV*** function in *Cell A2*, enter the information shown below, and then click **OK**.

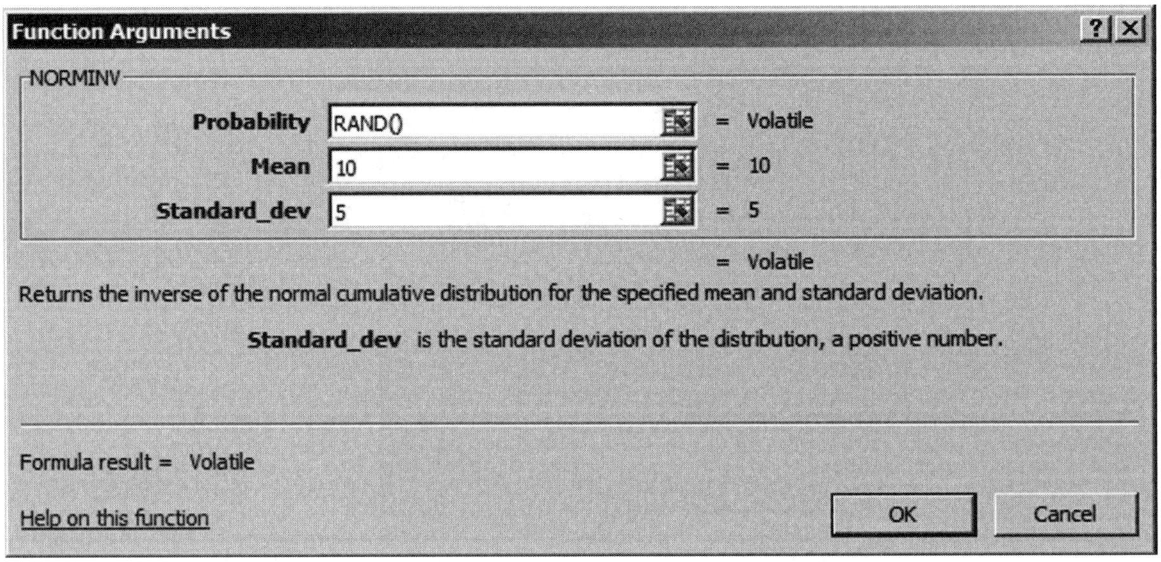


The first observation will appear in *Cell A2*. To generate the remaining observations, copy the formula from *Cell A2* into *Cells A3:A21*.

| Sample |
|---|
| 14.11869 |
| 8.380957 |
| 13.83364 |
| 16.27443 |
| 8.467802 |
| ⋮ |
| 3.765574 |
| 15.01212 |
| 8.878692 |
| 14.28119 |
| 5.108887 |

**Example 3:** Let $Y$ be a normal random variable with mean $\mu_Y = 800$ and standard deviation $\sigma_Y = 250$.

a. Use ***Random Number Generation*** to generate a set of 500 samples, each of size $n = 4$ observations of $Y$.

b. Compute the mean of each sample generated in Part a.

c. Recall that the sample mean, $\overline{y}$, for samples of size $n = 4$ observations of $Y$ is also a normal random variable. Compute the mean and standard deviation of $\overline{y}$.

d. Compute the mean and standard deviation of the 500 observations of $\overline{y}$ generated in Part a.

## SOLUTION

a. Open ***Random Number Generation***, enter the information shown below, and click **OK**.

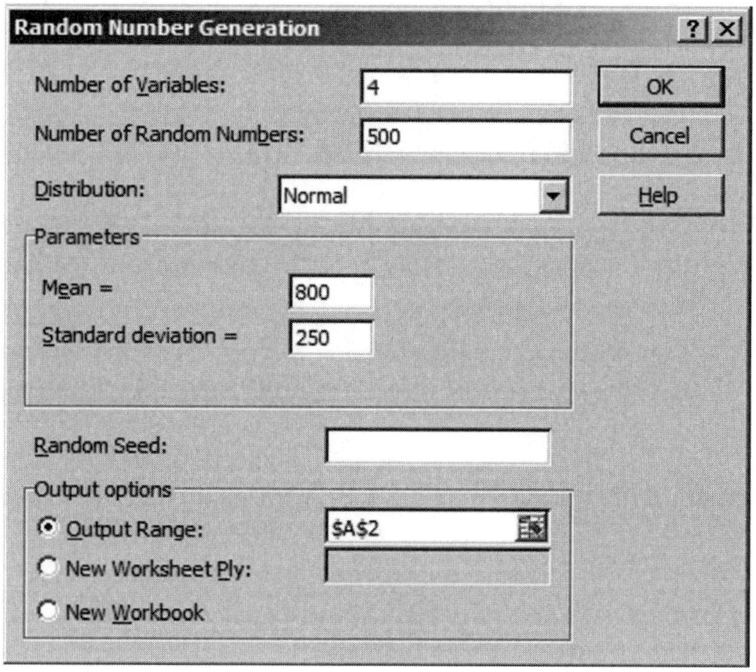


The results will vary; an example is shown below.

| | A | B | C | D |
|---|---|---|---|---|
| 1 | $y_1$ | $y_2$ | $y_3$ | $y_4$ |
| 2 | 1069.511 | 1049.611 | 883.2824 | 601.7631 |
| 3 | 893.6876 | 577.3386 | 737.5751 | 935.6003 |
| 4 | 427.9155 | 653.3877 | 1090.906 | 817.8105 |
| 5 | 682.6604 | 1239.873 | 978.7933 | 718.0707 |
| 6 | 750.831 | 720.6293 | 795.7828 | 1036.9 |
| 251 | ⋮ | ⋮ | ⋮ | ⋮ |
| 497 | 690.1669 | 814.8208 | 967.9206 | 148.4562 |
| 498 | 1113.792 | 732.5706 | 775.318 | 1061.247 |
| 499 | 840.7528 | 651.11 | 842.2847 | 116.0873 |
| 500 | 808.6939 | 990.6284 | 701.4113 | 1032.149 |
| 501 | 829.7996 | 654.997 | 974.7907 | 871.9427 |

b. Enter the formula =AVERAGE(A2:D2) in *Cell E2* and copy into *Cells E3:E501*. The results are given below.

| | A | B | C | D | E |
|---|---|---|---|---|---|
| 1 | $y_1$ | $y_2$ | $y_3$ | $y_4$ | $\bar{y}$ |
| 2 | 1069.511 | 1049.611 | 883.2824 | 601.7631 | 901.0417 |
| 3 | 893.6876 | 577.3386 | 737.5751 | 935.6003 | 786.0504 |
| 4 | 427.9155 | 653.3877 | 1090.906 | 817.8105 | 747.505 |
| 5 | 682.6604 | 1239.873 | 978.7933 | 718.0707 | 904.8492 |
| 6 | 750.831 | 720.6293 | 795.7828 | 1036.9 | 826.0358 |
| 251 | ⋮ | ⋮ | ⋮ | ⋮ | ⋮ |
| 497 | 690.1669 | 814.8208 | 967.9206 | 148.4562 | 655.3411 |
| 498 | 1113.792 | 732.5706 | 775.318 | 1061.247 | 920.7317 |
| 499 | 840.7528 | 651.11 | 842.2847 | 116.0873 | 612.5587 |
| 500 | 808.6939 | 990.6284 | 701.4113 | 1032.149 | 883.2205 |
| 501 | 829.7996 | 654.997 | 974.7907 | 871.9427 | 832.8825 |

c. The mean of $\bar{y}$ is given by $\mu_{\bar{y}} = \mu_Y = 800$ and the standard deviation is given by $\sigma_{\bar{y}} = \sigma_Y / \sqrt{n} = 250/\sqrt{4} = 125$.

d. Enter the formula =AVERAGE(E2:E501) in *Cell F2* and the formula =STDEV(E2:E501) in *Cell G2*. The results are given below.

| | F | G |
|---|---|---|
| 1 | **Mean of Means** | **Standard Deviation of Means** |
| 2 | 799.7573 | 114.0393 |

**Example 4:** The annual compensation, *X*, of a randomly selected CEO has a normal distribution with mean $\mu_X = \$634,961$ and standard deviation $\sigma_X = \$124,690$.

a. Use the ***NORMINV*** function to generate a set of 100 samples, each of size $n = 8$ observations of *X*.

b. Compute a 95% confidence interval for $\mu_X$ from each sample generated in Part a.

c. What fraction of the intervals actually contain $634,961?

**SOLUTION**

a. Insert the ***NORMINV*** function in *Cell A2*, enter the information shown below, and then click **OK**.

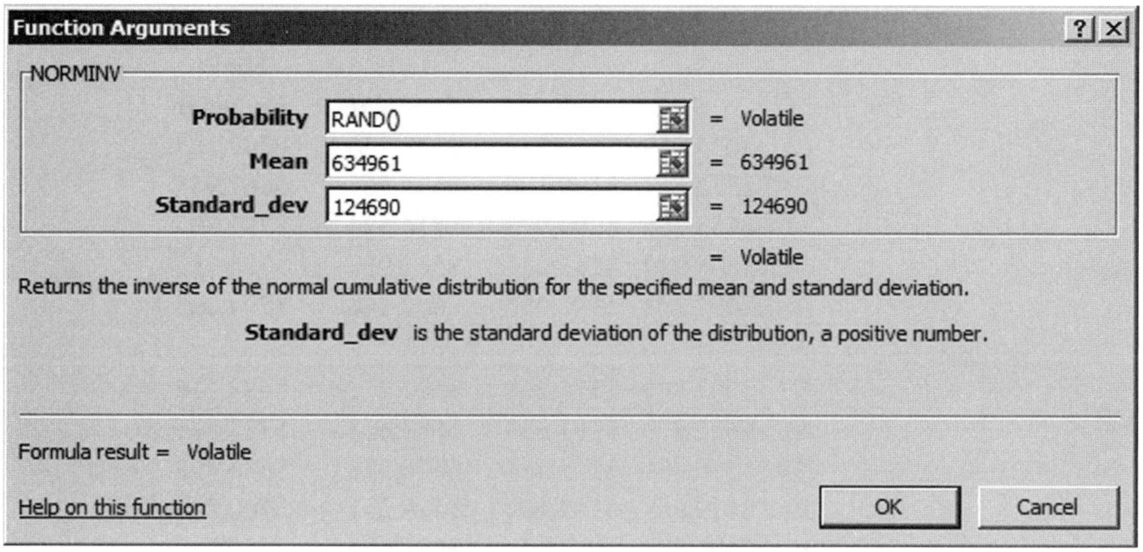


The first observation will appear in *Cell A2*. To generate the remaining observations, copy the formula from *Cell A2* into *Cells A2:H101*. The results will vary; an example is shown below.

| | A | B | C | D | E | F | G | H |
|---|---|---|---|---|---|---|---|---|
| 1 | $x_1$ | $x_2$ | $x_3$ | $x_4$ | $x_5$ | $x_6$ | $x_7$ | $x_8$ |
| 2 | \$690,221 | \$508,717 | \$481,142 | \$730,499 | \$777,696 | \$682,743 | \$818,847 | \$397,811 |
| 3 | \$439,182 | \$607,418 | \$544,749 | \$587,828 | \$821,923 | \$495,983 | \$387,750 | \$687,668 |
| 4 | \$766,894 | \$693,328 | \$404,848 | \$690,329 | \$816,061 | \$367,912 | \$545,354 | \$941,397 |
| 5 | \$686,866 | \$410,275 | \$941,533 | \$702,126 | \$537,102 | \$460,558 | \$607,909 | \$845,149 |
| 6 | \$679,462 | \$831,465 | \$799,313 | \$416,707 | \$647,146 | \$611,990 | \$777,206 | \$814,492 |
| 51 | ⋮ | ⋮ | ⋮ | ⋮ | ⋮ | ⋮ | ⋮ | ⋮ |
| 97 | \$584,519 | \$506,093 | \$609,613 | \$543,464 | \$619,496 | \$794,020 | \$437,695 | \$727,855 |
| 98 | \$596,415 | \$666,626 | \$709,842 | \$669,713 | \$518,780 | \$676,625 | \$558,225 | \$853,937 |
| 99 | \$472,663 | \$424,509 | \$656,151 | \$640,613 | \$473,071 | \$691,926 | \$431,311 | \$660,883 |
| 100 | \$723,791 | \$578,581 | \$526,712 | \$796,507 | \$624,562 | \$582,741 | \$702,195 | \$504,813 |
| 101 | \$742,575 | \$798,742 | \$638,281 | \$654,727 | \$681,865 | \$635,837 | \$805,929 | \$576,746 |

b. The confidence intervals are computed using the formula given below.

$$\left(\bar{x} - 1.96\frac{s}{\sqrt{n}},\ \bar{x} + 1.96\frac{s}{\sqrt{n}}\right)$$

| | I | J | K | L |
|---|---|---|---|---|
| 1 | $\bar{x}$ | $s$ | **Left Endpoint** | **Right Endpoint** |
| 2 | $635,959 | $153,289 | $529,736 | $742,183 |
| 3 | $571,563 | $139,106 | $475,167 | $667,958 |
| 4 | $653,265 | $200,070 | $514,624 | $791,907 |
| 5 | $648,940 | $183,231 | $521,967 | $775,912 |
| 6 | $697,223 | $140,086 | $600,148 | $794,298 |
| 51 | ⋮ | ⋮ | ⋮ | ⋮ |
| 97 | $602,844 | $115,261 | $522,972 | $682,716 |
| 98 | $656,270 | $103,333 | $584,664 | $727,876 |
| 99 | $556,391 | $115,464 | $476,378 | $636,403 |
| 100 | $629,988 | $102,107 | $559,231 | $700,744 |
| 101 | $691,838 | $82,589 | $634,606 | $749,069 |

c. Enter the formula =IF(AND(K2<=634961, 634961<=L2),"yes","no") in *Cell M2* and copy into *Cells M3:M101*. Then enter the formula =COUNTIF(M2:M3,"yes") in *Cell N2*. The results are given below.

| | M | N |
|---|---|---|
| 1 | **$634,961 in Interval?** | **Fraction Containing $634,961** |
| 2 | yes | 90% |
| 3 | yes | |
| 4 | yes | |
| 5 | yes | |
| 6 | yes | |
| 51 | ⋮ | |
| 97 | yes | |
| 98 | yes | |
| 99 | yes | |
| 100 | yes | |
| 101 | yes | |

Ninety percent of the intervals actually contain $634,961.

## *Exercises*

1. Let $X$ be a normal random variable with mean $\mu_X = 75$ and $\sigma_X = 15$. Use the ***NORMINV*** function to generate a sample of size $n = 10$ observations of $X$.

**SOLUTION**

The results will vary; an example is shown below.

| Sample |
|---|
| 70.49652 |
| 55.83475 |
| 78.66386 |
| 94.1471 |
| 92.97525 |
| 100.997 |
| 42.24619 |
| 71.48728 |
| 91.42534 |
| 58.69949 |

2. The rate of return, $R$, on the Standard & Poor's 500 Index in a randomly selected year has a normal distribution with mean $\mu_R = 12\%$ and standard deviation $\sigma_R = 16.5\%$. Use the ***NORMINV*** function to generate a sample of size $n = 25$ observations of $R$.

**SOLUTION**

The results will vary; an example is shown below.

| Sample |
|---|
| 10.1% |
| 27.5% |
| -13.9% |
| 40.0% |
| 16.2% |
| ⋮ |
| -1.8% |
| 18.3% |
| 7.1% |
| 37.4% |
| -2.0% |

3. The price, $P$, of electronic textbooks is a normal random variable with mean $\mu_P = \$400$ and standard deviation $\sigma_P = \$150$.

   a. Use ***Random Number Generation*** to generate a set of 200 samples, each of size $n = 9$ observations of $P$.

b. Compute the mean of each sample generated in Part a.

c. Recall that the sample mean, $\overline{p}$, for samples of size $n=9$ observations of $P$ is also a normal random variable. Compute the mean and standard deviation of $\overline{p}$.

d. Compute the mean and standard deviation of the 200 observations of $\overline{p}$ generated in Part a.

## SOLUTION

The results will vary; an example is shown below.

a.

| $p_1$ | $p_2$ | $p_3$ | $p_4$ | $p_5$ | $p_6$ | $p_7$ | $p_8$ | $p_9$ |
|---|---|---|---|---|---|---|---|---|
| \$426.98 | \$480.05 | \$317.95 | \$504.48 | \$442.80 | \$667.96 | \$463.63 | \$316.91 | \$503.79 |
| \$293.74 | \$612.70 | \$454.38 | \$258.49 | \$125.21 | \$67.29 | \$217.63 | \$115.63 | \$248.05 |
| \$201.06 | \$730.85 | \$491.81 | \$214.57 | \$771.82 | \$291.58 | \$378.74 | \$455.36 | \$264.91 |
| \$515.61 | \$623.18 | \$362.92 | \$391.82 | \$392.04 | \$281.09 | \$125.76 | \$317.48 | \$396.70 |
| \$213.45 | \$485.40 | \$484.87 | \$257.57 | \$249.95 | \$387.47 | \$271.50 | \$247.58 | \$86.20 |
| ⋮ | ⋮ | ⋮ | ⋮ | ⋮ | ⋮ | ⋮ | ⋮ | ⋮ |
| \$216.81 | \$201.28 | \$257.75 | \$302.47 | \$341.50 | \$428.39 | \$303.89 | \$99.19 | \$113.95 |
| \$360.54 | \$564.80 | \$168.46 | \$284.02 | \$336.88 | \$261.84 | \$463.31 | \$602.59 | \$347.21 |
| \$60.96 | \$523.57 | \$414.30 | \$467.05 | \$504.71 | \$525.35 | \$17.95 | \$296.82 | \$581.17 |
| \$414.75 | \$353.31 | \$365.32 | \$348.32 | \$516.48 | \$573.22 | \$385.56 | \$593.64 | \$180.23 |
| \$341.43 | \$587.07 | \$372.83 | \$606.34 | \$567.08 | \$281.69 | \$490.79 | \$409.92 | \$442.58 |

b.

| $\overline{p}$ |
|---|
| \$458.28 |
| \$265.90 |
| \$422.30 |
| \$378.51 |
| \$298.22 |
| ⋮ |
| \$251.69 |
| \$376.63 |
| \$376.88 |
| \$414.54 |
| \$455.53 |

c. $\mu_{\overline{p}} = \$400$; $\sigma_{\overline{p}} = \$50$

d.

| Mean of Means | Standard Deviation of Means |
|---|---|
| \$399.69 | \$57.38 |

4. Let $X$ be a normal random variable with mean $\mu_X = 120$ and $\sigma_X = 30$.
   a. Use the ***NORMINV*** function to generate a set of 400 samples, each of size $n = 5$ observations of $X$.
   b. Compute a 90% confidence interval for $\mu_X$ from each sample generated in Part a.
   c. What fraction of the intervals actually contain 120, the mean of $X$?

**SOLUTION**

The results will vary; an example is shown below.

a.

| $x_1$ | $x_2$ | $x_3$ | $x_4$ | $x_5$ |
|---|---|---|---|---|
| 124.2638 | 148.1593 | 143.4964 | 120.2233 | 153.7784 |
| 139.2829 | 158.2827 | 132.5595 | 104.9673 | 127.0745 |
| 88.52088 | 79.03872 | 118.5087 | 63.54104 | 73.01746 |
| 108.7963 | 146.8852 | 123.1008 | 171.3205 | 105.3307 |
| 125.8272 | 185.3711 | 93.95439 | 93.32361 | 93.80586 |
| ⋮ | ⋮ | ⋮ | ⋮ | ⋮ |
| 139.1814 | 82.25816 | 113.2187 | 124.462 | 104.6625 |
| 125.0905 | 109.3658 | 114.7014 | 103.674 | 100.7062 |
| 109.5152 | 141.3822 | 143.5599 | 127.2481 | 93.32465 |
| 102.7131 | 188.3896 | 80.70895 | 132.7279 | 115.1604 |
| 111.6652 | 63.61067 | 128.3347 | 111.4029 | 83.75384 |

b.

| $\bar{x}$ | $s$ | **Left Endpoint** | **Right Endpoint** |
|---|---|---|---|
| 137.9842 | 14.89189 | 127.0288 | 148.9397 |
| 132.4334 | 19.35342 | 118.1957 | 146.671 |
| 84.52536 | 21.05772 | 69.03391 | 100.0168 |
| 131.0867 | 27.8045 | 110.6319 | 151.5415 |
| 118.4564 | 39.91105 | 89.09522 | 147.8177 |
| ⋮ | ⋮ | ⋮ | ⋮ |
| 112.7565 | 21.39608 | 97.01617 | 128.4969 |
| 110.7076 | 9.671086 | 103.5929 | 117.8223 |
| 123.006 | 21.45418 | 107.2229 | 138.7891 |
| 123.94 | 40.70491 | 93.99476 | 153.8852 |
| 99.75348 | 25.77198 | 80.79391 | 118.7131 |

c. 80.75% of the intervals actually contain 120.